John Bateman, Janina Wildfeuer, Tuomo Hiippala
Multimodality

John Bateman, Janina Wildfeuer, Tuomo Hiippala

Multimodality

Foundations, Research and Analysis
A Problem-Oriented Introduction

DE GRUYTER
MOUTON

ISBN 978-3-11-047942-3
e-ISBN (PDF) 978-3-11-047989-8
e-ISBN (EPUB) 978-3-11-048004-7

Library of Congress Cataloging-in-Publication Data
A CIP catalog record for this book has been applied for at the Library of Congress.

Bibliographic information published by the Deutsche Nationalbibliothek
The Deutsche Nationalbibliothek lists this publication in the Deutsche Nationalbibliografie;
detailed bibliographic data are available on the Internet at http://dnb.dnb.de.

© 2017 Walter de Gruyter GmbH, Berlin/Boston
Printing and binding: CPI books GmbH, Leck
♾ Printed on acid-free paper
Printed in Germany

www.degruyter.com

Contents

How to use this book —— 1

Part I: Working your way into 'multimodality'

1   Introduction: the challenge of multimodality —— 7
1.1   First steps … a multimodal turn? —— 9
1.2   The journey ahead —— 15
1.3   What this chapter was about: the 'take-home message' —— 21

2   Recognising multimodality: origins and inspirations —— 22
2.1   The 'problem space' of multimodality as such —— 23
2.2   Materiality and the senses: sound —— 26
2.3   Materiality and the senses: vision and visuality —— 30
2.4   Language —— 38
2.5   Systems that signify: semiotics —— 51
2.6   Society, culture and media —— 64
2.7   What this chapter was about: the 'take-home message' —— 71

3   Where is multimodality? Communicative situations and their media —— 74
3.1   Stepping beyond the terminological jungle —— 74
3.2   Communicative situations —— 77
3.3   Medium (and media) —— 101
3.4   What this chapter was about: the 'take-home message' —— 110

4   What is multimodality? Semiotic modes and a new 'textuality' —— 112
4.1   Semiotic mode —— 112
4.2   Modes and media —— 123
4.3   Genre, text, discourse and multimodality —— 128
4.4   What this chapter was about: the 'take-home message' —— 135

Part II: Methods and analysis

5   The scope and diversity of empirical research methods for multimodality —— 139
5.1   What are methods? What methods are there? —— 139
5.2   Getting data? —— 141
5.3   Corpus-based methods to multimodality —— 152

5.4   Eye-tracking methods for multimodality —— 159
5.5   Computational methods in multimodality research —— 162
5.6   Summary and conclusions: selecting tools for the job —— 166
5.7   A word on good scientific practice: quoting multimodal artefacts —— 168

6   **Are your results saying anything? Some basics —— 169**
6.1   Why statistics?—and how does it work? —— 171
6.2   What is normal? —— 175
6.3   Assessing differences —— 178
6.4   Assessing similarities —— 184
6.5   How much is enough? —— 188
6.6   Refinements and restrictions —— 190
6.7   Inter-coder consistency and reliability —— 198
6.8   What affects what? Looking for dependencies —— 204
6.9   Summary: many types of tests and possibilities —— 209

7   **Multimodal navigator: how to plan your multimodal research —— 211**
7.1   Starting analysis —— 211
7.2   Undertaking multimodal investigations of phenomena —— 226
7.3   The basic steps in multimodal analysis reviewed —— 229
7.4   Conclusions and lessons for effective multimodal research —— 230

Part III: **Use cases**

Use case area 1: temporal, unscripted —— 238

8   **Gesture and face-to-face interaction —— 239**
8.1   Previous studies —— 240
8.2   Describing gesture and its functions —— 243
8.3   Conclusions —— 248

Use case area 2: temporal, scripted —— 250

9   **Performances and the performing arts —— 251**
9.1   Performance and scripted behaviour —— 252
9.2   Previous studies —— 253
9.3   Example analysis: theatre and its canvases —— 254
9.4   Example analysis: Berlin Philharmonic concerts 'live' —— 258
9.5   Conclusions —— 260

Use case area 3: spatial, static —— 262

10 **Layout space** ▬ 263
10.1 Perspectives from graphic design ▬ 263
10.2 Example analysis: school textbooks ▬ 268
10.3 Example analysis: posters ▬ 273
10.4 Summary ▬ 278

11 **Diagrams and infographics** ▬ 279
11.1 Aspects of the diagrammatic mode ▬ 281
11.2 Example analysis: assembly instructions ▬ 286
11.3 Example analysis: information graphics ▬ 289
11.4 Summary ▬ 294

12 **Comics and graphic novels** ▬ 295
12.1 Comics: basic ingredients ▬ 297
12.2 An aside on the notion of 'narrative' ▬ 314
12.3 Beyond narrative: comics as non-fiction and metacomics ▬ 315
12.4 Issues of literacy ▬ 321
12.5 Moving onwards: empirical multimodal comics research ▬ 322
12.6 Summary ▬ 324

Use case area 4: spatial, dynamic ▬ 326

13 **Film and the moving (audio-)visual image** ▬ 327
13.1 The technical details of the virtual canvas of film ▬ 329
13.2 Multimodal film analysis: an example ▬ 332
13.3 Films and comics: adaptation and convergence ▬ 336
13.4 Summary ▬ 338

14 **Audiovisual presentations** ▬ 340
14.1 Characterising the medium ▬ 340
14.2 Exploring the canvases ▬ 342
14.3 Summary ▬ 343

Use case area 5: spatiotemporal, interactive ▬ 346

15 **Webpages and dynamic visualisations** ▬ 347
15.1 Challenges and difficulties: determining objects of analysis ▬ 348
15.2 Example analysis: dynamic data visualisations ▬ 350
15.3 Summary ▬ 354

16 **Social media** ▬ 355
16.1 Previous studies ▬ 356

16.2   Communicative situations in social media —— **358**
16.3   Social media analyses and possible methods: Instagram —— **362**
16.4   Summary —— **365**

17   Computer and video games —— 366
17.1   Example analysis: turn-based strategy games —— **367**
17.2   Example analysis: first-person, real-time games —— **372**
17.3   Summary —— **377**

18   Final words —— 379
18.1   Lessons learned: the take-home messages —— **379**
18.2   Our goals in the book —— **381**
18.3   Be a multimodalist: have fun and explore! —— **382**

Bibliography —— 383

Index —— 411

How to use this book

There are several distinct kinds of introductions to 'multimodality' that you will find on the bookshelves. Some promise to provide what they call 'toolkits' for multimodal analysis. These toolkits include sets of ready-made distinctions to apply to images, films, advertisements and so on, perhaps suggesting 'visual grammars' or other kinds of 'grammars' for applying categories to examples of communication. Others work their way through various theoretical orientations that have traditionally been applied to the areas of 'multimodality' selected for concern. In both cases, the task for the reader is to learn to apply those frameworks to examples.

The problem with such approaches is the guidance (or lack of guidance) given when the tools do not fit your particular research questions or your particular objects of inquiry. A common experience is that a distinction that might appear straightforward to apply when confronted with the, often relatively simple, examples given in an introduction turns out to be difficult to apply reliably in many other cases encountered in investigation. The range and combinations of 'modes' already commonplace in our everyday world are extremely diverse and, what is more, *flexible*. One of the most often remarked features of modern communication is how we see new forms of combinations of existing media, which then quickly join the ranks of resources to be used and extended in their own right. Dealing with this kind of flexibility raises challenges for the toolkit view and requires a different kind of competence. In fact, we need to become tool-makers rather than just tool-users.

To meet this challenge we have to attempt something rather different in this book than to follow the toolkit view. We begin at a more foundational level so that we can approach any example of 'multimodal' signification—and we will characterise just what this is when we get started—no matter how complex. We set out theoretical distinctions and methodological steps that will enable you to tease apart complexity and to make that complexity amenable to productive analysis. This means that we offer not a ready-made toolkit but more a way of building the specific toolkits you will need for each job and for each particular kind of multimodal communication that you encounter and wish to analyse. We will see that this is a crucial basic competence and skill that needs to be learned when exploring multimodal analyses.

To use this book, therefore, we must begin by working over and reconfiguring some of the assumptions and preconceptions commonly found concerning just what multimodality is and how the phenomenon of multimodality is to be captured. This demands a more flexible openness to the properties of whatever it is that is being analysed and the situations in which it is used or occurs. We describe how to achieve this openness, the theoretical constructs that are necessary, and then offer a range of specific case studies that lead you through how very varied materials are to be approached.

Notation and layout conventions used in the book

Margin notes We make use of several distinct forms of information presentation in this book; these are generally marked out by distinctive layout environments so that you can easily recognise them. For example, we use margin notes so that you can easily identify the passages where some particular topic is introduced or discussed so that you can more easily find those passages again.

Quotations that play a role directly in the unfolding discussion are quoted either in the text or as visually set-off blocks as usual. In addition to these, we also offer additional quotations of a more background nature, opening up connections or other formulations of relevant points. These are shown in 'background' quotes, marked off from the main text by horizontal lines and indicated by a large opening or closing quotation mark in the margin.

> ...communication depends on some 'interpretative community' having decided that some aspect of the world has been articulated in order to be interpreted.
> — *Kress and van Leeuwen (2001: 8)*

We also use a range of additional information boxes, again with identifying symbols in the margin. ■ is used for additional information; ⚡ is used for a warning or point where you need to be careful; ❗ is used for a conclusion or summary point; ❓ is used for additional questions that you can beneficially think about to throw more light on a discussion or point being made.

Acknowledgements

The following people commented on the manuscript at various stages: Ylva Grufstedt, Björn Hochschild, Petteri Laihonen, Nick Moore, Rob Waller, Lukas Wilde, Christoph Wolf, and students at Bremen University who participated in relevant classes. We thank them all for their constructive criticism and feedback; all errors remain our own.

We are very grateful to the following who supplied material for analysis:
- Oliver Kutz for the photographs in Figures 1.1 and 5.4
- Raúl A. Mora (LSLP, Colombia), Chris Davis and James Lamb for the tweets in Section 1.1
- Austin Adams, Theo Boersema and Meijer Mijksenaar for Figure 4.2
- Mehul Bhatt for the heatmap in Figure 5.3
- Tracy Ducasse for making Figure 9.1 available with a CC BY 2.0 license
- Kaela Zhang for providing Figure 10.5
- John McLinden for making Figure 12.1 available with a CC BY 2.0 license
- Aaron Scott Humphrey for Figures 12.4 and 12.6
- Anssi Hiippala for providing the screenshots in Figures 17.4 and 17.5

In addition, the authors are grateful for permissions to reproduce the following copyrighted works:
- SAGE publications for Figure 5.1
- Thinkstock for Figure 7.1 (Item number: 78253847)
- Graphic News for Figures 11.4, 11.6 and 11.7
- DC Comics for Figure 12.2

All other images are reproduced for scientific discussion according to the principles of fair use, are in the public domain, or are our own. The authors offer their apologies should the rights of any further copyright holders have been infringed upon unknowingly.

Part I: **Working your way into 'multimodality'**

In this part of the book we meet the basic notion of 'multimodality', what it is, where it occurs and which fields have addressed it. We then introduce a suitable foundation for approaching multimodality as a basic phenomenon of communicative practice.

1 Introduction: the challenge of multimodality

Orientation

To get us started, we will immediately raise three related concerns: (i) *what* is multimodality? (ii) *who* is the intended audience of this book? and (iii) *why* might it be beneficial to know about multimodality anyway? We then give a hands-on feel for some typical cases of multimodality that have been addressed in the literature, pointing out the challenges both for the theory and practice of analysis.

If you are watching a TV news programme, where a presenter discusses some events backed up by textual overlays and recorded smartphone videos from the scene of those events, then you are interacting with a multimodal medium. If you are reading a book with diagrams and text, photographs and graphs, then you are also interacting with a multimodal medium. If you are talking to someone in a cafeteria, exchanging verbal utterances accompanied by facial expressions, gestures and variations in intonation, then you are, again, interacting in a multimodal medium. If you talk to a friend on WhatsApp and reply with an image instead of writing something, or when you draw or insert emojis on your Snapchat video, then you are, once again, communicating multimodally. And, if you are playing a computer video game (possibly in virtual reality), controlling the actions of your avatar while receiving instruction on your next mission, then this is also an example of multimodality at work. Given this, it is in fact difficult to find cases of communication and action that do *not* involve multimodality. So what then marks this area out as deserving a name at all? Is it not just the normal state of affairs, inherent to the nature of everyday experience in any case? Why have books on it?

Multimodality is a way of characterising communicative situations (considered very broadly) which rely upon combinations of different 'forms' of communication to be effective—the TV programme uses spoken language, pictures and texts; the book uses written language, pictures, diagrams, page composition and so on; talking in the cafeteria brings together spoken language with a host of bodily capabilities and postures; and the computer game might show representations of any of these things and include movement and actions as well.

Despite the fact that situations qualifying as 'multimodal' are everywhere, it may be surprising to discover just how little we know about how this fundamental human capability operates. We can observe that people succeed (often, but not always!) in making sense out of, and in, these multimodal communicative situations, but precisely how such varied and diverse forms of communication combine productively raises challenging issues at all levels. Moreover, the development of many traditional disciplines has focused attention in precisely the opposite direction: segmenting and compartmentalising rather than addressing how 'ensembles' of communicative practices work together. Thus, traditionally, linguistics deals with language, art history with paintings, etc., graphic design with page composition, architecture with buildings, film studies with film, theatre studies with theatre, and so on.

Focusing on particular areas has been immensely important for gaining deeper knowledge of the individual forms addressed. Nevertheless, as Jewitt et al. (2016) in another recent introduction to multimodality aptly put it:

> "These (sub)disciplines focus on the means of meaning making that fall within their 'remit'; they do not systematically investigate synergies between the modes that fall inside and outside that remit." (Jewitt et al. 2016: 2)

This situation is increasingly problematic for the growing number of students, practitioners, teachers and researchers who are being confronted with the need to say something sensible, and perhaps even useful, about these complex artefacts or performances. In fact, and as we will describe in Chapter 2 in more detail, there is now barely a discipline addressing any kind of communicative situation that is not feeling the need to extend beyond the confines of the focus originally adopted.

Linguistics, for example, now seeks to take in issues of gesture in spoken language and page composition for written language; art history makes forays into addressing film; film studies moves to consider TV and other audiovisual media; and so forth. In these and many other disciplines, the awareness is growing *that it is not sufficient to focus on individual 'forms of expression'* within a communicative situation as if these forms were occurring alone. Particular forms of (re-)presentation are always accompanied by other forms: their 'natural habitat', so to speak, is to appear in the context of others. As a consequence, the separation of concerns that has held sway over the past 100 years or more is placed under considerable pressure.

The mechanisms of combination involved when distinct forms of expression appear together often remain obscure. And individual methods or approaches to 'meaning making' rarely offer the flexibility to go outside of their principal areas of focus without losing considerable analytic precision. We need then to move beyond a superficial recognition that there appears to be 'something' being combined and ask what it means to 'combine' forms of expression at a more fundamental level. Strong foundations are essential for this—otherwise the danger is either that the complexity of multimodal 'meaning making' will remain beyond our grasp or that only certain areas of complexity become visible at all. For this, we will need to find points of contact and overlap both in the basic functioning of combinations of expressive resources and in the methods for analysing them. In short, we need to have a better understanding of just what the 'multi' in 'multimodality' might be referring to. Just *what* is being combined when we talk of combinations of diverse communicative forms? It is in this area that a more focused and what we term 'foundational' approach is necessary.

'Multimodality' for the purposes of this book addresses this concern directly. It is a research orientation in its own right that seeks to address what happens when diverse communicative forms combine in the service of 'making meanings'—however, and wherever, this is done. The claim inherent to such a position is that, when such ensembles of communicative forms appear, it is going to be possible to find similarities and parallels in the mechanisms and processes involved. This offers a complement to individual disciplinary orientations by providing more general understandings of what is going on when other forms of communication are considered. These understandings do not seek to replace existing disciplinary orientations, but rather *to add ways of dealing with the particular challenges and questions that combining diverse forms of meaning-making raises.*

This serves well to define the primary intended audience of this book. If you are reading this, you have probably encountered artefacts or performances that you want to study in more detail and which are made up of a variety of expressive resources. You may already have some theoretical or practical background with artefacts and performances where multimodality appears to be an issue and are interested either in taking that further, or in seeing how other approaches deal with the questions and challenges raised. What we will be developing in the chapters that follow is a framework that will allow you to pursue such investigations in a more systematic and well-founded fashion than is currently possible within individual disciplines on their own—that is: we set out a framework that places 'multimodality' at its core.

1.1 First steps ... a multimodal turn?

All 'multimodal' artefacts or performances pose significant and interesting challenges for systematic investigation. It is by no means obvious just which methods, which disciplines, which frameworks can help. The range of places where issues of multimodality arise is also expanding rapidly and so it is increasingly rare that knowledge about one area will suffice. This means that our journey will turn out to be quite an interdisciplinary endeavour, although one which is held together throughout by our central orientation to the phenomenon of multimodality as such.

In most multimodal situations, it has already become a genuine question as to just which forms of expression might be selected. Combinations of diverse forms must now be considered to be the norm in many of the media with which we regularly engage. This is partly due to the changing media landscape both in our daily life as well as in professional environments, where we regularly face the need to communicate via several channels and with the help of various media. It is also, however, certainly due to the increasing number of products and artefacts resulting and emerging from this ever-growing landscape and our habits of using these as well. Particularly problematic for traditional divisions of analytic interest is then the fact that it will often simply not be possible to have the particular forms of expression fixed beforehand. The sheer diversity of the current media situation consequently demands that reliable, systematic, but also extensible techniques and theories for analysing such combinations be found.

Multimodality in daily life

As an illustration, consider Figure 1.1. Here we see a situation no doubt very familiar to everyone reading this book. These photographs were taken in a restaurant without any prior planning and so are a little rough; they show how a group of diners have formed two pairs of interactants, each for the moment pursuing their own respective concerns. There is much that is happening in these images, and so we would generally need some guidance of how to focus in on just what is to be analysed for any particular purposes we select. Broadly we see four people sitting at a table; each pair that is interacting in the picture is, however, engaged in a typical face-to-face interaction with all of the usual expressive possibilities that such interaction provides—speech, gesture, facial expression, bodily position and more.

Fig. 1.1. Everyday contemporary restaurant interactions (photographs: Oliver Kutz; used by permission)

Characterising face-to-face interactions is an area addressed by several disciplines. In some approaches complex situations of this kind are considered to be 'layered', or 'laminated' (e.g., Stukenbrock 2014), in order to pull apart the various contributions being made. We will be suggesting something similar when we turn to methods in later chapters. Moreover, in this particular case, each pair is not only engaging in face-to-face interaction, they *also* happen to be discussing information (of very different kinds) that is displayed on a variety of electronic devices: two smartphones in the interaction in the foreground, and a tablet and smartphone in the interaction in the background. In order to understand such interactions we then have to push the layering concept even further—we must not only deal with spoken face-to-face interaction, but also with the various information offerings and interaction possibilities present on the electronic devices (which are also multimodal!) *and* the ways in which the interactants communicate with each other *about* and *with* the information displayed.

We can go on and vary aspects of this situation, each variation drawing on new disciplines and expertise but actually raising the same questions again and again. Lets move, for example, the interaction from a restaurant to a study room or university cafeteria and a group of students, similarly interacting around a table with a range of technical devices, discussing their term assignment, which may involve a range of diagrams and graphs: how do we analyse how the interaction in this group works (or does not work) in order to characterise their collective activity of 'meaning making'?

One (by now) traditional discipline would look at the spoken language interaction complete with gestures and gaze activity as the participants look at each other (as we shall mention in subsequent chapters, probably at the same time identifying 'multimodality' with just these particular facets of the situation). Another quite distinct community would look at the nature of diagrams, reasoning with diagrams, their construction and degrees of abstraction. The two kinds of knowledges—both essential to understanding the unfolding situation—have had virtually no interaction up until now and it is unclear how their individual descriptive frameworks would relate to one another. At the same time, another community from pedagogy and education might well have considered how people talk about diagrams and make sense of them – while drawing neither on the full range of what is known about face-to-face interaction nor on what is known about diagrams.

We can push this still further. Perhaps the students are also discussing the treatment of some controversial issue with relation to how a collection of newspapers and magazines present it (requiring knowledge of newspaper and magazine language, layout, typography, press photography), perhaps the students are discussing an artwork (requiring knowledge of art history, graphical forms and techniques), perhaps the students are not sitting around a table after all but are interacting via video (knowledge of the consequences of mediated communication for interaction), and so on.

Each of these diverse situations draws potentially on different disciplinary backgrounds, picking out facets of a situation that actually needs to be seen as a unified, multimodal activity in its own right. We see the understanding of how these activities become, or are, unified as one of the primary challenges for appropriate notions of multimodality. In subsequent chapters we will provide introductions both to the main ways in which this concern is being addressed currently—spanning cognitive, pragmatic/action and semiotic accounts—as well as suggesting further paths for integration.

We cannot predict the combinations of 'modes' which we will be confronted with. We therefore need methods and conceptual tools that will support us wherever we need to go. **!**

Our daily contact with websites, blogs or YouTube videos as well as printed magazines, leaflets, brochures or posters challenges more traditional perceptions and interpretations of the world. We use smartphones, tablets and computers as often as, or even more often than, pen and paper and produce digital texts, photos, videos or voice messages in hitherto unprecedented quantities. Such combinations can no longer be considered as exceptions, or odd cases, whose treatment may be postponed until we have better understandings of the expressive forms individually.

Indeed, problems arising from combinations may not necessarily be reduced by compartmentalisation in any case: it is often precisely the 'co-contextualisation' of forms that provides the key to their successful and effective use. For example, imagine you are analysing the speech of two people standing in a queue for coffee while discussing football results: at some point there will be language involved with ordering the coffee that will intermix with the language about football—these are different activities and so can most usefully be analysed separately. It is likely, for instance, that any accompanying gestures will need quite different interpretations for the football activity and for the coffee-ordering activity. Similarly, in a classroom situation, the teacher may at some point be describing something that has been written on the blackboard, at another point asking some pupils to pay attention: these are quite different and so combining them under some presumed mode of 'speech' can make the analysis more difficult (and confusing) than need be.

Multimodal
turnAwareness of the role of multimodality for everyday communication practices and within the broader media landscape as a whole is by no means limited to academic discussions. Some people then even talk of a 'multimodal turn', particularly in the areas of literacy and pedagogy (Goodling 2014), with the awareness of multimodality working its way not only into how investigations are structured and pursued but also into our everyday ways of thinking and acting. Consider, for example, the results of a quick Twitter search for the hashtag #multimodality shown below (undertaken in August 2016) which invoke several different mentions of the term.

Multimodality
in the
mediaAll of these tweets, being multimodal themselves by featuring a photo or image posted together with some verbal text and hashtags, mention multimodality as an aspect of daily life, as in an exhibition of pictures showing coffee mugs, as being part of a performance situation in the specific context of participatory theatre, as well as in a tourist situation where someone is taking a snapshot of some reflection in an object in the environment. None of these tweets further mentions particular research questions or analytical patterns to be examined for their specific situation and purpose. They just describe the specific moment they are sharing as explicitly multimodal, neither going into details about the modes included nor the particular connections one has to draw between their texts and hashtags and the image they use in order to understand the message. Nevertheless, multimodality plays an important role in all the tweets, functioning as the main motivation for their posting.

Given the kind of diversity of combinations of expressive resources and their uses we have now seen, it is unsurprising that almost equally varied groupings of interests and disciplinary methods have emerged to address facets of this overall concern. This means we need to consider the attention now being given to 'multimodality' from a variety of perspectives, or at a variety of 'scales'.

Media
convergenceAt the broadest level, we have the growing orientation to multimodality as a cultural phenomenon—across the board, in all kinds of contemporary media, there is an

 **LSLP Colombia**
@lslp_colombia ⚙ Following

Coffee as synaesthesia, coffee and photography as #multimodality and #literacies in #Caféidoscopio

Tweet by @LSLP Colombia, the account for the Literacies in Second Languages Project at Universidad Pontificia Bolivariana (https://twitter.com/lslp_colombia/status/760975732738785281, August 4, 2016 ©LSLP Colombia)

 **Chris Davis**
@chrisdaviscng

#multimodality of #ParticipatoryTheater becoming the river #COV #PreWriting #edchat #dtk12chat #makerEd #improv

Tweet by @chrisdaviscng (https://twitter.com/chrisdaviscng/status/749992198121385985, July 4, 2016 ©Chris Davis).

 **James Lamb**
@james858499 ⚙ Following

Multimodal dérive in Amsterdam with @j_k_knox #multimodality

Tweet by @james858499 (https://twitter.com/james858499/status/743015556815695872, June 15, 2016 ©James Lamb).

awareness, at least passive, sometimes quite active and incorporated as a central part of production, that expressive resources that might formerly have been restricted to particular media are now more typically combined. The emergence of 'super-media' (we will introduce more technical descriptions in subsequent chapters) that are capable of simulating or copying other media—as, for example, when an iPad shows a film or a newspaper, or a website plays music—has been referred to for some time in terms of *media convergence* (cf. Bolter and Grusin 2000; Jenkins 2008; Grant and Wilkinson 2009; Hassler-Forest and Nicklas 2015). The essential idea here is that we no longer have separate media: we have instead media that are capable of doing the jobs of many.

The prevalence of media convergence as a new norm for our everyday engagement with information is one supposed consequence of the ease with which digital technologies allow very different forms of expression to be combined and distributed. This can be argued to lead to new forms of communication and 'new media' emerging that precisely rely on such combinations in order to make their points, have their emotional impacts, raise interest and so on. Any webpage freely combining music videos, scrolling text and diagrams is already an example of this. In all such cases we see the growing importance of being able to engage productively and critically with combinations of quite different forms of expression. The study of popular culture and reception practices more generally can now never ignore the 'multimodal' aspect of the artefacts and performances being investigated.

Curricula   The need for such skills is consequently recognised at all levels and is already being worked even into secondary education. Many national school curricula now include as a matter of course the requirement that combinations of image material and text, of films, of other forms of combinations be addressed—generally without saying much about how this can be done. Consider, as examples, target competence areas formulated in the following two official school curricula, one from guidelines for art and design study programmes in Britain and another addressing media education in Bremen, Germany:

> **National curriculum in England: art and design programmes of study (Department of Education (GOV.UK) 2013)**
> Pupils should be taught:
> > to use a range of techniques to record their observations in sketchbooks, journals and other media as a basis for exploring their ideas
> > to use a range of techniques and media, including painting
> > to increase their proficiency in the handling of different materials
> > to analyse and evaluate their own work, and that of others, in order to strengthen the visual impact or applications of their work
> > about the history of art, craft, design and architecture, including periods, styles and major movements from ancient times up to the present day
>
> **Competence areas in the school curriculum for media education, Bremen, Germany (Landesinstitut für Schule Bremen 2012: 9–12)**
> > Understanding short sound and film sequences

Visual communication between individual and social interests
Discuss and evaluate different information sources, such as texts, graphics, statistics
Critical analysis of graphical illustrations; recognising manipulations

These points only roughly summarise the need for competences in dealing with several media techniques, genres and text types. All areas mentioned in these curricula represent typical artefacts and contexts in which the interplay of several materials and resources is relevant, such as in art, design, architecture as well as visual communication in texts or graphics in general. For the understanding, evaluation and critical analysis of these texts, knowledge about the resources involved as well as their multimodal interplay is indispensable, even though the curricula make no mention of the notion of multimodality among their target skills and competences.

The multimodal turn is then the willingness and, indeed, perceived need to examine combinations of expressive resources explicitly and systematically. This is fast working its way through all disciplines and practices. For some areas, such as graphic design or face-to-face spoken language studies, or theatre and film, this is hardly news, although general methods of analysis are still lacking. For others, such as theories of argument and sociology, it is rather innovative and may even still be greeted with some suspicion or doubt. Particularly in the latter cases, recent introductions may make the point that other forms of expression are relevant and need to be considered, but without offering much assistance as to how that could be done. Providing such assistance is another of the central aims of this book. No matter what discipline or background you may be starting from, the issues that arise when one has to take into consideration multiple, synergistically cooperating forms of expression show many similarities.

1.2 The journey ahead

The previous section gave us our first glimpses of the diversity of areas where multimodality and questions of how multimodality works (or does not work) are relevant. These opportunities and challenges are also accompanied by genuine problems concerning how such research is to be undertaken. Sorting out and moving beyond those problems is a further principal goal of this book. And to achieve this, we will need to move beyond and deviate from some of the methods and proposals existing in the literature.

Although often discussed in the context of 'new media', for example, the occurrence of such combinations of diverse expressive forms is no means a 'new' phenomenon historically. Multimodality also needs to be seen as *always having been the norm*. It is more the compartmentalisation of historically more recent scientific method that has led to very different disciplines, practices and theories for distinct forms of expression. As pointed out above, this has been very useful, indeed neces-

Multimodality as the norm

sary, to deepen our understanding of how particular forms of expression may operate, but now we also need to see how such meanings can function collectively, combining in the service of unified activities of communication.

Multiplying
meanings

One metaphor quite commonly employed at this point in work on multimodality is that of meaning *multiplication*. Suggested by the semiotician and educationalist Jay Lemke, this notion is used in order to emphasise that multimodality is more than just putting two (or more) modes 'next' to each other in what would then be an 'additive' relationship. When things go well, *more* seems to emerge from the combination than a simple sum of the parts and so a central issue for multimodal research and practice has been just what the nature of this 'more' might be (Lemke 1998).

Such 'multiplications' of possibilities were very evident in our examples above. At any point we might have needed to address aspects of spoken language, aspects of the design of various programs running on any devices being used or aspects of the interaction between speakers and such devices, or between their texts and images, and so on, and all of these would have been *interrelated* in interesting ways, playing off against each other rather than just existing 'side-by-side'. Look again closely, for example, at the pair interacting in the background in Figure 1.1: in the left-hand picture we see a hand gesture pointing to something on the screen of a smartphone, in the right-hand picture we then see that the hand is still placed similarly to before, but is now accompanied by a take-up of mutual gaze to make a verbal point about what had just been shown. This is what is meant by meaning 'multiplication'.

'Mode?' The multiplication metaphor has been a useful frame for pushing conceptualisations of multimodal situations further, but it also has some drawbacks. As is usually the case with metaphors, not all of the properties 'carried over' from source to target domain are equally apposite. Most problematic for our purposes here is the kind of methodological access to multimodal phenomena that the metaphor suggests. In order to engage in multiplication, we need things to multiply. And this, in turn, has led many to presuppose that the first step in analysis should be to identify these 'multipliers' and 'multiplicands', i.e., the things to multiply, as 'modes'. As a consequence, if called to do so, the vast majority of existing discussions simply list examples of the 'modes' they consider and proceed to the next step. Any of written text, spoken language, gestures, facial expressions, pictures, drawings, diagrams, music, moving images, comics, dance, typography, page layout, intonation, voice quality and many more may then informally come to be called *modes* of expression, and combinations from the list automatically raise issues of a 'multimodal' nature.

This does not, however, offer much guidance for how to take the investigation further. In fact, several central challenges are left unaddressed. One such challenge is that the various forms will generally have very different properties from each other— using forms *together* then requires that ways are found of making sense of *heterogeneous* ways of making meaning. Just how, for example, does one 'multiply' text and images, or apples and oranges? Informally, we seem to gain more information, but how, precisely, does this operate?

A further challenge, even more fundamental, is the presupposition that we can even find things to multiply in the first place. Empirical media researcher Hans-Jürgen Bucher sets out the problem here particularly clearly (Bucher 2010). Bucher argues that if multimodality is seen as a kind of multiplication of meaning, then it suggests that the meanings are somehow fixed and only need to be combined. When we multiply 3 by 4 to get 12, for example, we do not concern ourselves with the '3' possibly having some effect on the '4', or vice versa. To do so would be inappropriate for the operation of multiplication.

But meanings, and multimodal meanings in particular, are not like numbers in this regard. The contribution made in some modality can well depend on just which contributions it is combined with in *other* modes. For example, think about a picture of a group of people engaging in some activity accompanied by a text caption—a critical question then is how much of our analysis of the picture has *already* been influenced by the content of the text caption (and/or *vice versa*!). Are the people engaged in happy celebratory activities or striking a pose of defiant solidarity? Is someone walking along in front of a convoy of tanks waving a flag *supporting* that convoy or *protesting* its presence? Several examples of these kinds of interdependencies are discussed in the context of press reporting by Müller et al. (2012).

Interdependence is a difficult issue because the use of different modes together can have an influence on *what each is contributing*. This means that we need to be very careful that we can separate out in our analyses suggestions of what any particular mode is contributing 'on its own' and suggestions that only arise because of the combination with other modes. Modes presented together then need to be interpreted with respect to one another and so cannot be considered independently (cf. Baldry and Thibault 2006: 18 on 'resource integration' and composite products; and Kress 2010: 157 on 'orchestrations'); in this book, we often adopt 'multimodal ensembles' from Bezemer and Kress (2016: 28–30) as a relatively neutral term. Mechanisms for characterising such situations must, however, be built in from the start. The main challenge for multimodality research is then to find ways of characterising the nature of such interdependencies and to develop methodologies for investigating them empirically.

> Interdependence

Issues of this kind show how cautious we must be when suggesting that some contribution in some particular mode has some particular meaning. First, we must factor out the fact that perhaps our interpretation of that meaning has already been influenced by the other contributions it occurs with. Thus, in short, the common assertion that the meaning in multimodality is 'more than the sum' of its parts is no doubt true, but the metaphorical move from 'sums' to 'multiplication' is not quite enough, other kinds of operations will no doubt be necessary if accounts are to be made more precise and explanatory of actual signifying practices.

This fundamental property of multimodal 'meaning making' makes the task of identifying 'semiotic modes' considerably harder. Is a distinction that we draw motivated by what that individual mode is doing or because of the context (i.e., the other modes) that we happen to see that distinction operating with? Attempts to systematise

> Identifying modes

this complex area of what semiotic modes there are and what they do paint a less than clear picture, particularly when we move from general statements of what a mode may be and proceed instead to concrete practical analysis of actual cases.

!   **Modality quotes—what might a 'mode' be?**

"Mode is used to refer to a regularised organised set of resources for meaning-making, including, image, gesture, movement, music speech and sound-effect. Modes are broadly understood to be the effect of the work of culture in shaping material into resources for representation." (Jewitt and Kress 2003: 1–2)

"the use of two or more of the five senses for the exchange of information" (Granström et al. 2002: 1).

"[Communicative mode is a] heuristic unit that can be defined in various ways. We can say that layout is a mode, which would include furniture, pictures on a wall, walls, rooms, houses, streets, and so on. But we can also say that furniture is a mode. The precise definition of mode should be useful to the analysis. A mode has no clear boundaries." (Norris 2004a: 11)

"[Mode is] a socially shaped and culturally given resource for making meaning. Image, writing, layout, gesture, speech, moving image, soundtrack are examples of modes used in representation and communication." (Kress 2010: 79)

"we can identify three main modes apart from the coded verbal language. Probably the most important, given the attention it gets in scholarly circles, is the visual mode made up of still and moving images. Another set of meanings reach us through our ears: music, diegetic and extra-diegetic sound, paralinguistic features of voice. The third is made up of the very structure of the ad, which subsumes or informs all other levels, denotes and connotes meaning, that is, lecture-type ads, montage, mini-dramas." (Pennock-Speck and del Saz-Rubio 2013: 13–14)

"image, writing, gesture, gaze, speech, posture" (Jewitt 2014b: 1)

"There is, put simply, much variation in the meanings ascribed to mode and (semiotic) resource. Gesture and gaze, image and writing seem plausible candidates, but what about colour or layout? And is photography a separate mode? What about facial expression and body posture? Are action and movement modes? You will find different answers to these questions not only between different research publications but also within." (Jewitt et al. 2016: 12)

"[i]n short, it is at this stage impossible to give either a satisfactory definition of 'mode', or compile an exhaustive list of modes." (Forceville 2006: 382)

A list of definitions offered for the notion of 'mode' that we have selected from the multimodality literature is set out in our **Modality quotes** information box. It is interesting to consider these definitions in order to ask what kinds of guidance they offer for carrying out specific analyses. In fact, they turn out rather quickly to be of extremely limited practical help when analysing the modes we saw in action in the illustrative cases discussed above.

Reoccurring mentions are made, for example, of the perceptual senses involved, but this is rather difficult to reconcile with the ideas that gesture, speech, writing and so on are modes as well. As we will explain and illustrate in detail beginning in the next chapter, modes may cross-cut sensory distinctions rather freely. As a consequence, some approaches then see 'modes' more as a methodological decision for particular situations of analysis. This certainly has positive features with respect to its flexibility, but it also leaves the analyst somewhat too free. Just what is supposed to *follow from* such a decision? That is: what does it mean to include, perhaps, 'furniture' as a mode? Issues of demarcation are also rife: is there, for example, a 'visual mode' that includes *both* static and moving images? Films and painting certainly may have some overlaps in the kind of communicative work they can perform, but just as importantly they also have radically different capabilities.

Essentially, then, with such definitions almost anything that one considers as potentially contributing to a meaning-making situation may come to be treated as a 'mode'. Introductions to multimodal analysis that suggest that you then *start* by picking out the various 'semiotic modes' involved in your object of study—such as speech, gesture, music, painting or whatever—are consequently worrisome. Just when we would require a method and a theory to guide us and help organise our analytic activities, the opposite happens; we are left very much on our own. Countering this situation is then perhaps our most important motivation for this book over all.

Knowing how to go about investigating how such ensembles of expression work then becomes a fundamental skill for dealing with information presentation and representation in the modern age. And this moves beyond bare recognition that there *are* such ensembles: questions of *method* are going to become particularly important. Many issues at this time, for example, need to be seen as demanding empirical research—there is much that we simply do not know. What we *can* do at this stage, however, is to bring together robust starting points for analysis so that subsequent research can make progress—either for some practical task or research question that you need to find solutions or methods for, or as part of an educational process in some area of study where such questions have been found relevant.

This opens up many possibilities for new contributions. Consider, for example, the breadth of topics already being addressed in a host of BA and MA theses across institutions in the context of linguistics, literary theory, cognitive or media studies. Table 1.1 lists an opportunistically selected collection of such topics from just three institutions, which you can also use for inspiration or further thought, or just to get an impression of the diversity of relevant questions.

The projects vary with respect to whether they deal explicitly with the notion of multimodality, but they all address questions of interpreting all sorts of communicative artefacts as well as their various techniques and designs. We assume a similar orientation throughout this book. Finding out just what aspects of any combinations of

Table 1.1. Topics for BA and MA theses collected from recently submitted dissertations

Department of Communication and Information Sciences, Tilburg University, the Netherlands	– Creating and Interpreting Infographics – Differences in Pointing Gestures When Children Lie or Are Being Truthful – Comparative Advertising – Eye-Tracking Tinder Profiles – Aesthetics in Learning with Pictures
Faculty of Linguistics and Literary Science, Bremen University, Germany	– "We all go a little mad sometimes": A Social Semiotic Account of Mental Illness in the Movies – Villains in Animated Disney Movies – Intersemiosis between Accent and Color – An Analysis of Text-Image-Relationships and their Consequences for Focalisation in the BBC Series *Sherlock* – The representation of superhero(in)es in graphic novels
University of Helsinki, Finland	– The effects of visual mode, accent and attitude on EFL listening comprehension with authentic material (Modern Languages) – Integrated semantic processing of complex pictures and spoken sentences – Evidence from event-related potentials (Psychology)

expressive possibilities are really constitutive for what is going on and in what manner raises significant, challenging and interesting issues. And, because simple assumptions can easily turn out to be wrong, stronger theoretical and methodological frameworks are going to be crucial. This makes it rather urgent that approaches to communication achieve positions from which they are able to make useful statements and analyses of multimodal forms and styles: such analyses need to be able to reveal deficiencies and provide for deeper, more predictive theoretical understandings of precisely how multimodality operates.

Interaction not competition

The current lack of more inclusive accounts of the nature and operation of multimodal communicative activities dramatically compromises our abilities to deal with situations where multimodality is in play, despite the fact that these situations are already frequent and, in many contexts, even constitute the norm. Attempts to move beyond the confines of individual disciplines or communities on the basis of those disciplines' own approaches and techniques are commonly in danger of missing much of the complexity and sophistication of the areas they seek to incorporate.

 Multimodality in our view is *inherently and intrinsically an interdisciplinary cooperative enterprise*. Making this work is itself a major aspect of, and motivation for, the account.

But the relevant model of interaction here can only be one of cooperation, not one of replacement or colonisation. We need to see the approach in this book as *complementary* to such bodies of knowledge. At the same time, however, we will need to go further and to place many of those tools and frameworks within the context of a broader view of 'multimodality'. Making such interactions maximally productive is then one last goal for what we consider essential for appropriate multimodal research and practice.

1.3 What this chapter was about: the 'take-home message'

In the final sections of our chapters we will often summarise the preceding arguments, offering the final message that needs to be taken away from the reading. In the present chapter we have seen that there is a huge diversity of interesting cases of apparent combinations of ways of communicating and making meanings. Challenging combinations have been with us for a long time but now, with the help of new media and new production and access technologies, they appear to be proliferating into every area of communication and signification.

While opening up many opportunities for studies and explorations of just what combinations there are, how they work, how they can be effectively designed and taught and so on, we also learned that this is not straightforward and that there are various ways in which the complex interconnections can be seen. Questions arise at all levels: how to define the subject matter of multimodality, how to study it, and how to demarcate the objects of study in the first place. Research questions have tended to come first and investigations as a consequence may be driven more by what appears relevant for those questions rather than adequately reflecting properties of the multimodal configurations at issue. We suggest this situation graphically in Figure 1.2. Our analytic framework must now move beyond this state of affairs in order both to allow more principled studies and to guide investigations more effectively.

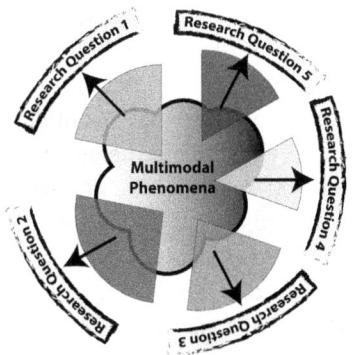

Fig. 1.2. The pre-theoretical approach to multimodality research: questions are posed and these are used to extract from the artefacts or performances under study some slices of material, using whatever theoretical apparatus the researcher may commit to. Some may look at websites, others at music videos, others at face-to-face interaction, others at films. There is little use of the regularities in the material under investigation and it is not straightforward to relate results across studies.

2 Recognising multimodality: origins and inspirations

Orientation

In this chapter, we begin the task of establishing just where and how multimodality can appear. We will also take our first look at some of the disciplinary responses to the challenges raised—independently of whether the term 'multimodality' is used or not. Several established disciplines address artefacts or performances that are inherently multimodal and so need to be considered regardless of whether they use the term themselves. There are also other relevant disciplinary orientations that address issues of multimodality while drawing on different terms, such as 'media' or 'mediality'. And, even among the disciplines that do adopt the actual term 'multimodality' as a cornerstone, it may be used in various ways. Our discussion will show that multimodality is a reoccurring challenge and so needs to be focused on in its own right.

Achieving a principled stance with respect to the broad phenomena of multimodality is challenging in its own right. Several introductions and overviews available in the literature begin by describing some selected set of approaches and disciplines where issues of a multimodal nature arise. We see this done very well in, for example, the *Routledge Handbook of Multimodal Analysis* (Jewitt 2014b), the Mouton de Gruyter *Handbook of Visual Communication* (Machin 2014), the Sage *Handbook of Visual Analysis* (van Leeuwen and Jewitt 2001) and, most recently, Routledge's *Introducing Multimodality* by Jewitt et al. (2016) as well as the de Gruyter *Handbook of Language in Multimodal Contexts* (Klug and Stöckl 2016). These books offer many useful illustrations or discussions of how multimodality may be addressed within particular approaches. There are also now several publications that adopt specific orientations and show these being used in detail; examples here include the systemic-functional approach adopted in Baldry and Thibault (2006), the discourse analytic approach of Gee (2015), the 'mediated action' approach to face-to-face interaction in Norris (2004a), the 'geosemiotics' approach of looking at language in real-world environments set out by Scollon and Scollon (2003), the 'interactional' approach descended from conversation analysis (cf. Deppermann 2013; Mondada 2014) and the sociosemiotic approach in Machin (2007) or van Leeuwen (2005a), again among several others—we will see several of these in more detail below and in the chapters to follow.

On the one hand, this breadth of literature is a good sign that we are dealing with a lively research area with significant ongoing activity and diverse opportunities for study. On the other hand, this diversity may also leave the beginner rather at a loss to just what methods may be appropriate for what tasks, and even just what approaches there are. All of the above publications have *selected* particular approaches to include, and so how do we know that there are not other approaches or methods that might be even more appropriate for some particular problem at hand?

Starting from particular disciplines and approaches does not yet, therefore, get to the heart of multimodality more foundationally—indeed, there is a positive danger that, as Deppermann notes: "'multimodality' is a label which is already worn out and has become most fuzzy by its use in various strands of semiotics, discourse and media analysis" (Deppermann 2013: 2). Moving beyond this situation is now our principal aim. And, for this, we need to develop a sense of what these various disciplines and approaches are *responding to*, i.e., what it is that gives rise to issues of multimodality at all? To move us in this direction, our approach here will have to be rather different. We will begin by laying foundation stones for *any* approach to multimodality that we subsequently try to apply or build, regardless of existing disciplinary orientations.

This has some interesting consequences. We cannot start, for example, as Jewitt et al. (2016: Chapter 1) put it, by first 'navigating the field' because we have doubts that there *is* yet a 'field' to characterise. It is precisely one of the aims of this book to help such a field emerge, driven by new generations of students, researchers and practitioners. For the present, we must start at a more foundational level and attempt not to navigate a 'field' but more to navigate the *problem space* of multimodal issues. That is, we must focus more on achieving clarity about just what constitutes problems and challenges of multimodality: how can such problems and challenges arise and what do they involve?

A 'field' of multimodality?

2.1 The 'problem space' of multimodality as such

The reality of much current research in multimodality is, as Jewitt et al. (2016) accurately describe it, that researchers largely select methods on the basis of personal preferences and circumstances of training. A 'socio-semiotician' will adopt a sociosemiotic approach; a psychologist will adopt a psychological approach; an art historian will adopt an art historical approach; and so on. While of course understandable, remaining within established traditions for an area as new as the systematic study of multimodality can also be problematic.

In fact, "choosing an approach" as the first step in a study (cf. Jewitt et al. 2016: 130) may close our eyes (ears, hands, mouth, …) to other perspectives necessary for getting maximal traction on our selected objects of analysis. It is a little like deciding to use a hammer before you have selected whether you are going to join two pieces of wood with a screw or a nail. Indeed, pushing this metaphor further, one needs to know why one is joining these pieces of wood together—and why one is using wood at all—before taking the step of a selection of tools.

What we need for a 'field' of multimodality is to emerge is therefore rather different. We need to be able to characterise just what problems are being posed by multimodality *before* we can go about sensibly selecting particular approaches or methods to take the investigation further. This is not to suggest, of course, that we can somehow reach some objective all-encompassing view that lies above and beyond the constructions of individual disciplines. When we reach the main foundational chapters at the end of this part of the book (Chapters 3 and 4 in particular), we will make some very specific proposals for a theoretical construction of the phenomena of multimodality. These proposals will attempt to build a *multimodal approach* that complements any other disciplines we may choose to link with. The motivations for the distinctions drawn will be anchored, as far as possible, in a characterisation of multimodality and what we currently know about how humans engage in signifying practices of any kind that involve diverse (i.e., multimodal) forms of expression. That is, we will construct a foundation built on the 'problem space' of multimodality as such.

We see this as providing a far more secure basis for subsequent theory and practice than, perhaps prematurely, settling on some specific methods or disciplines. In other words, and quite concretely, working foundationally is intended to help you pursue multimodal analysis *regardless* of disciplinary orientation and what the particular problem being addressed is. No existing discipline has a monopoly on methods relevant for all problems of multimodality—and so, in Chapters 5 and 6, we will also introduce the principal methods that may be employed to support empirical approaches to multimodal questions drawing from a number of disciplines. We will also emphasise there that mixing methods in judicious ways may be necessary for effective research, combining them anew in order to better answer the questions being posed.

⚡ A warning is in order here. You may not always receive words of support or praise when attempting to solve a multimodal problem in the cross-disciplinary fashion we propose! You will need to judge just how supportive your environment is for 'foreign' or 'external' methods before setting this out as your method. Not everyone finds the idea of perhaps using art historical methods for psychology, or linguistic methods for philosophy, etc. acceptable. Disciplinary boundaries are very real in the sense that borders often are. You may even have to 'go under cover' until you have results to show! Note that we do not mean by this to dispute the value and relevance of disciplinary orientations. Your results will always need to stand according to the standards and criteria of the discipline you are working in or with. It is more the methods that help you get to those results that may need creative (but always critical) extension.

Materiality and the senses
　　　We need here at the outset, then, to increase our sensitivity to, and awareness of, the 'problem space' of multimodality as such. And we begin this task by classifying some of what has to be there for *any* problem of multimodality to arise at all. At a foundational level, this is actually relatively simple: we need distinct materials—and materials not only in a physical sense, but also different materialities as they are taken up and organised by our senses in differing ways. Without some kind of 'externalised' trace, we have little need to discuss multimodality. This then steps back a bit from the examples we saw in the previous chapter. We pick up the issues of 'sensory channels' mentioned there but without yet saying anything about their relation to 'modes'. In fact, we will only reach this step in Chapter 4 below. For now, we need to clarify the properties, roles and functions of materiality and of the senses in providing the basis for multimodality as such.

This is in many respects in line with most approaches to multimodality that you will encounter today. Typically, strong statements are made about the central role of materiality and how this used to be overlooked in earlier accounts. Now, however, we need to undertake this *before* placing ourselves within any one approach to multimodality or another. Our concern is not with how any particular approach has extended itself to take in materiality, but with the possibilities for meaning-making that materiality, and moreover different *kinds* of materiality, offer at all. This means, in essence, that we are still at the stage of deciding whether to use pieces of wood or not: questions of nails, screws or hammers are still some way off!

We begin then with the senses, putting a certain focus on sound and vision since these are far more developed in humans than the other senses—although it is also necessary to emphasise that all senses would need to receive equal attention for any complete theory of multimodality. We use this as a launching point for saying something about the disciplines that have developed for addressing phenomena in each sensory realm.

We will then describe the particular use of certain 'slices' of material (→ §7.1.1 [p. 214]) in the service of *language*. This is not to state that language is distinguished as the most central or most important way of making meanings. Our concern is always with distinguishing the means that humans have for making meanings through externalised traces that have *properties sufficiently different from each other as to raise challenges of a multimodal nature*. Language certainly has a host of properties that single it out from other forms of expression, as we shall see. Language

The extent to which these properties are shared by other forms is still a matter of intense debate and research. Claims have been made, for example, that language and music share significant properties (e.g., Jackendoff and Lerdahl 2006), that language and action share significant properties (e.g., Steedman 2002), that language and static sequential visual representations in comic strips share significant properties (e.g., Cohn 2016), and so on; so here too, therefore, there is much exciting work to do. For our present purposes, we note simply that if we do not explicitly include language and the properties it provides for making meanings and signifying, then we will not be able to deal with many issues of central importance for multimodality, just as would be the case if we were to exclude 'visuality' or 'sound'.

For all uses of the senses in the service of making meaning and signifying— i.e., of responding to objects and activities that 'mean something' to their producers and consumers—there is the even more fundamental functionality of 'meaning' itself to consider. That is: how can it be that something comes to mean something for someone? There is a discipline that has attempted to address this particular foundational question in its own right: that is the field of *semiotics*. We will say something briefly about this endeavour below as well. Semiotics

Finally, working our way around the problem space, there is a further aspect that also appears just as constitutional as senses and materialities for making meaning and signifying. Semiotics, linguistics, visual studies and several branches of philosophy all point to the essential role of *society and culture* for human meaning-making. Without an assumed backdrop of society and culture, little of the necessary proper- Society ties for making meanings of any kind are available. We need to include this within our notion of the problem space as well, therefore. And, again, there are a range of disciplines that focus on this area as their main point of interest and which have had strong influences on other components of the entire picture—for example, seeing language as a social phenomenon, or addressing the (degree of) social conventionality of visual representations and perception and so on.

These four starting points—materiality, language, semiotics and society—give us much of what we need to begin distinguishing problems of multimodality. In each case, we will show that issues of multimodality are *always already present*. No matter what point of disciplinary departure is adopted, issues directly relating back to multimodal challenges can be identified. These issues often give rise to inherent tensions for any discipline that is still constructing its stated objects of interest in 'monomodal' terms. Disciplinary restrictions to particular points of access to the overall multimodal problem space in fact compromise the ability of those disciplines to deal with multimodal phenomena more broadly and to engage with those phenomena productively. Moving beyond limitations of this kind is then precisely what our explicit characterisation of the problem space is intended to achieve.

2.2 Materiality and the senses: sound

As with all of the areas we address, there are a host of disciplines that take this particular area as their point of departure. We can provide some additional organisation here and subsequently by considering the *levels of descriptive abstraction* involved. That is, are we dealing with something close to materiality, to actual physical events or objects that can be identified (with some reliability) in the world, or are we dealing more with objects or acts of interpretation, based perhaps on evidence or regularities in the world but essentially going beyond the merely physical?

2.2.1 Physical acoustics

Waves   A division of this kind lets us pick off relatively easily the contribution of the physics of sound and acoustics, for example, since the phenomena involved here are well understood. Sound is the result of a *wave* travelling through a medium, which means that the material of the medium is repeatedly 'displaced' in a regular pattern. Sound thus has all the usual properties of waves, such as frequency (i.e., the number of times per specified time unit that some single identified event occurs—such as a maximally high point, minimally low point, or 'zero' point), amplitude (i.e., how strongly the wave deviates from its zero point), and wavelength (i.e., how far apart are some single identifiable events spatially). Moreover, sound is a *longitudinal*, or *compression*, wave: this means that the dimension of the direction of displacement of the material carrying the wave is the same as the direction the wave is travelling in. Another way of describing this is as periodic increases and decreases in the pressure of a medium as the wave passes. These increases and decreases influence receptors in the ear and are rendered as 'sound'.

It is an indication of the considerable power of visuals (to which we return in the next section) that waves of this kind are *not* how we typically visualise waves at all! When we think of waves, the visual impression is usually that of a sine curve or of waves in the sea. These waves are *transverse* waves because the displacement of the medium is perpendicular, or transverse, to the direction of movement. They are also much easier to draw than compression waves: hence their dominance. Understanding something about such physical properties is always important for seeing why particular effects occur and how they might be controlled and perceived; exactly the same applies for vision and light waves as we will see below.

Whereas the bare physics of sound waves may be relatively straightforward (for our current purposes), the move to how sound is *perceived* immediately takes us into considerably more complex and uncertain territory. Again, we will see precisely the same situation when we turn to consider visual information. On the one hand, it is relatively unproblematic to give a 'monomodal' account of the physics of sound; but, on the other, it is substantially more difficult—probably even questionable—to attempt this when considering the interpretations and meanings that we may attribute to sound.

Sound perception gives access to physical frequency and amplitude in a fairly immediate fashion. Higher frequency is a higher tone; more amplitude is louder—precisely how this relates to perceptual properties is a discipline on its own, which we will not address further here. In addition to this information, however, natural sound perception also comes with indications of direction—that is we generally perceive stereophonic sound and relate this to direction rather than just 'sound' as such. We also generally perceive not sound in the abstract but 'things making sounds', i.e., sounds have *sources*. This is strongly related to evolution, as all perception is. When our ancestors heard sabre tooth tigers on the prowl, that is the information that was necessary, i.e., that there is a sabre tooth tiger in the vicinity making noises, rather than any immediate access to just the sounds 'themselves' that would then, somehow, be assigned to possible causes. In addition, we also receive information about the properties of the objects making sounds—i.e., whether hard objects are being struck together, or soft objects, or whether the contact is sharp or prolonged, and so on. And, finally for now, we also are given not only a sense of direction but of the properties of the space that we are in, whether it is large (producing echoes) or confined (being muffled).

From this brief characterisation it should be clear why talking of 'modes' in terms of 'sensory channels' can be quite misleading. Sound is not just sound or the hearing of tones; it also gives information about space, hardness, distance and direction. These 'meanings' of the sounds we hear are generally not negotiable—i.e., we cannot 'switch them off'. A high amplitude wave is a loud sound and that is that.

Sound
sources

When we turn to the use of sound in language, the situation is even more complex. We have all of the physical properties we have described so far plus indications of emotional state, physical state (age, health, tiredness) and much more—even without considering the particular phonetic details of speech. In general, whenever we have some physical material carrier that is used for further, more abstract purposes, all of the 'direct' meanings that the use of the medium 'commits to' come along as well, perhaps modified into particular paths of culturally-shaped interpretation, but present nonetheless. We return to the use of sound in language below, so for the purposes of the present section we will turn to some other disciplines that are crucially involved with sound, specifically the study of music and the practice of sound design.

2.2.2 Music, sound design and more

Music is arguably the purest 'time-based' form of expression (→ §7.2 [p. 227]). Ranging from the basic properties of sound waves, through the temporal duration, rhythms and sequences of its selected sounds, textures and timbres, through to larger temporal patterns and reoccurrences of motifs, themes and movements, music is particularly bound to time in a way that, for example, visual expression is not. The form of perception that accompanies music also has specific properties constitutive of the form. Whereas light of distinct frequencies is perceptually blended prior to perception—it is not possible, for example, to 'see' white light as made up of many component colours—sound perception by and large maintains the individual contributions of its contributing frequencies. It is for this reason that music can profit from combining notes, instruments and voices to form polyphonic wholes where individual contributions can still be heard and related to one another, reaching perhaps its culmination in full orchestral pieces.

Human perception also includes the capability both of remembering sequences of sounds arranged musically and of *predicting* particular outcomes of harmonic progressions. With this strong fundament in place the study of music is then concerned with the culturally conditioned manipulation of such sound combinations and sequences for aesthetic purposes. More important for our current concerns, however, is just how natural it has been throughout the existence of music and its study to make contact with *other* forms of expression.

Gesamtkunst-
werk

Combining melody and lyrics, for example, builds bridges not only between the musical form and representational contributions of the language involved but also between other common facets of the two forms of expression, including rhythm, amplitude and many other properties carried by sound as a physical medium (e.g., Andersson and Machin 2016). Moreover, many further complex forms of interconnection involving music already have long histories—resulting perhaps most significantly in the aesthetic notion of the *Gesamtkunstwerk* in relation to opera (cf. Trahndorff 1827; Millington 2001). Here instrumental music, libretto, performance, costume, lighting and staging were all seen as integral carriers of the work. In many respects, therefore, this is an archetypal illustration of multimodality in action (Hutcheon and Hutcheon 2010).

More recent connections of music with the visual are widespread, ranging from interactions between music and animation, between music and film in music videos, between music and film in quite a different form in music composed specifically for film and so on. All of these have substantial bodies of literature of their own and, moreover, raise issues for multimodality due to the reoccurring question of how distinct forms of expression may best be combined for common purposes. There are also uses of music for more 'abstract' aims, as in the case of *program music*, where particular musical elements are intended to be associated with specific narrative elements. This is also taken up in film music more directly since there the association can be *explicitly constructed* in the film itself, for example by making use of simultaneity of musical motifs and on-screen depiction.

A further dimension of multimodality arising with respect not only to music but to all uses of sound is due to the *physicality* of the medium. Generally sounds can have deep effects on emotion, mood and feeling. This is employed in the field of sound design, which attempts to influence affect by accompanying other expressive resources—most typically film—with sounds designed specifically to enhance the desired response (cf. Rowsell 2013: 31–43; Ward 2015) Sound design

There are also several further quite different types of multimodality to be observed 'around' music. Perhaps most obvious is the very existence of sheet music, i.e., notations for describing musical compositions in a visual form. Sheet music responds to very different requirements when compared, for example, with the relation of writing to spoken language, and consequently has many interesting properties of its own—not only *transcoding* (→ §5.2.3 [p. 148]) time into space but also the multi-layered voices of musical works into multiple spatial lines or tracks.

It is then no accident that the power of this form of expression has been considered for other modalities. In early work on film, the Russian film-maker Sergei Eisenstein explicitly likened film to music, drawing out notions of tonality and resonances as well as the need to synchronise the various contributions to film in a manner entirely reminiscent of multi-voiced orchestral works (Eisenstein 1943: 157–216). An example of Eisenstein's 'score' characterising the design of one of his films—*Alexander Nevsky* from 1938—is shown in Figure 2.1. Eisenstein worked closely on the music for the film with the composer Sergei Prokofiev and so the extension of the medium across all the modalities of the film shown here is logical.

Each of the varying forms of expression mobilised in the film are given a 'stave' in the overall plan, with the more familiar musical notation taking up the second main line, or track. The uppermost line shown in the figure depicts still images corresponding to the image track, the third shows forms and shapes characterising the visual composition more abstractly, and the fourth attempts by means of particular 'gestures' to show movement and changes of forms that would guide the eye across the frame—

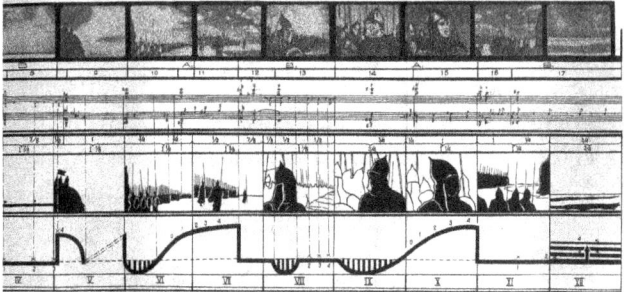

Fig. 2.1. The extended musically-derived notation used by Sergei M. Eisenstein in the late 1930s for describing film's multimodal construction (or, as Eisenstein theorised this for film: 'vertical montage')

note that the extent that this actually corresponds to how viewers look at the film can now be explored far more precisely as we describe in Section 5.4 in our chapter on methods.

This multiple track style of presentation is now the most common form for notating multimodal artefacts and performances of many kinds and has been incorporated quite directly within computational tools to cover a variety of forms of expression, including spoken language, intonation, gestures and much more; we discuss examples of this in Chapter 5 below and an example screen shot of such a tool is given in Figure 5.2.

Sound thus raises many potential challenges for multimodality and the disciplines engaging with sound have already produced some descriptive frameworks and notations of broader application now relevant for multimodality research and methods. A particularly useful characterisation of sound and its properties across a variety of expressive forms is given by van Leeuwen (1999). Reoccurring notions of rhythm, material sound qualities (and their emotional consequences), recognisable sequences and temporal synchronisation phenomena have implications of a multimodal nature far beyond the description of sound-based forms of expression considered more narrowly.

2.3 Materiality and the senses: vision and visuality

The distinction we drew above between the physical properties of sound and the perceptual qualities carried by that medium applies equally to visual phenomena. Some researchers thus make an explicit distinction between 'vision', which is anchored closely to light and the effect of light on the eye, and visuality, which goes on to consider all the meaning-making practices that can follow in its wake.

The sheer range of studies of aspects of visuality is enormous and so we will need here to be even more circumscribed than was the case with sound. A particularly useful overview of some of the more prominent approaches to visuality is offered by Gillian Rose's *Introduction to Visual Methodologies* (Rose 2012a); we will not repeat the individual characterisation of disciplines and methods here, therefore. We will also not go into detail concerning the equally diverse directions of study that centre them-selves on the 'visual': these range across journalism, art history, sociology, media stud-ies, education, visual perception, film studies, typography and document design, sci-entific visualisation and many more. You may already be anchored in one or more of these fields.

Visual approaches are also now being proposed increasingly as *methods* for carry-ing out research—for example by producing video recordings of interactions in natural situations, examining photographs from earlier times for historical research, produ-cing visual materials for 'thinking about' some topic and so on. A good sense of this breadth can be found in handbooks such as David Machin's *Visual Communication* (Machin 2014) or Margolis and Pauwels's (2011) overview of visual research methods. Also useful are more philosophically (e.g., Sachs-Hombach 2001, 2003; Schirra and Sachs-Hombach 2007) or culturally (e.g., Grau 2003) oriented accounts of 'image sci-ence' that specifically engage with the forms, meanings and uses of images.

What we need to point to immediately for current purposes, however, is a tend-ency of many researchers on visuality to emphasise the visual as if it were a *monomodal* phenomenon. Consider, for example, Manghani's (2013) opening to his introduction to *Image Studies*:

> "We are surrounded by images. In our day-to-day living we come into contact with all man-ner of images from advertising, newspapers, the Internet, television, films and computer games. We leaf through junk mail, display birthday cards, ponder graphs and charts, deliberate over what to wear and manipulate digital photographs. The public domain is framed by architec-tural design, interior décor, landscaping, sculptures, shop fronts, traffic signals, and a plethora of video screens and cameras." (Manghani 2013: xxi)

It is clear from this list (which we have in fact truncated) that artefacts and perform-ances relying on the visual are immensely important. It is also clear, however, should we care to be more precise, that *not a single example that Manghani offers is only concerned with 'images'!* The 'image'-only advertisement is something of a rarity; the 'image'-only newspaper does not exist. Films and TV combine image, sound, music, language in wild profusion. Architecture is experienced through movement, not only by looking at pictures in books. And charts and diagrams make use of a considerable range of properties that the naturalistic picture does not (and *vice versa*).

None of this would be denied by those working in visual studies and many of Manghani's examples already explicitly consider images together with other modalit-ies. Moreover, in an influential article, the visual theoretician and art historian W.J.T. Mitchell has prominently claimed that "there are no visual media" (Mitchell 2005), which is in many respects comparable to the position we will be following here. Never-theless, it remains the case in visual studies that considerations of the consequences of an intrinsic and deep-running multimodality in itself are quite rare or explicitly sidelined.

"There are no visual media!"

This reoccurring de-emphasis of how meaning is made by *combinations* of signifying practices is then what we mean by a 'monomodal view'. We propose that it is largely due to such views that Manghani is compelled to continue:

> "Yet, for all the profusion of images, or maybe precisely because of their variety and ubiquity, we remain unable to say definitely what images are and what significance they hold." (Manghani 2013: xxi)

Characterising the properties that we know to hold of the profusion of very different kinds of visually-carried artefacts and performances is certainly important in its own right—we therefore recommend engaging with introductions such as Manghani's and Rose's as preliminary reading whenever visual materials are to be explored. But it is equally important, we argue, to address just what happens when visual materials are *used* since this will always involve issues of multimodality. Satisfactory accounts will therefore need to build in multimodal phenomena and mechanisms at a far deeper level than has typically been the case until now.

2.3.1 Properties of the visual and visuality

Returning then to our main task in this chapter of stretching out the 'problem space' of multimodality, we must be concerned with just what capabilities visuality brings to the multimodal mix, rather than arguing, unfortunately as often done, for or against its relative centrality. Just as with the other areas we address, visuality and visual artefacts and performances have significant contributions of their own.

Visual associations. Visual depictions, representations, images and the like are generally seen to operate in terms of *associations* to a far greater extent than representations such as those of language. Our visual processing system is able to recognise similarities in images with extreme speed and flexibility. Shown a collection of images, people are generally able to say with high accuracy (within certain limits that have been probed by more precise psychological experimentation) whether they have seen the images before.

This capacity of the visual system is used to considerable effect in several semiotic modes and media. Film, for example, relies on our ready recognition by association of similar images in order to construct complex constructions such as shot-reverse shot sequences (e.g., where two participants in a face-to-face interaction are shown in alternating shots) and visual homage (i.e., where the composition of a shot in a film deliberately resembles images in another film or medium).

Iconography
The associative functioning of visual material also plays a role in many accounts of how to analyse the visual. It is central, for example, to the French philosopher Roland Barthes' well known adaption of the Danish linguist Louis Hjelmslev's notions of *denotation* (what is 'actually' shown) and *connotation* (what it might mean by associ-

ation in some culture), as well as to the broader accounts of visual meaning pursued in *Iconography* developed by Aby Warburg, Erwin Panofsky and others in the early 20th century (Panofsky 1967 [1939]). The basic idea of iconography is that a significant area of visual meaning needs to be picked out by tracking how particular representations reoccur across the history of a culture, thereby accruing particular associations relied upon by their producers (and knowledgeable recipients) to mean more than what appears to be shown. Analyses of this kind can be extremely insightful and demonstrate just how important it is to 'know' your material; this approach has been extended to all areas of visual representation including modern media visuals (Rose 2012*a*).

Compositionality. Visual depictions are often distinguished from verbal language by virtue of the fact that they do not work according to principles of *compositionality*. Compositionality is where the meaning of a more complex whole is built up out of the meaning of its constitutive parts. One of the clearest and compelling statements of this has been made by Ernst H. Gombrich, one of the foremost art historians and visual theoreticians of the 20th century, as follows:

> "I well remember that the power and magic of image making was revealed to me ... by a simple drawing game I found in my primer. A little rhyme explained how you could first draw a circle to represent a loaf of bread (for loaves were round in my native Vienna); a curve added on top would turn the loaf into a shopping bag; two little squiggles on its handle would make it shrink into a purse; and now by adding a tail, here was a cat. What intrigued me, as I learned the trick, was the power of metamorphosis: the tail destroyed the purse and created the cat; you cannot see the one without obliterating the other." (Gombrich 1960: 7)

Although there have also been many attempts to explain visual processing and understanding in terms that *do* appeal to compositionality—suggesting various basic visual components from which larger meanings can be made—we consider it more beneficial here to accept the fact that visuality simply does not need compositionality of the linguistic kind. This will save a considerable range of rather strained and probably inaccurate descriptions when we turn to more complex objects of analysis below. Attempting to force the visual into the verbal mode almost always distorts the account.

A further related aspect of the non-compositionality of images is addressed in Gestalt approaches to perception (Koffka 1935; Köhler 1947). Gestalt psychology reverses the notion of compositionality: we recognise a particular form or motif not by picking out parts and building these into a whole, but instead by recognising a whole and using this as a bridge to determine parts. Many good examples of Gestalt principles at work can be found in visual illusions.

In Figure 2.2, for example, we show the classic vase-face composition: here it is possible either to see a face or a vase depending on one's selection of figure (what stands out) and ground (what is taken as the background for what stands out). It is not possible, however, to see *both* face and vase at the same time; our perceptual system

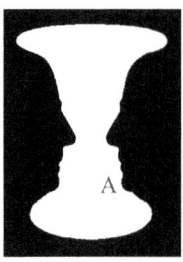

Fig. 2.2. Classic face-vase illusion developed around 1915 by the Danish psychologist Edgar Rubin

must select one Gestalt and then work with this. Now consider the black-white contour in the figure marked with an 'A', is this a part of the chin or of the beginnings of the base of a vase? Building up an interpretation of the whole in terms of the parts is particularly problematic in this case: there do not even appear to be stable parts available!

The combined issues of non-compositionality and Gestalt processing resurface in many discussions. Elkins (1999) offers a very interesting argument worthy of close attention that the study of images is marked by the desire to have both a certain compositionality in images—thereby allowing determinate 'readings' and the possibility of interpretation—*and* a lack of compositionality—thereby preserving the mystique of the visual arts and openness of interpretation. The former orientation has workers in the field suggesting various forms of structural elements (ranging from the early Wittgenstein to more mainstream 'visual semiotics' applying models derived quite directly from structural linguistics); the latter orientation rejects any such possibility. Elkins argues that one needs to engage (somehow) with both, which is actually what our own approach to the workings of semiotic modes in general in Chapter 4 undertakes.

Show and tell **Resemblance.** Probably the most well-known distinction drawn between images and verbal languages is the claim that images 'show' and language 'tells', sometimes characterised as the mimetic/diegetic distinction from classical Greek times (Schirra and Sachs-Hombach 2007). In short, an image of a dog looks like a dog, whereas the word 'dog' clearly does not. There has been substantial discussion in the literature on the nature of 'resemblance' that such positions presuppose. Some have argued that resemblance is only conventional, and so there is perhaps not such a difference on this dimension to language as often suggested.

The usual example adopted in this line of discussion is the emergence of the representation of perspective in Western art. Since there exist many cultural traditions where perspective is not used, its occurrence may be seen as a convention that is established rather than a necessary consequence of pictorial depiction as such. Counter to this argument, however, is the observation that pictorial depiction is not 'free': there are many styles of depiction which do not occur and so there appear to be limits on convention. These limits might then be found in properties of our perceptual system and so lie 'outside' of cultural influences—a 'universalising' statement that many in the fields of aesthetics and cultural studies feel duty bound to reject.

The phenomenon of resemblance is then itself philosophically far from straightforward. Stating that a picture of a dog resembles an actual dog, for example, begs the question of just how, or in what respect, that resemblance is to be found. After all, what a postcard showing a dog most resembles is other postcards, *not* any real dogs in the world with their three-dimensional properties, hair, smell, movements and textures. Resemblance is thus necessarily restricted and the main question is then just how restricted, and perhaps transformative, can that resemblance be and still qualify as 'resemblance'—a question picked up again in visual semiotics (→ §2.5.2 [p. 59]).

Various positions between the extreme poles of 'nature' and 'convention' have been pursued (cf., e.g., Gombrich 1982b: 70, 78, 100, 278–297) and this complex of issues is certainly important for those engaging with pictorial artefacts of all kinds. However, much existing discussion shows relatively under-developed understandings and models of how semiotic modes support meaning-making, including visual meaning-making. As a consequence, if this is of interest as a research topic, we recommend working through the account of meaning-making set out in this book before engaging, or engaging again, with the resemblance debate! The keywords to direct literature searches are then 'pictures', 'depictions', and 'visual conventions'.

Visual propositions. Several arguments are commonly pursued with respect to the kinds of 'speech acts' that images can and cannot perform when compared with language. Many argue, for example, that images 'cannot assert'. Whereas a statement such as 'it is raining' asserts the truth of its content, i.e., it really should be raining otherwise the statement is false, an image of rain, even a photograph, *does not make such a claim*. What such an image does is 'simply' show a state of raining, possibly with the implication, if the image is recognisable as a photograph, that it was raining whenever and wherever the photograph was taken (→ §2.5.2 [p. 59]). Similarly, many argue that images can only show what is there, they cannot show what is *not* there (Worth 1982), nor can they 'lie' (Nöth 1997)—mostly by virtue of not being able to assert.

These arguments play out in various ways: some argue that they show that images are restricted because they cannot perform the broad range of 'acts' that language can; others point out that verbal utterances also need to be embedded in some context in order to function unambiguously. It is now most often accepted that one needs to address visual *pragmatics*, i.e., the relation of images to their use in context, in order to characterise what precisely might be being done with an image more appropriately (e.g., Sachs-Hombach 2001). We can also take the discussion further below after more background information has been provided on how any semiotic mode may be operating: we will therefore return to this issue only when that background, and particularly 'discourse semantics', has been introduced in Chapter 4.

Pragmatics

Visual experience. We mentioned above that there have been many attempts over the years to apply linguistic models to visual materials. One further, particularly powerful argument against such positions is that such models fail to get at the 'experience' of engaging with the visual, an aspect that is particularly important for aesthetics and art criticism. Manghani (2013: 3) uses this angle to suggest that it is necessary to move "beyond semiotics", which, in its traditional forms, has often been argued to be a linguistically dominated approach to signs which is therefore inappropriate for visuality (Mitchell 1994: 16).

We sketch some of the basic tenets of semiotics in a more neutral fashion in Section 2.5 below—for the purposes of the present discussion, however, we do need to note that calling for moves 'beyond' or 'post'-semiotics (cf. Mitchell 1994: 16) is a little like calling Einstein's theory of relativity 'post-physics'! Relativity is, of course, not post-physics, it *is* physics, albeit physics drawing on new frameworks and assumptions. What is wrong and needs to be corrected for visuality, therefore, is various standpoints voiced within the general study of semiotics, not the appropriateness of the field as such.

In many respects, this is precisely what we undertake in this book—and we do this by taking in the entire problem space of multimodality, including those aspects specific to visuality, rather than allowing any particular area to dominate. One corollary of the above is that one needs to be very cautious when reading partial accounts of 'other' disciplines offered in the literature—the critiques offered can often be more political and strategic in orientation rather than engaging with the foundations required.

2.3.2 Expanding the visual

Adopting a broader perspective will also be a useful methodological step to take for the many cases where visual artefacts and performances are already being brought together with other kinds of information and signifying practices. This is the case already in classical painting and related arts, where the images depicted were often related to stories, narratives, biblical events, mythology and so on. Even broader cultural connections are explored within *iconography* as we mentioned above, now taken further for many distinct kinds of visual artefacts (Mitchell 1986, 1994; Müller 2011). Visual depictions have also been important for all forms of *scientific* discourse, providing what Manovich (2001: 167–168) has termed *image-instruments*, i.e., images for thinking with (Robin 1993). And Elkins (1999) argues that the methods and questions of art history can equally be pursued for such visualisations, again showing a broadening and interconnection of concerns.

Haptic Visuality

Moves beyond considerations of the 'static' visual arts are also taken up when addressing film and the moving image more generally, even though the addition of movement substantially changes, and enlarges, the kinds of meanings that can be made. Particularly in work on film, for example, a further property of moving visual images that has more recently come under study is their ability to engage other senses and responses. This is discussed in most detail under labels such as *haptic visuality* (Marks 2000): imagine watching a film without sound where someone licks a very rough wall or drags their fingers against a blackboard—the fact that you probably have reactions to these that are not visual at all is what haptic visuality describes. This suggests once again the need to seriously engage with cross-modal, at least in the sensory sense, of any objects of study. Visual images are, whatever they may be, not simply objects of vision and so addressing their role in terms of multimodality is clearly an important issue.

Moves beyond images as such to explicitly take in, for example, image and language as composite communicative artefacts are also increasingly central. This is pursued in studies of comics, graphic novels and picturebooks, as we set out in one of our use case chapters below (Chapter 12). As David Lewis describes his 'ecological perspective' on meaning in picturebooks:

> "words and pictures in picture books act upon each other reciprocally, each one becoming the environment within which the other lives and thrives." (Lewis 2001: 54)

which closely echoes our points on interdependence of modes in Chapter 1 (→ §1.2 [p. 17]). This is a position which therefore needs to be considered for almost all occurrences of images, not just picturebooks. Characterising such reciprocal relationships is yet another way of describing multimodality at work.

Finally, as a way of assessing the state of affairs in research stemming from a concern with visuality, it is useful to consider some of the concluding remarks of Gillian Rose in her introduction to visual methods as these also draw attention to some problems and ways forward particularly resonant with the call for multimodality we are making here.

After setting out a scaffold of questions that may be raised with respect to images for framing research on images (construed broadly), she notes that that list is actually "very eclectic" (Rose 2012*a*: 348) and turns instead to the theoretical perspectives that she usefully overviews throughout her book—including art history, 'semiology', psychoanalysis and discourse studies. She then points out in addition that, although these perspectives are certainly important, they nevertheless exhibit many 'gaps' that arise from traditions in which particular areas address particular kinds of images with particular kinds of methods. Thus:

> "Since many of the methods discussed here are related to specific arguments about how images become significant, it is not surprising that many of them produce quite specific empirical foci when they are used, as well as implying their own conceptual understanding of imagery. …[I]n some cases these foci are more a matter of what has been done so far by those researchers interested in visual matters than what the method itself might allow." (Rose 2012*a*: 348)

To counter this, Rose suggests that adopting multiple or mixed methods is probably a better approach.

She argues that all images should be subject to questions that cover their situations of production and reception as well as the properties of the images themselves. And, in order to pose those questions, a variety of methods are going to be relevant. This is precisely the position that this book is suggesting that you take—not only to images, but to *all* cases of multimodality. Moreover, we argue that it is essential that more robust foundations be achieved that align the posing of questions with theoretical understandings of how multimodality works. This then avoids the danger of methods devolving to eclectic collections.

In short, a rich array of approaches to the visual are available and these certainly need to be engaged with by anyone studying cases of multimodality that include visual aspects. However, the methods existing are still in need of more appropriate foundations that allow the very diverse range of questions that need to be posed to be more effectively integrated and interrelated. Moreover, the focus on visuality has often *backgrounded* the fact that visuals never occur alone—there is always at least a *pragmatic* context, since any particularly image will be used in some concrete situation or other, and most often there will be a *co-context* of simultaneously presented other modes as well. Thus multimodality itself, and the role that this plays in guiding the production and interpretation of visual materials, again needs to be added to the issues under study.

2.4 Language

Language, restricted for current purposes to *verbal natural languages*, such as English, German, Finnish and so on, also brings some interesting and rather specific properties to the multimodal table. As mentioned above, the extent to which some of these properties overlap with those of other 'modalities' is still an ongoing matter of research.

Language design features

There is also equally active research concerning the extent to which these properties are specific to *human* language: during the late 1950s and early 1960s suggestions were made about particular 'design features' that distinguished human language from the communication systems of other species (Hockett 1958)—most typically, those of birds, ants, bees and apes. Few of those design features now remain as *uniquely* the province of human language, although, of course, their level of species-specific development varies considerably. Viewed evolutionarily, it would be a mystery indeed if human language were completely disjoint to all other forms of communication (cf. Tomasello 1999, 2005).

Convention-ality

Several features of language nevertheless can be picked out that contribute significantly to our developing articulation of the problem space of multimodality. The first is the role played by *convention*. As noted above, whereas at least some claim can be made that there may be a non-conventional, i.e., motivated, connection between a picture of a dog and an actual dog by virtue of resemblance, for language this is clearly not the case. Moreover, the fact that different languages may use quite different forms to pick out the same or similar objects and events shows that convention is at work. Convention, according to the philosopher David Lewis, is where a number of options could have been taken with equal success but only one is regularly employed by some community (Lewis 1969). Conventions can be adopted in other forms of expression, but without convention human language as we know it would certainly not exist.

A further central feature of language is the converse of the situation already described for visuals above concerning *compositionality*. In the case of language, compositionality is constitutive: i.e., in verbal languages the meaning of complex wholes *can* generally be constructed from the meanings of the parts of those wholes and this marks out an essential mechanism that permits language to operate so flexibly. Thus, in order to understand "the girl chased the dog" we need to know the meaning of 'girl', 'chased' and 'dog' and can then compose these meanings following the instructions given by the syntax, or grammar, of the language being used. This property plays out very differently for visual images.

Composition-ality

A correlate of compositionality is *substitutivity*. Given some collection of elements making up a verbal unit, in verbal language one can generally substitute any of those elements by other elements without effecting those elements that were *not* substituted. If, for example, in the sentence above, we replace 'girl' by 'boy', we still have a sentence about chasing dogs. Compositionality thus relies on any changes in parts only having *local* effects. In contrast, elements within a picture readily have 'non-local' consequences and so, as Aaron Sloman, an early figure in the development of Artificial Intelligence, observed with respect to 'representations', pictorial depictions generally *violate* substitutivity (Sloman 1985). Consider, for example, a picture of the African veldt showing an elephant. If we now replace the elephant with a giraffe, this will *also change how much of the background can be seen*, simply because the giraffe and the elephant are different shapes and perhaps sizes. This is a further death knell for the simple notion of semiotic multiplication that we introduced in Chapter 1, re-emphasising yet again that interdependence between contributions is going to be a central issue for multimodality.

Substitutivity

Languages even make use of compositionality at different levels of abstraction. Most importantly for distinguishing language from other communicative systems is the principle of *double articulation*, or 'duality of patterning' (Martinet 1960): this refers to the fact that languages not only compose (meaningful) words to form meaningful sentences, they also compose *meaningless* sounds to form words. Thus the minimally meaningful item 'dog' is itself composed of the (abstract) sounds, or phonemes, 'd', 'o' and 'g', which are in themselves meaningless. The letters are used here as a shorthand for a proper phonemic description in terms of how particular sounds are distinguished; we will not discuss this as it belongs properly within linguistics and can be found in any introductory text. In short, however, 'd' and so on do not 'mean' anything by themselves, they are just sounds. Double articulation is often proposed as a unique property of verbal language, although there is, as always, discussion of the applicability of the concept to other forms of communication. Certainly for language it is essential as it provides the mechanisms by which any language can create an unlimited number of lexical items with only a few distinguished sounds.

Double articulation

The next feature to be considered is *functionality*. All forms of communication can be said to perform communicative functions of various kinds, but language is generally seen as going further with regard to how those functions *become inscribed* in the

Functionality

basic structures of a language system itself. That is: not only does a sentence perform some communicative functions, but grammar itself appears to be *organised internally* according to those functions. Different linguistic theories place their weight of description on different aspects of functionality, but nowadays functionality will always be present in some form or other. Below we will describe the particular approach to functionality adopted within the linguistic theory of systemic-functional linguistics because this has found very broad application in multimodality studies as well.

As was the case for visual studies, there is naturally a very broad range of disciplines concerned centrally with 'language' in some way. These include linguistics, literary studies and studies of the verbal arts, as well as more applied areas such as literacy, rhetoric, journalism and so on. Many of these include in turn orientations both to spoken language and to written language, sometimes differentiated and sometimes not. Our main concern here, as with all our descriptions in this chapter, will be restricted to where any approach engaging with language has begun to consider language in relation to other forms of expression. This turns out to be very common, even though often not yet thematised by the areas themselves. The starting points adopted within separate disciplines again contribute to the fact that largely disjoint communities arise working in parallel but independently on some very similar issues.

We now briefly run through some of the main orientations combining 'language' and moves towards multimodality, picking out challenges, consequences and opportunities that each has for multimodal research and method.

2.4.1 Dialogue and face-to-face conversation

Disciplines addressing spoken language as their prime interest are often situated within linguistics, albeit quite varied forms of linguistics, each with their own sets of methods and histories. There are consequently several main 'schools' or 'paradigms' and we will meet some of these again at later points in our discussion and particularly when we address specific use cases in Part III of the book. Concern with spoken language has in most cases led quite directly to 'multimodally' relevant issues—and it is certainly with respect to spoken language that the longest traditions of what might be termed 'multimodal research' can be found.

Transcription   This engagement with multimodality manifests itself in two principal respects. First, there is the basic problem of even describing the data that one is analysing. When analysing spoken language, for example, it was long standard practice not to work with sound or video recordings directly—indeed, the technology required for this is still some way off—but to work with 'transcriptions' of that data in forms more amenable to reflection: i.e., *textual* renditions of properties such as pauses, intonation and so on which are generally expunged from written language. We describe this entire area of 'transcription', as it is called, in more detail in Chapter 5 below on methods.

This has had the most direct influence on all areas of multimodal research because the same basic problem recurs regardless of the modes being investigated.

Second, multimodality is also involved by virtue of the inherent richness of this type of data itself. Spoken language does not, after all, consist solely of simply pronounced renderings of the written forms—stress, timing, rhythm, intonational contours are in many respects additional, but constitutive, properties of spoken language that need to be addressed and which are not covered by noting that some word is made up of some sequence of sounds. In addition, and most obviously, speakers will make various gestures with their hands, arms and entire body that evidently support or augment what is being presented verbally. There is also variation in aspects such as gaze, i.e., where interactants are looking, in the distances taken up between interactants (i.e., 'proxemics': Hall 1968) and so on. Despite recognition that such properties were important (Birdwhistell 1970), earlier research often found them difficult to include—mostly because of the technological problems of gaining access to and recording such natural data. Nowadays, these aspects are considered essential for understanding how spoken face-to-face interaction apparently operates so seamlessly (Norris 2004*b*; Harrigan 2008).

These components of spoken language are thus strongly multimodal. All approaches to spoken language are faced with the challenge of addressing data which is intrinsically made up of coordinated strands of communicative behaviour of quite different kinds. Several techniques for dealing with such data, as well as theoretical considerations of how best to describe what is going on, have therefore been developed that are both relevant and influential for multimodality in general. We provide more from this perspective in Chapter 8 below.

2.4.2 Human-computer interaction

Depending on your background, you may not have expected human-computer interaction (hereafter HCI) to appear among the approaches to multimodality. However, conversely, if you work within HCI, you may well also have been surprised at just how much *other* work there is on multimodality! This is the usual problem of compartmentalisation that we refer to repeatedly across the entire book. In fact, in HCI we find some of the most detailed and at the same time exploratory investigations of how modes may be combined in the service of communication at all. HCI also includes some of the earliest focused studies on multimodality as a communicative resource, drawing on yet earlier work on graphic design (e.g, Bertin 1983; Tufte 1983), speech and interaction. There were also connections drawn here with cognitive science and the question of how distinct modalities might combine (cf. Paivio 1986; Stenning and Oberlander 1995). It should, therefore, actually be considered quite impossible to address multimodality without also being aware of how this is being actively pursued in HCI.

Interaction We place HCI at this point in our discussion primarily because of HCI's 'interaction' component. In other respects HCI has long moved far beyond language in any narrow sense; many more current texts talk directly of *interaction design* instead (for introductions and overviews, see: Rogers et al. 2011; Benyon 2014). Current HCI research consequently explores novel tactile, or haptic, spatial, visual, and gestural forms of communication, is engaging with virtual environments, with embodied interaction, with combinations of language, gaze and gesture in real, augmented and virtual environments, with the detailed problems of 'telepresence' and much more besides. Many of the contributions in HCI therefore address concerns that overlap directly with specific issues of the theory, description and practice of multimodality.

Some of the earliest attempts to provide systematic overviews and definitions of the various communicative possibilities of modalities grew out of the need to characterise under what conditions it made sense to present information to a computer user in one form rather than another (cf. Arens and Hovy 1990). This has led to some very extensive classifications, such as Bernsen's (2002) characterisation of 48 distinct modalities organised around basic properties such as linguistic/non-linguistic, analogue/non-analogue, arbitrary/non-arbitrary, static/dynamic as well as distinct information channels such as graphics, acoustic, haptics. Distinctive about such approaches was their aim to draw also on empirical physiological and behavioural evidence, on the one hand, and to predict useful combinations of single modalities, on the other. There is very much to draw on from these frameworks and to relate with other accounts of multimodality being discussed.

Implementation Moreover, the need for implementations—that is, running systems that actually do what is intended—naturally pushes in the direction of explicitness and detailed modelling, often with respect to larger bodies of collected data of people actually interacting in the manners intended and with the devices developed (Kipp et al. 2009). Indeed, as soon as technological support for new styles of interaction appear, they are explored within HCI for their potential to enhance and extend the interactive possibilities of people and so clearly make important contributions to our broader understanding of multimodality (Wahlster 2006; Oviatt and Cohen 2015). Much of this work has led to sophisticated tools and schemes for analysing multimodal interaction that generally go beyond what is found in more humanistic research projects; we return to this in our chapter on methods in Section 5.3 below; it is, for example, in the field of HCI that we currently find the development of some of the most usable tools for detailed multimodal research (⟩ §5.3.2.1 [p. 156]).

This will become increasingly central for many areas of multimodal research in the future. A more traditional view of HCI as involving someone sitting at a computer screen clicking their way through some computer program is already obsolete. Instead, or rather in addition, we have the physical interactions of touching, swiping, pinching, dragging, etc. on smartphones, while the ability to sense physical gestures means that 'entire body' interfaces supporting commands by hand movements and body posture are moving towards the mainstream—often, as now usual, beginning in gaming environments. On the side of the information being offered there is a similar explosion—moving from relatively low resolution screens first to cinematographic quality (itself a steadily rising moving target from 2K to 4K to 8K, i.e., 7680 × 4320 pixel). When this is combined with immersive virtual reality (VR) presentations, we enter another domain of HCI raising a multitude of multimodally relevant questions and challenges as well as opportunities. To talk of HCI as a purely technical affair is thus to underestimate what is currently happening with this form of interaction considerably and several authors have now offered more appropriate conceptualisations (cf. Murray 2012; → §15.1 [p. 348]).

Another partially related area here is *computer-mediated communication*, where human-human communication is carried via some computational platform, as in social media, blogs, telepresence and so on (cf. Herring et al. 2013). Much of this work has been strongly language-based however. We will see some of these areas in the discussions of our use cases, although their operation and analysis is very much dependent on the capabilities of the interfaces on offer and so in many respects needs to build more on HCI results.

2.4.3 Literary studies

Since our most general orientation in this book is geared towards the possibility of empirical research (cf. Chapter 5), the concerns of literary studies are probably the most distantly related to the core theoretical and methodological issues we are pursuing. Most literary work does not see itself as 'empirical' in this sense. Nevertheless, literary studies have also engaged with a very broad range of artefacts and performances that are extremely challenging multimodally. In fact, art is always concerned with breaking and stretching boundaries, and this applies equally to 'verbal art'. As a consequence, there is a tremendous diversity of experimentation which has also broached, and continues to broach, themes of relevance to multimodality—both in practice and in theory. These are then very useful for the purposes of evaluating whether one's theories of multimodality are up to the task of engaging with such objects of inquiry also.

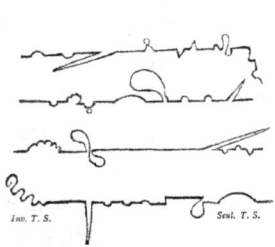

Challenging boundaries has, for example, regularly led even those working with, or in, verbal language to explore crossovers into markedly non-language-based forms. Consider, for example, the fragment taken from Laurence Sterne's *Tristram Shandy* shown in Figure 2.3. This novel, written in the latter half of the 18th century, employs a very high degree of self-reflexivity and at one point characterises potential courses that a plot (or life) might take *graphically* instead of in prose. Now, although it might be claimed that one could 'just as well' have described these twists and turns in words, it

Fig. 2.3. Plot lines depicted in *Tristram Shandy* by Laurence Sterne (1759–1767)

should be clear that this would lose a considerable amount of information. It would, moreover, be far more difficult to *compare* alternatives 'at a glance'. We see here, therefore, a direct use of the 'affordances' of the media used: i.e., language is good at some things, but other things may be better covered in other forms of expression. There will always be information that is lost when moving across modalities, and this is one of the major reasons why we need to study multimodality at all.

Lessing   Attempts to place such differences in the expressive capabilities of contrasting forms of expression on a sound footing have a long history. Within literary studies and art history more broadly, discussions of the contrast between words and image, for example, stretch back to classical times. The position that *ut pictura poesis*—'as is painting, so is poetry'—held that the 'sister arts' have similar concerns and say similar things, but do this in different ways. A poem might consequently be 'a speaking picture', a picture 'a silent poem', and so forth. The poet and philosopher Gotthold Ephraim Lessing (1729–1781), for example, argued against such a free-for-all, offering more rigorous distinctions between media that are still discussed and used today, even though he himself eventually came to the conclusion that the distinctions drawn were less than tenable. Quite independently of the use of the term 'multimodality', therefore, there has been considerable discussion of the capabilities of different forms of expression and their combinations (cf. Bateman 2014*b*).

Ekphrasis   Not all works with multimodal leanings need to include such emphatic moves beyond the language system as those employed by Sterne. A more constrained connection that has received extensive study is the rhetorical technique of *ekphrasis*. This traditional poetic term denotes attempts to evoke an intensively visual or otherwise sensual experience of some work of art or situation by employing a detailed *verbal* description. From a literary perspective it has been interesting to consider just how it is possible for a piece of verbal language to function in this way at all. Current results in brain studies showing just how tightly language comprehension is anchored into other aspects of perception and action, including embodiment, makes this phenomenon perhaps less mysterious, but it is still a hallmark of explicit attempts to move across expressive modes.

Ekphrasis is currently enjoying something of a renaissance, although primarily within 'inter-art' discussions (Azcárate and Sukla 2015). We can see this as a further general increased awareness of the potential of relating differing forms of expressions—i.e., precisely the same very general configurations and social movements that have led to the development of multimodality. It is now common to encounter discussions of ekphrasis involving other media, such as painting and music, architecture and poetry and so on. Within discussions of film, the term is being taken up particularly actively (cf., e.g., Pethő 2011), primarily because within film it is a genuine *aesthetic choice* whether to describe, or have some character describe, an event or object in words or to show that event or object. As we will see below in more detail when we turn to semiotics (→ §2.5.1 [p. 53]), the fact that there *is* a choice means that a variety of meanings may hinge on the selection. It then again becomes a multimodal issue as to just what those meaning differences might be in any particular context of use. In literary studies, such movements are most often referred to as *trans-medial* rather than multimodal—we will clarify these relations considerably below and particularly in Chapter 4.

Another very broad area of potential interaction between literary studies and multimodality has its roots in experimental literature in the 1960s that tried to move beyond 'linear narrative'. Writers such as Raymond Quenau, Georges Perec and others of the time constructed texts that required progressively more 'work' on the part of their readers to uncover the sequences of related events traditional for narrative.

This direction found a natural further development in literary *hypertext* (Landow 1994; Aarseth 1997). The first hypertext fictions were also produced around this time—the most well-known and, arguably most successful, being Michael Joyce's *afternoon, a story* (1987). As discussed by Rettberg (2015), however, the use of hypertext for serious literature fell into decline rather quickly. Hypertext

More recently, the development of new media and more powerful computational techniques at all levels has led to something of a revival of experimental works engaging with the application of computational technology for literary purposes. Such endeavours are rarely limited to text and have begun to employ the full range of capabilities that modern media and the tools available for their creation offer. This then re-engages with issues of multimodality at rather profound levels. Indeed, creating 'literature' that draws on rich media, including their capabilities for interaction and content that is produced programmatically, i.e, by computer programs rather than directly by an author or creator (cf. Gendolla and Schäfer 2007; Simanowski et al. 2010), raises the need for a more thorough understanding of how multimodality works with a new urgency. An overview of some of these interactions between literature and the extended possibilities of new media is offered by Gibbons (2011), while Koenitz et al. (2015) provide a, often still programmatic, discussion of one very active area within this general field, *interactive digital narrative*.

Computer games, 'video games', etc. form another related area which has tended to be addressed on a theoretical level primarily by literary theorists rather than linguists or others concerned with multimodality. Several researchers who formerly addressed literary hypertext now look at games and gaming (cf. Aarseth 2004; Bogost 2006). This restriction will certainly need to change because the multimodal properties of these artefacts are absolutely constitutive for their functionality. Properly multimodal analyses of computer games and their use will therefore be necessary; we address this further in our use case in Chapter 17 below.

2.4.4 Literacy studies and multiliteracies

With the much vaunted explosion of combinations of forms of expression in all forms of communication, it has also been natural that attention should turn to questions of how people learn to understand them. The issue of whether multimodal understanding needs to be explicitly *taught* at all in the educational context is an important component of such considerations. Bill Cope and Mary Kalantzis of the New London Group are often credited with bringing the effective use of diverse forms of expression into the spotlight of *literacy*, which traditionally was concerned only with verbal language (Cope and Kalantzis 2000). This movement considers that a prerequisite of being literate in today's society is the ability to have a command of a range of diverse and complex modes of expression and their technologies, or in short, to be *multi-literate*.

Awareness of this issue has now moved through most education systems and, as we saw in the previous chapter (→ §1.1 [p. 14]), many national curricula now explicitly include references to forms of expression other than language. As a consequence:

> "In education, for instance, the question of what theories are needed to deal with learning and assessment in a multimodal constituted world of meaning is becoming newly and insistently urgent." (Kress 2010: 174)

This opens up discussions of using more varied media combinations in education more broadly, moving beyond the traditional use of film extracts to illustrate literary or historical points.

Although the positive value of picturebooks for pre-school children has long been accepted, subsequent schooling has traditionally been aimed at achieving competence in purely textual styles of presentation. Now attention has shifted to considering explicitly ever more diverse forms of expression—particularly visuals, diagrams, graphs, films, comics and so on—both more analytically in their own right and in combination with other forms. However, as Serafini (2011: 349) observes, doing this in an effective fashion demands that teachers and other educators are sufficiently familiar with both the media involved and methods for characterising how these media products are operating. This is rarely the case, which marks out another reason why more foundational views of the nature of multimodality are becoming increasingly necessary.

A very real danger is that insufficient theoretical understandings of what is involved in multimodal artefacts and performances will leave the would-be educator trailing behind the competence pupils are already bringing to their engagement with diverse media. Serafini outlines several approaches he considers useful for this process, ranging across the linguistically-inflected approach of Theo van Leeuwen and Gunther Kress, central figures in the development of the sociosemiotic theory of multimodality that we will hear more of below (→ §2.4.5 [p. 48]), the art historical and iconographic approach of Panofsky that we mentioned above (→ §2.3.2 [p. 36]) and media literacy more generally (Iedema et al. 1995), for which the *Journal of Media Literacy* offers a natural outlet.

Visual literacy

Although most approaches to literacy have taken their starting points in language, as suggested by the term itself, there is now also a growing body of literature that is centred in the visual from the outset. Approaches to such *visual literary* include those of Messaris (1998), Brill et al. (2007), Jacobs (2013) and others. Elkins (2003) engages specifically with visual literacy, arguing that it should involve a far broader conception of its scope than, for example, more traditional limitations to art appreciation. For Elkins, considerations of scientific images, of images used for distinct purposes in distinct cultures, as well as the creation of images are all relevant areas of investigation, study and education—an orientation entirely in tune with the orientation we develop in this book.

One further relatively well-known position in education relating to multimodality is the 'multimedia literacy' account of Mayer (2009). Mayer proposes specific design principles that should be adhered to when constructing communication offerings made up of varying forms of expression. These have the benefit that they are relatively straightforward to understand and seem plausible; they remain, however, at a very general level: for example, pictures and text about the same thing should be positioned close to one another, etc. As we will shall see in many of our example analyses below, such generic guidelines are quickly overwhelmed when we attempt to deal with more complex combinations, leading to less clear or even conflicting criteria for evaluation. There is often just so much 'going on' in a piece of multimodal communication that multiple components, in multiple expressive forms, may well be dealing with various aspects of both the 'same' thing and other matters. More stringent and methodologically secure forms of analysis are then required for tighter conclusions to be possible.

Siegel (2006) offers a further review of the literature for 'multiliteracies'. In general, however, in work on multiliteracies the basic notions of forms of expression, or 'semiotic modalities' as we are developing the idea here, remain pre-theoretical. Traditional distinctions among language, image, colour, and so on, augmented by the properties of particular material objects, may be employed without deeper critical or theoretical reflection. Where theoretical positions are appealed to, they are most commonly the accounts developed within social semiotics (cf. Archer and Breuer 2015). Thinking about these possibilities explicitly is no doubt a step forward, but at some point questions need to be raised in more detail concerning just what is being distinguished in order to provide sharper tools for the related tasks of both analysing and teaching multimodality.

2.4.5 Linguistics and 'language as such'

We mentioned approaches to spoken language above. Here we turn to other approaches to language within linguistics that have not specifically seen themselves as addressing spoken language. While usually not explicitly ruling spoken language out of scope, they still nevertheless have formed distinct communities of practice and, most important for us here, have run into multimodality in rather different contexts to those occurring with regard to spoken language.

Even general linguistics has naturally come into contact with other communicative forms right from its beginning. Saussure, one of the founding fathers of the modern science of linguistics and who we will return to below, drew attention to the different ways in which meanings could be made and distinguished the largely conventionalised forms of verbal language from more 'similarity'-based forms whereby a sound might receive a meaning because of its resemblance to the noises made by the thing referred to—such *onomatopoeic* forms, such as 'moo', 'woof', 'tick tock' and so on, were seen by Saussure as exceptions to the general rule that language works by convention and so did not receive close attention. As we saw above, resemblance is often thought to play a role in pictorial forms of communication as well and so this entire class of communicative forms were also sidelined as far as linguistic methods were concerned.

In stark contrast to the long history of combined use of text and images and aesthetic discussion mentioned above, linguists only came to address the issues raised by such combinations in the 1960s and 1970s. This was driven in part by the move to consider increasingly natural *texts* rather than words or sentences in isolation; it then became clear that in many contexts attempting to explain how some communicative artefact operated by focusing on the language alone would be fruitless. The classic case here is that of advertisements: when linguists attempted to describe and explain the workings of language in advertisements, ignoring the visual component generally made the task impossible or pointless. Since then there has been considerable work on how verbal language in the form of text and visual aspects, such as typography, diagrams and pictorial material, can all function together in the service of a successful advertisement. An overview of this kind of research can be found in Bateman (2014*b*); many early text linguists offered proposals for how the combination of text and image might best be captured.

Reading Images

By the beginning of the 1990s, acceptance of the need to deal with aspects of visual information alongside verbal language had taken hold and several groundbreaking publications appeared in which notions developed for linguistic description were extended and applied to other forms of communication, particularly visual forms. Among these Gunther Kress and Theo van Leeuwen's *Reading Images: The Grammar of Visual Design* has been the most influential by far. Published first in Australia in 1990 and then with far broader circulation as Kress and van Leeuwen (2006 [1996]), the approach proposed still structures the work of many multimodal researchers, particularly those beginning in a more linguistic context, to this day. Around the same time, an in many respects similar application of linguistic descriptive tools was proposed by Michael (O'Toole 1989, 2011 [1994]), focusing more on art works and including architecture and sculpture in addition to the visual forms of painting.

SFL

Although there had been other attempts to apply linguistic methods to more diverse communicative forms previously, the extremely broad reach of Kress and van

Leeuwen's approach was made possible by the linguistic methods that they drew upon—this involved the treatment of language as *a resource for making meanings* rather than as a formal system of rules. This orientation was provided by systemic-functional linguistics (SFL), an approach under development since the early 1960s and originating in the work of the linguist Michael A.K. Halliday; Bateman (2017*b*) sets out a broad introduction to the principles of systemic-functional linguistics seen against the general backdrop of linguistics today.

Within systemic-functional linguistics, not only is language always seen as func- **Metafunctions** tioning within a *social context* but it is also considered to have been fundamentally 'shaped' by this task. The particular view on functionality (→ §2.4 [p. 39]) taken within systemic-functional linguistics shows this principle in action. The theory posits three *generalised* communicative functions that always need to be addressed. These functions are termed *metafunctions* in order to capture their generalisation away from specific communicative functions, such as persuasion or describing, etc.—that is, they are functions that are to be considered 'meta' to any particular instance of communication.

The three metafunctions are the *ideational*, which picks out the representational, or world-describing, role of language, the *interpersonal*, which indicates the role of language in enacting social relationships, evaluations and interactions between participants in some language situation, and the *textual*, which characterises the way in which languages provide mechanisms for combining individual contributions, such as sentences, into larger-scale texts that are coherent both internally and with respect to the contexts in which they are used. Systemic-functional linguistics argues that these functions are so central that they have also shaped the internal form of the linguistic system; thus, different areas of grammar with different properties can be found that are 'responsible' for the various metafunctions (Martin 1992).

The tight relationship between language and social context can then also be used in the 'other' direction. For example, since the metafunctions are actually taken as characterising social situations and the *social* work that communication is achieving rather than language *per se*, the framework they offer for analysis may apply equally regardless of the communicative form in question. Gunther Kress, as a sociosemiotician, has been particularly influential in arguing this position, suggesting further that metafunctional organisation may be considered definitional for 'semiotic mode'. A semiotic mode is then a use of some material for the achievement of the three metafunctions; Jewitt et al. (2016) provide a good introduction to several accounts of multimodality that take this position and we will see further illustrations in some of our use case discussions in Part III of the book.

We will not, however, follow this position in the approach and methods set out in this book—not because the idea is inherently wrong or false, but because we still consider it a broadly empirical question as to what extent these categories can be applied to forms of expressions other than language. Note that, because of their generality, it is always possible to *describe* any particular communicative situation in terms scaffolded by the metafunctions—e.g., what does this situation represent, what relation does it assume or construct between those communicating, and what is the situation's organisation 'as a message'? This may, moreover, be a useful methodological stance to employ. But this does not yet mean that the forms of communication *themselves*, i.e., what we will later come to define as semiotic modes proper, are internally organised along these functional lines, which is the central claim of the original idea of metafunctional organisation (cf. Halliday 1978). We thus remain agnostic on this point until more evidence is in and introduce instead the tools necessary for addressing this question empirically.

Visual grammar

The general view of a bidirectional relation between language and social organisation pursued within systemic-functional linguistics has made it logical to explore possible applications of distinctions found in grammar to other expressive forms. This is the sense in which Kress and van Leeuwen mean a 'grammar of visual design' in the title of their influential work mentioned above. 'Grammar' is, again, seen as a characterisation of ways of achieving communicative functions and *not* as a set of rules determining 'grammaticality' or formal 'acceptability'. It is only in this somewhat weaker sense that notions of grammar are then applied to non-verbal materials. Following this, Kress and van Leeuwen set up the following research agenda:

> "to provide inventories of the major compositional structures which have become established as conventions in the course of the history of visual semiotics, and to analyse how they are used to produce meaning by contemporary image-makers." (Kress and van Leeuwen 2006 [1996]: 1)

This approach is still being pursued for multimodal analyses by many in the systemic-functional tradition today. Kress and van Leeuwen themselves together with colleagues have, for example, proposed inventories in this style of transitions in film (van Leeuwen 1991), of colour (Kress and van Leeuwen 2002), of voice quality (van Leeuwen 2009), of lighting in film (van Leeuwen and Boeriis 2017) and more.

Discourse analysis

One of the primary benefits of carrying out analyses of this kind for multimodality is that a common language is established between the visual representations and the linguistic representations; these and other examples are discussed further in Bateman (2014*b*). But several open questions remain, some of which we pick up in Part III of the book. Such an approach also opens up visual depictions to the same kinds of *ideological* critiques well-known within this style of linguistic analysis (cf. Kress and Trew 1978) and pursued within *Critical Discourse Analysis* (cf. Wodak and Meyer 2015). Multimodal critical discourse analysis drawing on this connection is described, for example, in Machin and Mayr (2012). This then overlaps and complements approaches to visual interpretation that address visuals as communicative acts, asking what those visuals are achieving for the producers and what effects they may be having on their recipients.

To conclude, it is difficult these days, especially for linguists of a discourse persuasion—however this may be defined (cf. Blommaert 2005; Renkema 2006)—to avoid making moves into multimodal areas. This is increasingly done quite deliberately: as mentioned above, for example, James Paul Gee after a series of introductions to various aspects of linguistic discourse analysis now offers in Gee (2015) a common approach to language, interaction and video games; while, from a very different perspective, Spitzmüller and Warnke (2011) include aspects of text appearance and its environmental placement in their overall framework for analysing linguistic discourse. In contrast to this level of interest, however, there is still considerable work to be done to achieve theoretical underpinnings sufficient for engaging with both language and other forms of expression on an equal footing. The recent handbook on language in multimodal contexts (partially in German) from Klug and Stöckl (2016) offers a detailed and very broad overview of the state of the art from a linguistic perspective.

2.5 Systems that signify: semiotics

The field of *semiotics*, the study of signs in general, has had a bad press. In principle, a discipline that considers how meanings—of any kind whatsoever—can be made through the use and exchange of specific meaning-bearing vehicles—again of any kind whatsoever—should sound very relevant for multimodality. This is a rather fundamental set of questions and so semiotics should actually be considered whenever questions of meaning are raised. And, indeed, particularly in the 1960s, semiotics was heralded as a major new discipline capable of providing insight into all kinds of communicative activities.

Unfortunately, since then, the field has been subject to considerable criticism—some justified, some less so. Many writers begin, as we do here, with a note saying that reading semiotics might prove to be a masochistic exercise due to semiotics' love of complex terms and opaque distinctions. Grudgingly it may then be admitted that semiotics may say something useful, although the general tone can remain negative. In contrast to such positions, the brief introduction we offer in this section will do its best to be neither complex nor opaque—we will try and cut through a lot of the red tape that has arisen around the field, keeping out the unwary. And we will also make it clear that many of the distinctions uncovered are just as important and fundamental as the broad definition of the field would lead one to expect.

Several reasons can be noted for semiotics' often rather complicated way of describing what it is about. One reason for complicated terms is that what is being talked about is genuinely complex—and there are indeed many area of semiotics where this is the case. But there are also some basic distinctions that need not tax us too much and which will still be beneficial when discussing multimodality both theoretically and practically in subsequent chapters. This is what we focus on here.

One further area of suspicion sometimes voiced against semiotics in the context of multimodality—particularly by those working on visual media—is that semiotics is

Saussure

too closely aligned with linguistics for comfort. By fixating on language, the concern is that other forms of expression may be being distorted to fit. There is much to this argument because the histories of linguistics and semiotics are indeed closely intertwined. The most direct relationship is due to the work of Ferdinand de Saussure that we mentioned above. Saussure's work in the early years of the 20th century is seen as one of the most important foundations for *both* linguistics *and* semiotics (or 'semiology', as the Saussurean tradition still sometimes calls it).

General template

Saussure was a linguist and so what he wrote about signs in general must usually be seen as working out from 'language'; for Saussure, the kinds of signs one finds in language constituted the most important signifying practice; other kinds were given less attention. Saussure even went as far as to suggest that language could be seen as a kind of 'template' for describing all kinds of signification—a statement that has caused considerable dispute and which we have already explicitly rejected in this chapter.

Text

A further methodologically less beneficial development in semiotics has been the progressive extension of just what it considers a 'text' to be. As semiotics as a field addressed an ever broadening range of phenomena, the objects of those analyses tended to be labelled as 'texts', thereby steadily stretching the term. Semiotic analysis has even then been called 'text analysis' in some cases, with the consequence that anything subjected to semiotic analysis is thereby a 'text'. Problematic with this is the fact that 'text' already had some properties associated with it and these then continued to colour both the analyses performed and *critiques* of those analyses. Many of these properties reflect the model of 'text' as conceived within linguistics in the 1960s and 1970s and this is simply inappropriate for many of the more extended forms of expression that we currently want to deal with. As we have argued above, a piece of music or a painting have very different properties to linguistic texts and this must be respected in any satisfactory account. We return to this in Chapter 4 when we set out a theoretical foundation for 'text' in the context of multimodal meaning-making in detail.

Peirce

Nevertheless, this weakness has understandably formed a rallying cry for those attempting to move beyond traditional semiotics. And, although Saussure was not responsible for the extensions made to the notion of 'text', many of the subsequent problems are attributed to 'Saussurean semiotics'. Semiotics also has another founding father, however, and that is Charles Sanders Peirce (pronounced 'purse'), the originator of the philosophical direction called Pragmatism. In stark contrast to Saussure, the work of Peirce was never limited to language—many of his innovations were precisely intended to move beyond purely verbal signifying practices. Occasional accusations that you may find in the literature that Peirce was also linguistically-limited are therefore quite inaccurate.

In fact, Peirce's position was already far removed from any particular sensory modalities, and so naturally offers a more supportive frame for multimodality than Saussure's account. Peirce sets out foundations that can be used for discussing signs *of all kinds*. Particularly when combined with more recent developments in semiotics, the Peircean framework offers much to guide our understanding of multimodality and so it is useful to see some of its basic constructs, even if only very briefly; we take this up below.

We should also know some of the basics of semiotics (and their implications) for one further reason. Although one cannot go far in reading the literature in any of the areas pursued in this book without encountering 'semiotics' in one guise or another, the descriptions offered of semiotics in those areas are often misleading or even inaccurate. This means that we need to be suitably prepared in case strange things are said about semiotics, or semiotics' assertions, for the more ideological purposes of building barriers between approaches as subject matters. For multimodality, such positions are particularly unhelpful. Semiotics, just like any other area of research, is a dynamic growing enterprise and it is not always easy from outside the field to keep up with newer results. Moreover, old prejudices die hard.

Our account here therefore steers clear of many of the presumptions still circulating when semiotics is discussed so as to place us in a stronger position for conducting multimodal research. We begin by sketching some basic aspects of Saussure's position and then go on to contrast Saussure and Peirce in order to emphasise the points of difference between them that are relevant for the task of multimodality.

2.5.1 The two-part sign: Saussure

Saussure is well known to linguists because he proposed several organisational principles that subsequently came to form the basis of all linguistic investigation, even spreading into neighbouring areas in the form of *structuralism*. The essential idea here is that meaning-making systems rely on a more or less complex set of structural relationships: the concrete elements organised by that structure are *not* important, just the structures.

For language, this means, for example, that it is not the fact that we have a word 'tree' that is important but rather that we have a set of contrasting words that stand in a discriminating relationship, such as: grass, bush, tree, forest. If a language broke up the realm of growing plantlike things differently, then we would be dealing with different meanings. This situation is well known in translation because languages may well divide up areas of experience differently. Thus, following Saussure, to make any meaning is also at the same time to indicate the meanings that we are *not* making.

An analogy that Saussure gives is the game of chess: it does not really matter what form the chess pieces have or what they are made of, it is their structural placement on the chess board and the rules of the game that give them their meaning. Signs, for Saussure, are then what gets placed into these structural relationships and we can only know what they mean by virtue of what they stand in contrast to.

This then grounds a fundamental principle of semiotics: for meaning to be possible, one must have *choice*—if there is no choice, then no meanings can be made. The 'choice' involved does not have to be a conscious choice, or one that is entertained as such by some sign-user, but the 'semiotic system' being described must support contrasting alternatives that can each be weighed differently and thereby construct differences in meaning.

The example with trees and bushes above then also leads us directly to an aspect axiomatic to Saussure's approach, also mentioned above (→ §2.3.1 [p. 34]) with respect to language and visuals. English, as a natural language, makes these divisions of trees and bushes one way, while another language might do it quite differently. Such sets of contrasts are not then given by nature, i.e., they are not in the world waiting to be picked up by the language-learning child: they are, crucially, a matter of *convention*. The structural patterns making up the distinctions that a language makes are patterns made and 'agreed' (by common usage) within a language community.

Arbitrariness

Thus, for Saussure, language was essentially a social fact, even though he also emphasised that signs were mental constructs. Language is therefore an internal reflection of a social organisation. Moreover, just as the structures are conventional, the elements demarcating those structures, e.g., the words of a language, are also conventional: 'tree' in English, 'arbre' in French, 'Baum' in German and so on. This is the famous tenet of the *arbitrariness of the sign*. Figure 2.4 shows the usual graphical depiction of this notion of a sign according to Saussure.

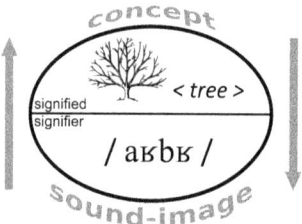

Fig. 2.4. Saussure's model of the sign. The 'tree' in angle brackets is intended as some label for a 'conceptual' correlate of 'arbre'.

Saussure put forward several methodological and theoretical constructs for working out what signs there are in a language and how these might be related. We already introduced one of the most important of these above—the notion of choice, which Saussure termed the 'associative' relation, later renamed as *paradigmatic*. Saussure employed the paradigmatic relation as one way of characterising how alternative items could be brought together—for example, we might say that 'bush', 'tree', 'forest' are all 'associated' by virtue of being something to do with plants. This is, of course, rather vague in that anything might turn out to be associated with something else: thus the more restricted and precise notion of a paradigmatic relation as *structural contrast* is now more generally preferred. Its origins can be found in Saussure's original descriptions, however.

Another basic distinction established by Saussure allowed him to 'fix' the object of attention in linguistics in a way that proved highly beneficial for many years following. What people actually say in any given situation of language use may appear to be rather unpredictable. Consequently, Saussure wanted to make sure that he could uncover the general principles of language as social fact, untainted by potential vagaries of memory, slips of the tongue, changes in what one wanted to say in mid-sentence and so on. He considered such variation to be potentially chaotic and not indicative of the social phenomenon that was his target. To deal with this situation, Saussure introduced a basic distinction that has remained fundamental to linguistic research ever since: that between *langue* and *parole*. Here the original French terms are used as the preferred technical terms in English as well.

Langue is the 'abstract' system of language, the internalised organisation that makes up what an adult speaker of the language knows as a manifestation of the social phenomenon. *Parole* is actual speech behaviour, i.e., what comes out of people's mouths when they speak in real situations of language use and are subject to all the contingencies of situation and cognitive distractions that might occur. Saussure considered the purpose of linguistics as revealing the properties and organisation of *langue*. *Parole* and its linking to actual behaviour and contexts of use was, on the one hand, considered too unruly to be part of the domain of linguistics and, on the other hand, best taken up by other fields of study in any case.

This distinction is also echoed in the well known difference drawn between *competence* and *performance* from the linguist Noam Chomsky—Chomsky, as Saussure, saw the 'proper' area of concern for linguistics to lie in the abstract competence constituting knowledge of a language, rather than how that knowledge might be used. Although this restriction is useful for some purposes, we also now know that 'parole' is equally worthy of linguistic investigation; this has been made clear by the immense advances that have been made in sociolinguistics and accounts of *variation* since the 1960s (cf. Labov 1994, 2001) as well as by detailed studies of regularities in interactional behaviour (e.g., Heritage and Atkinson 1984). Indeed, the entire mutual dependency relationship between *parole* and *langue* is now far better understood and forms a central building block for signification in general.

More immediately relevant for our present concerns, however, is Saussure's discussion of the nature of the sign and arbitrariness. One of the most argued points of distinction between discussions of images and of verbal language is that images cannot be arbitrary in this way and so any notion of semiotics based on language must be irrelevant to how images work. This fundamental point raises many problems. If images are not conventional and language is conventional, then what happens when we mix the two in text-image combinations? Does the free-floating, because conventional, text become fixed by the more anchored, because non-conventional, image? Or perhaps *vice versa*? Although there are many proposals concerning how to apply semiotic principles to visuals, these are nowhere near as developed as those for language. There are even still many who would argue that semiotics as such, with its talk of signs referring to things, is something that is language-centred and so is inherently of less value for visuals and so should be avoided.

Langue/Parole

2.5.2 Three-part signs and semiosis: Peirce

One response to this situation is to turn to the other main source of semiotics at the turn of the 20th century, Peirce. Suggestions of this kind are primarily based on the fact that Peirce's account puts much less weight on arbitrariness and so might provide a more congenial starting point —one that is less centred on language and so more likely to help us with images and other medial forms (e.g., Iversen 1986; Jappy 2013). As we shall see in a moment, arbitrariness arises with just one of the rather diverse ways of being a sign that Peirce discusses.

Perhaps even stronger support for the Peircean position, however, can be found in the fact that Peircean semiotics casts its net far more broadly than the notions of 'representation' assumed for language and accepts all manner of responses to signifying material. This is particularly important when discussing artefacts and performances such as film, music and the like, where it is difficult to talk of 'signifieds' in any clear sense. For Peirce, an emotional response to a piece of music is just as relevant for semiotics as any other response to a 'sign'.

Charles S. Peirce on 'signs':

"a sign is something by knowing which we know something more." (Letter to Lady Welby, October 12th. 1904, Peirce 1977 [1904]: 31–32)

"A sign ... is something which stands to somebody for something in some respect or capacity. It addresses somebody, that is, creates in the mind of that person an equivalent sign, or perhaps a more developed sign." (Peirce 1998 [1893–1913]: 228)

"I define a sign as anything which is so determined by something else, called its Object, and so determines an effect upon a person, which effect I call its interpretant, that the later is thereby mediately determined by the former." (Peirce 1998 [1893–1913]: 478)

Most of Peirce's work is also much less known to linguists precisely because Peirce's concern was not language, but 'knowledge' in general. Peirce wanted to know how it was that we could know more *about anything at all*, how a society could gain knowledge and so know more about the world and our experience in it. He took signs to be crucial for this: a sign, for Peirce, is a way of knowing more than what the sign by itself 'says'. 'Says' is in quotes here because Peirce did not make any restrictions to language: for Peirce anything can be a sign. As long as someone, an interpreter, is using something to know more about something else in any way whatsoever, then we are dealing with a sign.

Although we will see in a moment that we actually have to be a bit more careful, the most commonly cited difference between Saussure and Peirce is that Saussure's model of the sign is a binary model—i.e., made up of two indivisible parts as we saw above—while Peirce's model is a *three-way* model. Saussure focuses almost exclusively on language as a cognitive or mental instantiation of a 'social fact'; Peirce brings an additional component into the equation: an 'external' object that is referred to or brought to attention by the sign.

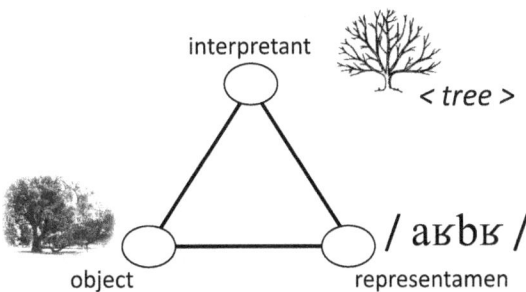

interpretant

< *tree* >

object

representamen

/ aʁbʁ /

Fig. 2.5. Peirce's model of the sign. The 'tree' in angle brackets is in this case intended as a suggestion that there is some notion to do with trees at issue; this could be anything, however, including the sound of the leaves blowing in the wind or the smell of the bark—it is *not* to be understood in only a linguistic sense.

The triadic framework and the terms Peirce chose to refer to them are summarised graphically in Figure 2.5. Although the terms are, as often with Peirce, somewhat unusual, this should not get in the way of understanding what is meant, at least loosely at this stage. This should also be helped by comparison with Saussure: the figure shows the same example of a linguistic unit, again the sequence of sounds used in French for picking out 'trees', and the relationship drawn with its 'meaning', but this time embedded within the Peircean scheme.

Thus, the sound-image from Saussure is the thing that some user of a sign actually encounters or produces: Peirce calls this the *representamen*; other names for this that have been used are 'sign-vehicle', i.e., the thing actually carrying the sign or, a little confusingly, just 'sign'. We will generally use 'sign vehicle' in this book in order to distinguish this clearly from the more inclusive notion of 'sign' if we need to. The interpretation that someone produces of this sound-image is then the *interpretant*. As before, this might be something conceptual or a mental image or some other internal construction. But it may also be any of a broad variety of further possible entities as well; Peirce in fact allows this to become very broad indeed, including feelings and dispositions to act in particular ways rather than others. Finally, the interpretant that a sign interpreter produces in response to the representamen leads the interpreter to pick out some external referent or response to the sign: this is the *object*.

The object is the new additional component which is not mentioned in Saussure's model—and, in certain respects, quite rightly so since Saussure and Peirce were concerned with very different tasks. On the one hand, Saussure is quite right in excluding the referent of a sign from the sign itself in *his* sense, since such a referent is obviously outside of language—our language does not include the external world. Peirce, on the other hand, is also quite right in including the referent for *his* aims, which were to describe, not language, but rather what a sign user actually *does* with signs. Thus, from this perspective, a model that leaves out the fact that someone who understands French might interpret the sequence of sounds given to actually pick out real trees is missing something.

This concern with what sign-users do with signs (and what signs do to their users) is a central aspect of Peirce's philosophy and one of the reasons he considered his

Pragmatism

approach to be a philosophy of *pragmatism*. So, these quite different concerns on the part of Saussure and Peirce lead naturally to different models. Suggestions that are sometimes made that they simply be combined should therefore be considered with caution.

Both Saussure and Peirce are in agreement, however, on the fact that what is being described in both cases is a *relation*. As Saussure writes:

> "I call the combination of a concept and a sound-image a *sign*, but in current usage the term generally designates only a sound-image, a word, for example (*arbre*, etc.). One tends to forget that *arbre* is called a sign only because it carried the concept 'tree', with the result that the idea of the sensory part implies the idea of the whole." (Saussure 1959 [1915]: 67)

Thus the two sides of the sign in Saussure's account are not really separable; they are like two sides of a coin or, in a further metaphor used by Saussure, like two sides of a piece of paper. This means that drawing a shape on one side—which is like choosing a particular word and its *denotation* (→ §2.3.1 [p. 32]) and then cutting along its edges necessarily effects *both* sides. Peirce emphasises a very similar point but within his three-way system: a representamen is not a type of object that exists independently, it is instead only a *role* taken on by something *when some sign-interpreter uses it* to build some interpretant. Nevertheless, comparison needs to be very careful here, since for Saussure it is the composition of signifier and signified that constitutes the sign, whereas for Peirce the sign is one pole *within* the relation itself!

Saussure's focus on language also naturally makes his account rather more fixed with respect to the kinds of entities that can be considered to be related in a sign. Peirce deliberately does *not* do this. Whereas Saussure only describes in any detail signs whose signifying part are sound-images, Peirce is quite willing to accept *anything* as a 'sign'—that is, if someone uses something as a source of further meaning, then that is sufficient. A word (or rather sound-image) can thus stand in the role of being a representamen as in Figure 2.5, but equally a stick or a painting or an entire book or art gallery can take on this role as long as a sign-interpreter chooses to treat it so. Peirce's account then tries to characterise the various ways in which this might work. The result is potentially very flexible and is another of the reasons why Peirce's account has sometimes been preferred for work on visual semiotics.

Signs and the process of sign-making can, therefore, be considered from various perspectives, and Peirce spent much of his life probing just which perspectives made sense given the internal logic of his overall account. The most well known result of these considerations is the three-way division into icons, indexes and symbols—this characterisation is found in several disciplines and is generally reported in introductions to the semiotics endeavour as a whole. Because these terms are used so frequently and pave the way for several distinctions commonly drawn in multimodality research, we will briefly characterise them here as well. This is also made necessary because the ways they are described are so often, in certain basic respects, in error. Essentially the division is concerned with explaining the ways (or, in yet another sense to that which is employed in multimodality, the 'modes') in which the *connection between signs and their objects* can be constructed—that is, just what is it about some particular sign-vehicle and its object that supports that usage as a sign.

Peirce offers three 'modes', or manners of support: (i) the sign may either 'resemble' its object, in which case we have an *iconic* relation, or (ii) the sign may rely on some 'caused' (more on this in a moment) connection of sign and object in a shared situation, in which case we have an *indexical* relation, or (iii) the sign may be conventionally associated with some object, in which case we have a *symbolic* relationship. We should emphasise at this point that Peirce also developed *other* important perspectives on ways of being signs which are necessary to understand just why Peirce drew the categories and their distinctions in the way he did; we cannot pursue this here, however—Jappy (2013) provides one of the most readable introductions to this broader picture to date, working primarily with visual materials.

Resemblance and 'iconicity' is, in principle, perhaps the most straightforward to understand, although here too there is more than immediately might 'meet the eye'. A sign-vehicle works iconically if it possesses particular properties which are sufficient by themselves to allow an interpreter to recover some other object or event. That is, a footprint in the sand has a particular shape and that shape by itself is sufficient to call to mind the interpretation 'footprint'. Pictures are the most prototypical cases of iconic signs—they are pictures of what they are pictures of because they resemble their objects *in some particular sense*. Peirce developed this basic notion of 'resemblance' far more deeply and proposed three further subclasses of icons which would take us too far afield for current purposes. Icon

More important for the current discussion is Peirce's essential idea of a sign functioning in order to 'know more'. By virtue of an iconic relationship, one can draw out more knowledge about an object on the basis of the iconic properties that hold—the sign relationship thus contributes to a 'growth in knowledge' of a particular kind. For example, considering the footprint again, one can draw conclusions such as the number of toes, the relative sizes of those toes, the distance from toe to heel, the width of the foot and so on. It is the iconic relationship that lets an interpreter transfer properties of the representamen, or sign-vehicle, to the object itself.

The indexical sign is quite different in its 'mode' of being a sign. Here there must be a relationship of causality or 'co-location' within some situation that brings the sign-vehicle and the object or event indicated together. The most cited example is that of the relation between smoke and fire. If we can see smoke, then this may mean that there is a fire. The smoke does not 'resemble' the fire—and hence we do not have a case of iconicity—but it nevertheless 'means' that there is a fire. The existence of the smoke is conditioned by the existence of the fire. The other most cited examples of indexicality make the picture a little more complex, however: 'indexicality' is also commonly related to *pointing*—in fact, in linguistics, that is just what 'indexicals' are, terms that point to something in the context. Index

A typical Peircean example is a weathervane which shows which way the wind is blowing by pointing. The direction indicated by the weathervane is conditioned by the existence of wind blowing in that direction and so the relationship is indexical. Similarly, if someone points in a particular direction, that is also indexical, because the pointing is conditioned by the direction that needs to be indicated. More complex still, if one has a map of some country, then the outline of the country on the map is also *indexical of* the outline of the country in the world—precisely because the shape of that outline on the map is conditioned by the shape of the country in the world. The most prototypical example of an indexical sign of this kind is the photograph: this is indexical (or used to be before the advent of photorealistic computer-generated graphics) because the image is actually caused by the reflection of light off objects in front of the camera.

Symbol   And, lastly, the symbolic sign is the one fixed by convention. That is, there is no necessary resemblance (the word "dog" does not look like a dog) and there is no notion of the existence of the sign-vehicle being conditioned by what is being referred to (the dog did not 'cause' the word "dog" to be used for it). It is only because a community of language users 'agreed' (implicitly by shared usage over time) to employ this phonetic form when attempting to talk about dogs that the association exists. This is then the single case—the *arbitrary* sign—treated most prominently by Saussure.

Various interpretations and usages of these basic categories can be found in the literature and this can be confusing. Some reasons for confusion can be found in the fact that Peirce's model was already at the leading edge of accounts of how signs work and there was much still to be clarified. Other reasons for confusion are more systematic: whenever focus is placed just on the icon-index-symbol triple, there is the danger of forgetting the far more extensive framework within which Peirce placed these categories. The most common confusion lies in the relationships often claimed *between* the categories. Some prominent commentators make the suggestion that somehow these properties may be considered as 'ingredients' of a sign, i.e., a sign may have some 'mixtures' of these types contributing to its make up. Unfortunately, this undermines much of the precision of Peirce's original conception; and, as Short (2007: 226–227) sets out in far more detail than we can address here, it has not helped that Peirce himself occasionally wrote of 'mixed' or 'blended' signs as well! This should, however, be seen as a shortcut that is incompatible with the broader sweep of his account. When any particular sign can be any of the three categories, far more subjectivity than is actually licensed creeps in and the analyses produced can become quite muddled.

As an example of how these categories and their interrelationships can be teased out more accurately, we consider a little more of the semiotic complexity that our footprint in the sand above in fact involves. First, the footprint can be considered an indexical sign because it is related directly to the foot that caused the print to be there; just as 'smoke means fire', seeing a footprint may mean 'someone was here'. What then lets us know that this particular imprint is a *foot*print and not, say, the tracks of some bird (which would also be indexical), is provided by the iconic relationship we mentioned above holding between the *form of the imprint* and the originating person's foot. This does *not* mean that there is one single sign that is both indexical and iconic; what it means is that there is some isolatable aspect of a situation that is working (i.e., is being isolated by an interpreter) indexically and that there are some *other* isolatable aspects of the situation that are working iconically. Because quite different 'facets' of the situation are being picked out in each case, this involves *two* signs. Indexicality and iconicity are not thereby 'mixed'. They are orchestrated together as two responses to a single 'encounter with the world' needing interpretation.

Although one reads quite often of 'iconic symbols', 'indexical icons' and the like, we consider this confused. Imagine the following. We know from chemistry that water is made up of hydrogen and oxygen; and water, fortunately, has very different properties to both hydrogen and oxygen. Nowhere in this scenario, however, does one have a 'hydrogen' version of oxygen or a 'oxygen' type of hydrogen! We are concerned with completely different 'levels'. So it is with signs. Failing to make this distinction raises considerable problems when the objects of analysis are anything other than trivial.

As a consequence, we will always take the three ways of being signs as strictly separate—it may be the case that something we are analysing makes use of more than one of these separate ways of signifying, but this means that we have several actual sign relationships occurring together, not that the ways of being signs are somehow confused. Peirce's categories then pick out *different facets* of a signifying situation. His framework as a whole then goes on to provide considerable detail about how these facets are related to one another. This is beyond our scope here but will underlie much of our account and analyses of 'semiotic modes' in subsequent chapters.

To conclude this brief characterisation we need to mention one further contribu- Abduction
tion of Peirce's account that will also be crucial for the definition of 'semiotic modes' that we give in Chapter 4 below. Peirce was very much concerned not only with 'signs' considered as entities of study but also with the process of working out interpretations for any things being considered as signs. This led to his development of a further form of logical reasoning alongside the more traditional principles of deduction and induction known since classical times. This further form of reasoning is called *abduction* (Fann 1970; Wirth 2005; Magnani 2001); abduction is currently held to be a basic mechanism operative in linguistic discourse interpretation (Lascarides and Asher 1993; Redeker 1996; Asher and Lascarides 2003) and in several other processes of sign interpretation as well (Moriarty 1996; Bateman 2007; Wildfeuer 2014).

The process of abduction can be described as 'reasoning to the best explanation'. That is, rather than starting from a premise and deducing logically what follows, one attempts to *explain* a relationship by hypothesising a further fact or situation that would make the relationship follow as a matter of course. An important feature of abduction is that it allows things to go wrong: that is, an interpreter might come up with a *wrong* interpretation. This is then different from a traditional logical deduction: if A entails B, and A holds, then B is true—end of story.

Abduction instead provides a formal account for what are sometimes called 'natural' deductions. An example would be the reasoning process behind 'birds usually fly', 'Opus is a bird' therefore Opus probably can fly. Now, if we find out that Opus is a penguin, then we have an exception to the rule and the reasoning process must 'backtrack' due to the new knowledge that we have. This cannot happen with traditional logical inferences, such as deduction, but is fine with abduction. Abduction also covers the kind of reasoning we see in 'X is flying' therefore 'X is probably a bird': Peirce saw abduction as a fundamental way of moving beyond strict logical conclusions to explanatory hypotheses. In short, one tries to find an explanation that fits the facts, but that explanation may be superseded when more is known—for example, that X turns out in fact to be Superman and was not a bird after all.

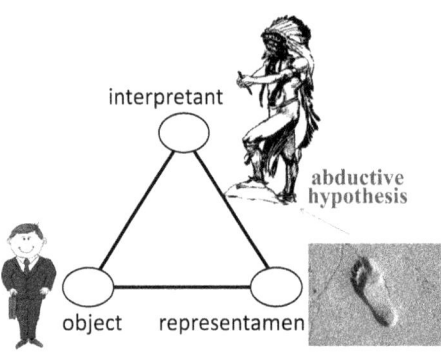

Fig. 2.6. Peirce's semiotic model and 'collateral' knowledge: an indexical sign going wrong

Figure 2.6 shows this at work in the Peircean semiotic triangle. Imagine some sign-interpreter sees our footprint in the sand; this is then interpreted. By virtue of being treated as a sign, the footprint then occupies the role of representamen and the interpreter comes up with an interpretant for it. This interpretant may, however, turn out to be wrong, due perhaps to the interpreter's over-active imagination: if the footprint was actually made by a businessman on holiday then this would be the actual object that the presence of the footprint points towards (indexically). Now, it may take some time to work out who the footprint actually belonged to, and perhaps the interpreter never manages this, but in principle, with some detective work, this would be possible. Peirce saw detective work of this kind as standing in for the process of scientific exploration and discovery in general: that is, finding out the objects actually referred to by signs may involve genuine increases in knowledge and the involvement of an entire community or society.

Discourse interpretation

In Chapter 4 we will relate the processes of abduction, sign-interpretation and semiotic modes more closely via a multimodal characterisation of *discourse interpretations*. Discourse interpretation is a form of abductive reasoning where the sign-user, or person interpreting some signs, makes hypotheses in order to 'explain' the coherence of what they are interpreting. This crucial facet of sign use has been neglected for a long time but is now beginning to reassert itself as the actual *process* of making meanings comes back into the foreground. The notion of interpretation as making discourse hypotheses can be usefully applied in linguistics in relating parts of a text to one another; but it can equally well be applied in visual studies to combining contributing parts of an image (Bateman and Wildfeuer 2014*a*) or in relating text and image (Bateman 2014*b*). In Chapter 4 we show how this then applies to *all* semiotic modes.

2.5.3 Summary: semiotics for the future

In previous sections, we already touched on some of the main areas where any semiotics that is going to aid our study of multimodality must be expanded. For example, we have drawn attention to the gradual acceptance of the meaningfulness of *materiality*. Many earlier semiotic accounts, particularly those drawing on linguistics and approaches to language, have backgrounded this important aspect of how signs work. This has also been one of the reasons that accounts of the visual, particularly of art, have tended to avoid or denigrate semiotic approaches—if the approach does not take the particularities of the medium and the actual properties of concrete instances into account, then it would appear to have little to offer. Semiotics has now moved on, however, and there is considerable attention given to the role of materiality.

A semiotics which is intended to be adequate for the description of the multimodal world will need to be conscious of forms of meaning-making which are founded as much on the physiology of humans as bodily beings, and on the meaning potentials of the materials drawn into culturally produced semiosis, as on humans as social actors. All aspects of materiality and all the modes deployed in a multimodal object/phenomenon/text contribute to meaning.

— *Kress and van Leeuwen (2001: 28)*

The omission of materiality represents a particular historical stage in the study of communicative situations rather than any inherent lack in semiotics as such. When we go back to basics and consider the account offered by Peirce, we can readily see that the model he proposes is not one that divorces signs from their material carriers at all.

Nevertheless, there is still substantial work to do to achieve a conception of semiotics maximally useful for the challenges of multimodality. We have seen that, on the one hand, because Saussure was a linguist and really only examined language, his accounts are inherently limited when we move towards other areas of multimodality. Peirce, on the other hand, made no restrictions on what could be considered a sign and so it is sometimes suggested that Peircean semiotics should be more suitable. However, the very generality of Peirce's position comes at a cost: it often remains unclear just how the framework can be applied to specific cases *systematically* and *reliably*—particularly when we turn to the more complex combinations of expressive resources, often themselves interactive, manipulable and dynamic (cf., e.g., Andersen 1997: 216–241), that we need to address today. This then demands far closer attention to the crucial notion of *method* that we return to in detail in Chapter 5.

In contrast, the writings of Saussure set the scene for the rapid rise of linguistics during the 20th century and this involved the establishment of strong methodological guidelines for carrying out research on language. The principles of analysis that resulted can now be drawn upon whenever we need to engage with complex artefacts or performances. No such guidelines have been forthcoming in the case of Peirce. This leaves a gap that needs to be filled, and several more recent semiotic accounts take on this task directly. This entire book can in fact also be seen in this light, focusing on possible methodological frameworks for pursuing multimodal research in a fashion that is also compatible with semiotic foundations—for without methods, our analyses will never advance beyond the speculative.

2.6 Society, culture and media

The starting point for the last perspective to be addressed in this chapter is 'sociological' in orientation—either indirectly via the study of social groups and cultural institutions or directly as contributions to the discipline of sociology itself. As we shall see, the path from sociology to questions relevant for achieving a strong foundation for multimodality is surprisingly short.

One of the central questions faced by theories of sociology, rising to a head particularly in the early to mid-20th century, was how it is possible for something like 'social order' to exist at all. Given that what we actually 'have' in the world is only the actions of individuals, how can it be that such a large-scale sense of social order and coherence (relatively speaking) can result? On the one hand, we appear to have these large-scale, determining social structures while, on the other, it is only individuals who actually 'exist'. Moreover, on the much smaller scale of individuals—how is it that individuals are enculturated into a complex of structures and values that is, again, apparently supra-individual? Note that any assumption that the 'social order' has the same kind of direct existence and objective reality that observable physical actions have would have significant philosophical consequences that many would nowadays be unwilling to accept—this then leaves many foundational questions in need of answers.

Naturally, then, a broad range of theoretical positions have attempted to address these concerns with greater or lesser degrees of success. Significant for us here is that most proposed answers to this dilemma draw on some account of *communication* for mediating between the micro and the macro: communication is then a fundamental 'enabler' capable of reconstruing individual action as social action (cf. Schutz 1962; Habermas 1981; Berger and Luckmann 1966; Giddens 1984; Searle 1995; Latour 2005).

> [S]ocial order is a human product. Or, more precisely, an ongoing human production. ... Social order
> exists only as a product of human activity. No other ontological status may be ascribed to it without
> hopelessly obfuscating its empirical manifestations. Both in its genesis (social order is the result of
> past human activity) and its existence in any instant of time (social order exists only and insofar as
> human activity continues to produce it) it is a human product.
> — *Berger and Luckmann (1966: 51)*

Certainly, given that the existence of language appears to be the major difference between humans and other species, the kinds of flexible but still coordinated responses to the world that language provides would appear good candidates for marking out how human societies can exist in the way they do. This is a rich and complex area in its own right of course that we cannot do justice to here. We will, however, need to extract some notions and theoretical considerations that are of immediate relevance for current considerations of multimodality.

2.6.1 Society as communication

A good point of departure is offered by the emergence of *communicative action* as a central sociological construct. Communicative actions are seen as providing much of the 'glue' that holds the complex constructions of society and culture together. And here it is particularly interesting how this concept, originally conceived more narrowly with reference to verbal language, has since been progressively extended to areas very relevant for multimodality. As, for example, communication researcher Hubert Knoblauch observes in relation to the original characterisations of communication action by the prominent sociology theorist Jürgen Habermas:

Communicative action

> "One has to acknowledge the merits of Habermas in being the first to develop a systematic theory of communicative action. Nevertheless, his concept suffers from various major shortcomings. In focusing on the role of language as the basis of communicative rationality (and, ultimately, utopian justice), he not only underestimated the role of non-linguistic communication but also neglects the bodily forms of 'expressive' communication and visual communication, such as diagrams, charts, or pictures (used even in 'rational' scientific and legal discourse)." (Knoblauch 2013: 301)

Here we see quite explicitly what should now be a rather familiar move: notions of communication are pushed beyond verbal language to become broader, more inclusive, taking in increasingly diverse expressive forms and practices. This then opens up a prospective channel of interaction between multimodality on the one hand, and sociologically-oriented work on communication and communicative situations on the other.

Research questions here focus on the 'nature' of communication in society and its structural consequences for the development and maintenance of societal configura-

Principle of communication

tions of specific kinds. This may, or may not, offer analyses of the content and form of individual interactions. Sociology more generally consequently sees communication in quite an abstract manner, orienting to sociological concerns of social groupings, power relations, social difference and so on and less so towards detailed analyses of individual instances of communication: it is the *principle* of communication that is at issue. And, indeed, certain basic properties characteristic of communication already appear to go a long way towards suggesting how social order can emerge, at least in theory.

One prominent development in this direction is set out by Couldry and Hepp (2013); here attention moves away from considering powerful institutions as more or less given and towards an examination of the processes by which power is "reproduced everywhere in a huge network of linkages, apparatuses, and habits within everyday life" (Couldry and Hepp 2013: 194). Within such a view the role of 'the media' in facilitating and structuring communication becomes central, particularly as the use of media must now be seen as interpenetrating every aspect of our everyday life, as our first example in Chapter 1 illustrated. This is seen as the primary consequence of the technological developments that have occurred in communication and new media since the 1990s and the realisation of McLuhan's often repeated *bon mot* "the medium is the message" (McLuhan 2002 [1964]). The French philosopher Michel Foucault's view of discourse as 'constitutive of society' has also been particularly influential here (Foucault 1969), although also overlapping with several prominent thinkers that have concerned themselves with how ideologies and cultural structures and practices emerge, how they are maintained, how they are contested and so on.

Media-
tization

Several authors, including particularly Krotz (2009) and Hepp (2012), consequently suggest **mediatization** as a suitable term for characterising the central role played by the media in this situation, marking it out as a distinct perspective on communication and the role of communication for society. This perspective brings to the fore the realisation that:

> "something is going on with media in our lives, and it is deep enough not to be reached simply by accumulating more and more specific studies that analyze this newspaper, describe how that program was produced, or trace how particular audiences make sense of that film on a particular occasion." (Couldry and Hepp 2013: 191)

Mediatization thus looks closely at the use and embedding of media in everyday practice and plays a constitutive role in the construction, maintenance and change of societal configurations. The methods employed in related research are generally those of media and communication studies, often with a focus on more sociological considerations and questions of evolving communication networks involving a broad and differentiated set of potential acteurs. A further detailed introduction to this approach to the social and the role of communicative media practices is offered by Couldry and Hepp (2017).

[M]ediatization is a concept used to analyze critically the interrelation between changes in media and
communications on the one hand, and changes in culture and society on the other. At this general level,
mediatization has quantitative as well as qualitative dimensions. With regard to quantitative aspects,
mediatization refers to the increasing temporal, spatial and social spread of mediated communication.
...With regard to qualitative aspects, mediatization refers to the specificity of certain media within
sociocultural change. It matters what kind of media is used for what kind of communication.
— *Couldry and Hepp (2013: 197; emphasis in original)*

This view on the role and consequences of media use will be accepted as part of a
general background for much of the framework that we present in this book. Tracking
the changes in media use is relevant for all studies of multimodal communication and
surfaces constantly in discussions of media convergence and explorations of 'transmedia' phenomena—such as, for example, the question of whether games can narrate
or in what ways narration in audiovisual media such as film is necessarily different to
that found in more traditional narrative forms such as written text. Also central are
questions concerning the differences between particular media uses: what effects, for
example, does the distribution of films on portable devices such as tablets have for
both the consumption and production of films? Or: what effect does the capability of
readily creating audiovisual materials and distributing these openly across social networks and dedicated websites have on that creation? There is clearly much here to
explore.

Discussions of the mutual relationships between society and the media, and the
mechanisms by which society is enacted through patterns of media use, then mark out
one scale of engagement with the phenomena at issue. As Couldry and Hepp set out in
their introduction to the field: "we will give less emphasis to specific media texts, representations and imaginative forms ..." in order to focus more on "the achieved sense
of a social world to which media practices, on the largest scale, contribute" (Couldry
and Hepp 2017: 3). Much research on media products from this perspective naturally
then focuses on questions of the political-economic conditions of production for any
media artefact or performance without necessarily considering the internal details of
those artefacts or performances beyond broad questions of their respective 'content'.

While this broader orientation serves very well to open up the kinds of communicative actions considered to include embodied meaningful action in general, regardless of which communicative forms or modes are mobilised, there is clearly still far to
go to articulate the methodological constraint and guidance necessary for exploring
the mechanisms of multimodal practice and signification in detail. This is still by and
large left to the individual finer disciplinary positions that engage with specific kinds
of behaviour.

Several broadly sociological approaches operate at such finer scales, attempting   Transmediation
to characterise the specifics of individual media use more tightly. This may be attempted for various reasons. Moving from general media use back to audience and recep-

tion studies, for example, requires that specific communicative options taken can be characterised sufficiently that consequences for the reception of design choices can be tracked (cf., e.g., Schnettler and Knoblauch 2007; Bucher 2007); this is then related to methods, several of which we discuss in Chapter 5 below. For our present purposes, such a focus is clearly going to be crucial; our pursuit of multimodality therefore aligns far more closely with this latter, finer-grained engagement with media and the social. It is then relatively straightforward to draw several interesting lines of connection. Sociological work of this kind includes, for example, workplace studies, where the forms of interaction in particular work-related situations takes centre stage, health communication studies, where the consequences of differing forms of interaction for the effectiveness or not of certain procedures, interventions and education is explored (e.g., Heath et al. 2000; Slade et al. 2008; Rider et al. 2014; Graaf et al. 2016), and political discourse analysis (cf. Dunmire 2012).

A further area exhibiting many overlaps particularly with studies of verbal language is media linguistics, a particular area situating itself within journalism, publication studies and media studies. Media research of this kind has a long history, particularly addressing concerns that arose about the effects of the rise of the mass media in the mid-20th century. Many studies at that time and since have examined just what kinds of programming decisions are made by different mass media outlets, including TV, radio and the press. A variety of methods were developed for characterising the range of materials distributed as well as particular slants or biases that particular media or media outlets might have been creating; we will see some of these when we discuss methods in Chapter 5 as they are relevant today for a much broader range of media offerings.

Visual communication

This is also one of the more well trodden paths by which students and researchers with a social and media orientation come to address issues of direct relevance to multimodality, since the media addressed are rarely, if ever, 'purely' verbal. Studies here therefore include some of the earliest multimodal work concerned with the classical 'text-image divide' at all. In the late 1980s and 1990s, for example, this became a particular focus of interest within German text linguistics and combinations of linguistic and journalistic media analyses continue as a thriving field to this day (cf. Muckenhaupt 1986; Doelker 1988; Harms 1990; Titzmann 1990; Renner 2001; Straßner 2002; Meier 2014). Particular attention here has also fallen on issues of *visual communication* as well, which now forms a strong area of research in its own right. This field generally adopts orienting questions in part from sociology, while also continuing with many of the disciplinary trajectories established in art history and iconography (cf. Müller 2007).

Other approaches with strong connections both to sociological orientations and to individual instances of social interaction include any that employ *ethnographic methods*, studies of 'place', *anthropology, conversation analysis* (→ §8.1 [p. 240]) and many more. As usual, there are many points of overlap between such approaches and studies of linguistic interaction such as those we introduced above, as well as with explicitly socially-oriented forms of linguistics such as critical discourse analysis (cf. Wodak and Meyer 2015; Flowerdew and Richardson 2017).

There are also theory-oriented approaches that seek to be more precise about what occurs when communication moves 'across' media (cf., e.g., Elleström 2010, 2014): i.e., what exactly is 'transferred' and how? However, whenever more fine-grained analyses of communicative options are the main focus of attention, it becomes natural to consider issues of 'multimodality' rather than, or in addition to, issues of 'media'.

Drawing on multimodal analyses with respect to media use is beneficial for several reasons. On the one hand, the accounts of modality (in the sense of multimodality) that have so far emerged within media and communication studies are, as we shall see in more detail in theory in Chapter 3 and in practice in some of our use cases in Part III of the book, relatively weak. Conversely, the accounts of media and their social embedding that have been produced in existing multimodality research are also weak. Dividing the necessary work more appropriately makes the best of both worlds as we will see below.

2.6.2 Relating the macro and the micro: an example

Although the extent to which a detailed account of the deployment of particular modalities no doubt depends on the particular research questions posed, there is little reason for rejecting *a priori* the idea that the analysis of specific examples of communication may be relevant for revealing issues of broader social import. We finish this section, therefore, with a brief illustration.

Imagine walking through a park by a lake or some other public place and coming across several apparently randomly spaced groups of people sitting on any benches or spaces available. Each group has between four and eight people in it, all with ages approximately between 17 and 25, mostly male but with a fair scattering of women as well. Prior to the summer of 2016, we might have made various guesses about what was going on—if we add in the information, however, that they all looking quite intently into their smartphones then the likely cause becomes almost a certainty. They were, of course, all engaged in stocking up their experience points and fighting in arenas in the augmented reality game Pokémon Go.

This social phenomenon, and the dramatic spread of the game following its introduction in the early summer of 2016, offers a good indication that knowing more about the actual communicative modes that are employed within some media can characterise the situations of its use far more finely than discussion of the properties of the medium 'alone'. That is, although analyses of the media platforms, their connectivity, the social networks supported and so on can tell us much about 'who' may be participating (and to what extent), it does not provide much information about *how* the communication is proceeding, about what strategies may or may not be being used, or *why* the situations are occurring at all. This information is important because it directly shapes the *experience* of use of the medium—Pokémon Go with an unusable interface, for example, would no doubt have compromised much of the promise of the game.

After all, Pokémon Go was by no means the first augmented reality game supporting play interleaved with others in the real world—both *Parallel Kingdom* (PerBlue, 2008–) and *Shadow Cities* (Grey Area, 2010–2013), for example, relied similarly on location-based smartphone capabilities to support so-called MMORPGs (massively multiplayer online role playing games), but they are by no means so well known. The take-up and use of some media product is thus crucially dependent on how its multimodal resources are being deployed *in detail*, rather than, or in addition to, the overall platform or network of actors involved. The earlier games evidently did not reach the 'multimodal sweet spot' necessary for broader acceptance; for further discussion of mixed reality 'games', see: Haahr (2015) and Rieser (2015).

Similar points related to the precise content and forms of communication can be made in all areas where one is concerned with effectiveness (or not) on finer-scales, ranging down even to individual outcomes (as in health communication, for example). Gillian Rose stresses this also with respect to the 'site of the image' in addition to sites of technological production and social use of visual materials:

> "...many writers on visual culture argue that an image may have its own effects that exceed the constraints of its production (and reception). Some would argue, for example, that it is the particular qualities of the photographic image that make us understand its technology in particular ways, rather than the reverse; or it is that is those qualities that shape the social modality in which it is embedded rather than the other way round." (Rose 2012a: 27)

We generalise this point here, and throughout this book, to the entire multimodal ensemble. The take-up of Pokémon Go is not sufficiently characterised, and even less explained, by the simple availability of certain medial configurations. *What* is communicated *how* in which modes makes a difference; and the further question is what *can* be communicated in specific modes. In the case of Pokémon Go, these 'modes' included live video recording and simultaneous playback giving framed views of the environment, animations representing a range of characters and those characters' abilities, additional 'state of game' data display and the possibility for action on the part of the player that combined movement in the world with interaction with the animations. This is not then simply a description of a combination of media, but of a very specific profile of multimodal capabilities.

2.6.3 Summary: multimodal communicative forms

Describing the modalities, their combinations, their histories of uptake in other media and the affordances of those combinations then offers much of value for explaining the take-up and spread of particular media practices at particular socio-historical points in time. What is more, the ability to tease apart the distinct communicative 'modes' that are involved becomes increasingly important the more capable the 'medium' is of carrying distinct modes. This does not apply only to devices such as smartphones—even paper can carry a very broad range of potential modes, including written language, diagrams, pictures, typography, layout and so on. Each of these modes may be employed in combinations with other media and exhibits its own history of development.

This should clarify further that what is proposed here is not that we 'replace' accounts of media by modes, or vice versa, since both contribute rather different parts of the overall puzzle. Communicative forms are still seen as the basic building blocks of society, but should now be construed with multimodal breadth. In many areas of sociological study it is now increasingly accepted that communication involves many forms of communication apart from verbal language. This is already driving theories of the social to engage with issues of multimodality more directly. As Knoblauch describes it:

> "the structures of society are constructed by, and differ with respect to, communicative forms—be it the specific linguistic code, the materiality of the action or the technicality of its implementation" (Knoblauch 2013: 306).

To pursue this more closely, we see the theories and methods of a sociology-aware multimodality as making an essential contribution.

Media need to be seen as particular 'bundles' of semiotic modes. Whereas the specific media offer an interface to considerations of society, institutions, distribution and technology, the 'modes' within any specific media describe just what they are capable of doing in terms of meaning-making. This then very intimate relationship between media and modes will be further articulated and theorised in Chapter 4. For the present, we just note that for multimodality, we must always see both the modes and their contexts of use within sociohistorically-situated medial practices.

2.7 What this chapter was about: the 'take-home message'

The central idea of 'multimodality' that it is possible to beneficially combine different forms of expression is very old—indeed, we can find such combinations going back in history almost as long as there are traces of designed artefacts at all. Notions that multimodality is a new phenomenon are therefore inaccurate. It is actually rather more appropriate to ask the question the other way around—i.e., when did people come up with the idea that different forms of expression could *not* be combined?! This tendency to separate out different forms of communication into distinct areas of concern is, historically speaking, a rather recent 'achievement' of a variety of disciplines and

Media and modes

institutional practices. As Gunther Kress and Theo van Leeuwen succinctly characterise the situation:

> "For some time now, there has been, in Western culture, a distinct preference for monomodality.
> The most highly valued genres of writing (literary novels, academic treatises, official documents
> and reports, etc.) came entirely without illustration, and had graphically uniform, dense pages of
> print. Paintings nearly all used the same support (canvas) and the same medium (oils), whatever
> their type or subject. In concert performances all musicians dressed identically and only con-
> ductor and soloists were allowed a modicum of bodily expression." (Kress and van Leeuwen 2001:
> 1)

Multimodality now has the task of challenging these divisions and of providing sufficiently robust methods for coping with the results.

In this chapter, therefore, we have started setting out a foundational scaffold
for the 'space' within which multimodal issues can arise. We used this to support
a necessarily brief overview of disciplines where concerns of multimodality have
asserted themselves in one form or another—sometimes more centrally, sometimes
peripherally—leading to many different positions with respect to the problems to be
addressed.

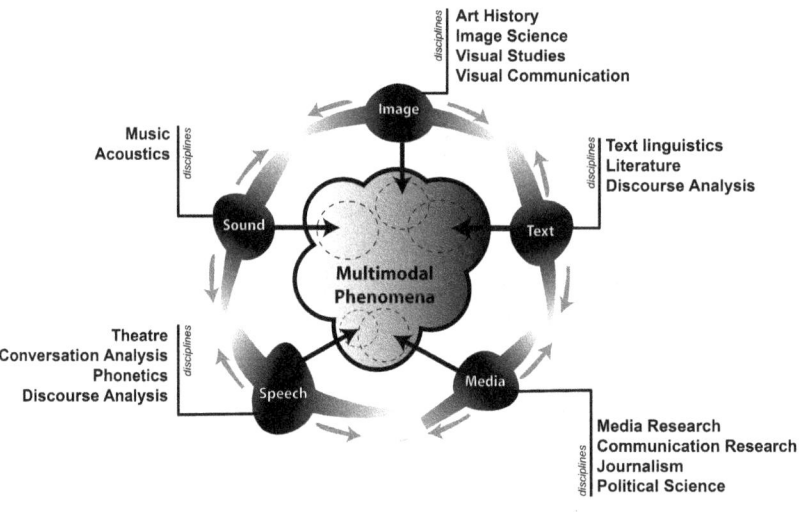

Fig. 2.7. The pre-existing discipline-based approach to multimodality research: disciplines pick out
phenomenal domains for study that are each limited multimodally. These are then 'stretched' to
attempt to take in a broader multimodal reach. Extensions preserve the original disciplinary orientation and so their move into broader domains may not always be appropriate.

In each case we have seen that individual disciplines are becoming subject to some rather similar pressures. An original focus on some domain of phenomena, often loosely anchored in some area or areas of sensory perception, is pushed to include other phenomena that appear in some way related, or to function in similar ways. We suggest this situation graphically in Figure 2.7, extending the view discussed at the end of the previous chapter.

While expansion is generally to be welcomed and is no doubt well motivated, there are also accompanying dangers. If one attempts to construct a general model of signification based on any one form of expression, this will always tend to compromise descriptions of semiotic modes which are *not* organised in that way. The over-application of linguistic principles has had the longest history in this respect, where it is termed 'linguistic imperialism'. It is equally important, however, to avoid this mistake in other domains as well. For example, if one attempts to construct a general model that assumes the static pictorial image as its preferred template, this is equally likely to distort what is being examined. No single approach that we have seen so far can therefore claim to be an account of 'multimodality' as such, even when useful examples of how to go about multimodal analyses in particular areas are offered.

As a consequence, then, although any reader of this book may already be placed within one or more of the disciplines we have mentioned, it is useful to go beyond such individual disciplinary boundaries in order to draw out more of the commonalities in both approaches and difficulties faced. Restricting attention to specific disciplinary orientations readily obscures many of the basic mechanisms that make multimodality so effective and prevents the much needed import and export of results, insights and experiences from other domains.

All of these individual discipline developments offer important parts of the overall puzzle of multimodality, but we also need more of a foundational contribution for bringing out the constitutive properties of multimodality 'as such', as a phenomenon in its own right. This means that we need to uncover more of the mechanisms and principles of multimodality in general in order draw maximally effective lines of connection between, and 'triangulation points' across, disciplinary approaches. This is the task that we take up in the rest of the book.

3 Where is multimodality? Communicative situations and their media

Orientation

The main purpose of this chapter is to provide definitions of several key terms necessary for pursuing multimodal research. The definitions can be considered complementary to definitions that may exist in other disciplines, formulated here with the express goal of supporting multimodal research. Taken together, they offer a solid foundation for considering all and any kinds of multimodality-in-use. We begin with a multimodally-appropriate construction of the notion of 'communication', and then move on to define the 'media' of multimodal communication with respect to this. These definitions have to be flexible enough to cover new media as they evolve and to provide a classification within which both established and new media can be related to one another. Only then can we achieve a field of multimodality that will be able to keep abreast of the diverse and diversifying range of multimodality *as it happens*.

3.1 Stepping beyond the terminological jungle

In the first chapter of the book we saw something of the sheer diversity of places where there are questions to be raised concerning multimodality. Then, in the previous chapter we saw just how widespread the need to respond to multimodal phenomena is. However, when we look across the ways in which disciplines have responded to these challenges, it is clear not only that the *scope* of what is included under 'multimodality' varies considerably but their respective capabilities with respect to multimodality vary as well. What different researchers take multimodality to be—and even whether they name it 'multimodality' or something else—usually depends on the research traditions being worked within and the research questions those traditions define to be of interest. Similarly, what different practitioners take multimodality to be depends by and large on the practical questions they need to deal with in their line of work.

This sheer diversity of theoretical positions can be a substantial hindrance when considering how to explore multimodality further. Different disciplines may pick out different aspects of multimodality that are by no means straightforward to relate. Such divergences or gaps in interconnections, often at quite foundational levels, are particularly problematic when we need to address a research question that appears to require drawing on several of the diverse possibilities all at once. This, as we have already suggested at several points, is increasingly the 'new normal' as far as multimodal research is concerned. Increased and increasing 'media convergence' is already making many of the multimodal discussions of the past 10 years appear restricted in their outlook.

On the one hand, this is positive in that it is now drawing an ever increasing number of researchers and practitioners into the multimodal arena. On the other hand, this demands that stronger techniques and methodologies be practised that are fully capable of engaging with the ever richer multimodality of the artefacts and performances that constitute much of our daily lives. These challenges will only grow as even more challenging developments, such as interactive full resolution virtual reality, augmented reality, ubiquitous media platforms and many more, work their way from research labs into the hands of consumers.

If we are to succeed in placing multimodal analyses on a firmer, and also broader, footing, we will require stronger foundations concerning just what is involved when multimodal phenomena are addressed. Offering a framework within which such foundations might be profitably articulated is the primary goal of this chapter. In pursuit of this aim, we will set out particular ways of defining the 'modes' out of which 'multimodality' emerges, the 'media' by means of which such multimodal situations are constructed, and the 'communicative situations' in which multimodality can be an issue.

To achieve this, we will need to reverse the direction of discussion in contrast to the overviews of the previous two chapters. Rather than seeing how particular research questions or disciplines have each attempted to take their various 'cuts' out of the broader phenomenon of multimodality, in this chapter we begin with multimodality and, more specifically, the situations in which multimodality is made to function for meaning making. In short, this means that we will be working through a somewhat simplified *ontology* of multimodality—by this is meant that we ask precisely just what is multimodality and what other constructs and interrelationships are going to be necessary to make multimodality work. This will be done independently of the questions or needs of any particular discipline: the goal will be to characterise certain principles of multimodality that may then be picked up by *any* discipline or research question where multimodality is a concern.

Adopting multimodality as our central point of departure will result in a certain degree of questioning of existing notions. Consider, for example, the role of 'new media' in media studies where discussions often take the properties of those media as motivations for rejecting or extending 'traditional notions' of communication. Such 'traditional notions' include what we will term below the 'post office model' of communication whereby messages are transported from senders to receivers. Our reorientation to multimodality will show such models to be inappropriate for *any* form of communication and has nothing particularly to do with 'new' media. We thus redesign the concepts from the inside out so that what results naturally include multimodal 'new media' just as well, and as centrally, as any other form of communication. Properties often discussed in the context of 'new media'—such as interactivity, virtuality, materiality and the like—are shown to be, and to always have been, properties of all communication, and so corresponding models must be prepared for this.

A reorientation of this kind is essential if we are to achieve a foundation for a field of multimodality capable of addressing broadly diversified modal developments 'as they happen' rather than running behind, chasing the newest combinations of modalities with terms that already hardly fit the old ones. And, for this, we need to step back and get some distance on just what is going on whenever some notion of multimodality is at work. That is, we need to go into multimodal communicative situations prepared for their complexity and with strong analytic frameworks available for segmenting and decomposing situations in ways that make sense—i.e., in ways that make sure that we do not make more problems for ourselves than we solve.

This is not to be done by choosing specific analytic frameworks before looking at what we are analysing. That is, analysis needs to be responsive to the challenges raised by the communicative situation we are engaging with. And, for that, we need to be able to approach such situations analytically so as to allow them to draw out the kinds of methods and frameworks for analysis that will be needed. In short, this means that we need to prepare ourselves for approaching new combinations of expressive resources in ways that respond to the needs of any particular data to be addressed, and not by taking some particular theoretical framework—which has in any case probably been developed only for some subset of (other) modalities—for granted.

The temptation is always great, of course, to apply and extend what one knows. Thus we see approaches to narrative in comics extended to film and even non-narrative representations; we see approaches to gesture and intonation extended to multimodality as such; approaches to linguistic texts applied to buildings and architecture; and so on. In contrast to such approaches, here we step back and set out how to pursue a far closer examination of data with respect to that data's multimodality before deciding on just how it is to be analysed. The result of our considerations will be a reusable methodological and theoretical framework for thinking about multimodality as such, wherever and whenever it might occur.

Comm-
unication

To reduce confusions and inconsistencies as far as possible, it is especially important to get the 'semiotic angle' right. This is often not done, or not done enough, and it is this that results in accounts that quickly become difficult to follow and apply. Under such circumstances, analyses are generally prone to be less than incisive. We begin, then, by asking the following question: if there is such a thing as multimodality, what other constructs are necessarily present in order for it to be coherently defined? One of these constructs that is particularly central is, of course, *communication* as such. There are many models in the literature from various disciplines that say things about 'communication'—but these have rarely been constructed with the demands of multimodality already taken into consideration. So we will spend some time at the outset pulling apart just what needs to be considered part of communication if we are to move to analysis of multimodality. For this, we will introduce rather carefully the notions of communication, communicative situations and the roles of different kinds of expressive resources within those situations.

3.2 Communicative situations

One of the most significant problems found in particular disciplinary accounts of 'communication' is that of missing 'levels' of abstraction or organisation. Failing to separate out levels of representation sufficiently is the single most common cause of descriptive confusion or lack of specificity. Phenomena that are at heart not particularly complex suddenly take on far reaching and contorted ramifications, which researchers and practitioners alike have great difficulty in extracting themselves from. We will see a few of these cases discussed in the use cases sections of this book. For now, though, we want to make sure that we avoid such difficulties as far as possible. And we do this by sorting out necessary distinctions from the outset so they do not come round and surprise us (or, more often, confuse us) later.

To get started on this, let us consider the variety of different kinds of situations in which 'communication', considered very generally, might be said to occur. In particular, we will want to focus in on the kinds of objects, materials and phenomena that 'carry' the communication in those situations. This is because our concern will always be on providing analyses of such 'objects' of investigation. Being more precise about this, and the range of possibilities that there are, gives a useful framework for selecting methods and approaches for undertaking analysis. That is, a framework of this kind will help you structure your investigations of how communication, and particularly of course multimodal communication, operates.

We will need to see communication very generally when constructing a foundation of this kind. We cannot be concerned solely with situations where it is fairly easy to pick out some specific message that is being deliberately sent to some receiver. This sender-message-receiver view, often called the 'postal model', is still quite commonplace in many theories of how communication works, despite equally common criticisms. The postal model can be quite misleading if it is used to describe situations that lie outside its scope. What, for example, is the 'message' of a Hollywood film, a website, a computer game or an entire face-to-face conversation? Talking of a 'message' in such circumstances often distorts the questions that one asks and, worse, the answers that one gets. Nevertheless, criticising the model is not yet enough—workable alternatives also need to be provided so that analysis can proceed. Postal model

More broadly, then, for our current purposes we will need to consider *any* situations where some recipient or group of recipients 'gains something' by virtue of interaction with something that they are giving 'meaning' to in that interaction. This is, of course, still very vague—or rather *under-specified*. When trying to cover all cases, the results will always seem to lack specifics. We need to make a connection with specifics, however, in order to get any particular analysis done—guiding how to do this is the task of methodology. So we will need to get a more refined understanding, not only of what our very general statement means, but also how we can apply it to particular cases. To do this, we will proceed very cautiously—we do not want to pre-judge the kinds of 'multimodality' that may be at issue by making perhaps unwarranted as- Method sumptions too early. Similarly, we will initially leave open just what 'is being gained' from the communication: this also needs to emerge from the discussion.

Signs and
Semiotics

In many respects, an extremely general notion of communication of this kind echoes what we saw in the previous chapter under Peircean semiotics. It always helps to keep this foundation in mind when working multimodally, both to provide some structure for thinking about what is going on and to avoid unwarranted 'overcommitment' about particular forms of meaning-making. Such overcommitment is commonly found when starting from too narrow disciplinary starting points, for example. Consequently, communicative situations should always be seen as involving the use of *signs*. And signs, considered most generally, are simply entities (again construed as broadly as possible) that lead an interpreter from one thing (the thing that is being interpreted as a sign) to another (the interpretation) by virtue of some properties exhibited by, or attributed to, the mediating sign itself (→ §2.5 [p. 51]).

Abstract statements of this kind are quite accurate, in that they cover all cases, but they manage this accuracy only at the cost of becoming less straightforward to understand and, more importantly for us here, much less straightforward to actually *apply*. That is: the very abstractness of the statement means that it does not particularly help us with concrete practical analysis. That is: we have to refine the general position to include enough detail as to know *how to do some actual analysis with it*. As our general statement stands, and regardless of the fact that it is a very good definition of what a sign is, it leaves open far too much freedom of interpretation. So our task now will be to show as we proceed just how such abstract notions can be successively filled in to meet the purpose of supporting practical multimodal analysis.

3.2.1 Basic conditions for the existence of communicative situations

Our line of attack will be to move progressively from the simplest of situations where some kind of 'communication' might be said to be present, to the kinds of complex combinations that are increasingly commonplace today. Taking this development very slowly will be beneficial because it will avoid skipping over important steps and qualitative changes on the way. One problem often encountered in descriptions of methods for multimodal analysis is that the discussion goes too quickly. By being very explicit about just what properties are added by each increase in complexity, we can better ensure that we can repeat the procedure for *any* situation that we might want to consider.

! Moving too quickly from simple to complex descriptions often means that important properties are being skipped over, and these may come back to haunt us later if a different communicative situation demands attention to be paid to just those properties that were formerly ignored.

Encountering difficulties when facing new communicative situations is very common for those beginning multimodal analysis. Suddenly it becomes very unclear how to proceed, the possibility of simple descriptions evaporates and one is confronted with complexity that does not appear to fit how one was thinking of the situation beforehand. When this occurs, it is likely that analyses become less well motivated, or even arbitrary, because the material being studied appears different to what was expected and it is unclear how to relate it to previously known distinctions.

To avoid this—or rather to prepare for it appropriately so that one can always 'back up' and start unpicking the complexities of a communicative situation—we will start off deliberately *too* simply. On the one hand, this will help place some boundaries on what we are considering as communicative situations at all; and, on the other, it gets our foundations in place for building a powerful framework capable of addressing communicative situations of all kinds.

3.2.2 Not (quite yet) communicative situations

First, then, imagine picking up a pebble on the beach and examining it, turning it around in your hand. We can learn or surmise much about this physical object by such an investigation—we might think of the erosion of wind and water, of the action of the waves, of its 'use' as a possible ecological niche for plants to grow. In each case we are interpreting the physical traces that these possible interactions have left on the object under investigation.

Now, since we are 'gaining something', i.e., information about the history of the   Traces
object, from our examination, some people, in some contexts, would be quite happy to call these traces 'signs' as mentioned very generally above, that is, they allow an interpretation of something that is not actually physically present. For theoreticians of this kind, such material traces are fair game for semiotics, since semiotics is the study of signs of all kinds—the material traces are, after all, signs of how the pebble got to be in the state that it is. They are *symptoms* of other processes, activities, states of affairs—just as red spots and a fever may be symptoms of measles.

Critical for our discussion here, however, will be our use of the word 'meaning' above—that is, we are looking for situations where what is gained is an *assignment of meaning* to the object or event interacted with. There are many different ways of defining these slippery terms in the various disciplines that use them, but for current purposes we will draw some sharp boundaries that will help us later when the going gets tough and multimodality starts being complex.

We will not then call the kind of interpretations that we just made of the physical traces left on our pebble 'meanings' at all. We do not construct a 'meaning' for the traces, for the seaweed and scratches, and so on, we construct an *explanation* for their presence. This leads to very different kinds of considerations and has very different kinds of properties as a systematic endeavour. Although interesting, it is just not our task as multimodal analysts.

It is important to be perfectly explicit and conscious of this fact: we will *not* be addressing such 'natural' signs precisely because they are not *communicative* in the sense of being designed and produced so that someone, an interpreter, 'gains' anything by interacting with the pebble. This is the sense of 'meaning' that we will be following throughout. Even if there are signs of earlier stages of life in the pebble, such as fossil traces, it is unlikely that we are seeing a communicative act. And, when we are not dealing with a communicative act, the techniques and mechanisms that we are learning about here cannot be expected to be sensibly and usefully applicable. This is a beneficial line to draw because it means that there is a considerable range of 'things in the world' that are not going to be good candidates for multimodal analysis. Placing something outside our scope will help us be more incisive about the things that remain and also supports stronger methodological decisions about what to address and what not to address.

We need to be precise about the status of this boundary, however. It is always necessary to be aware that it is, in the last instance, *an analytic decision*. We might not always be able to know whether or not some material trace on the pebble was deliberately made to communicate or not—as may well be the case in archaeology. This could just as easily be the case with our pebble: is some criss-cross marking a result of natural erosion or the indication of a long lost indigenous population? These kinds of situations always require a return to the theoretical premises being identified here. As we proceed, we will see that this often turns out to be more a theoretical concern than a practical one—the kinds of traces that we will primarily be concerned with throughout this book are overwhelmingly cases where there *cannot* be any reasonable doubt. And when there is doubt, the lack of further information about the potential origins and intent of the markings being examined will mostly already have precluded much of the detailed methods of analysis we set out below.

3.2.3 Almost communicative situations

 Let us now make our object under investigation a little more (semiotically) complex and see if we can identify the points where we cross over the border into situations that seem to be more appropriately characterised as communicative. Say we pick up a different pebble and discover that this one has a cross scratched into it. There are now several further possibilities distinct from those that apply to either natural objects or simply non-communicative artefacts.

Intentions We may, for example, well assume that there has been an *intention to communicate*: a sign has been selected for communication. Thus, the fact that we see some marks that have apparently been made deliberately gives us a strong cue that some kind of communication at some time was intended. We saw in the previous chapter how Peirce gave this a particular semiotic status: that of the *indexical* sign (→ §2.5.2 [p. 59]). This is something that points us towards something else by virtue of a 'causal connection'. The cross on our pebble is consequently 'indexical' of a communicative situation having held. But, and this is a substantial but, *we do not yet know just what kind of situation that was*. We are not yet *in* a communicative situation, we only have some cues that someone else was!

To make a mark or trace a single line on a surface immediately transforms that surface, energizes its neutrality; the graphic imposition turns the actual flatness of the ground into virtual space, translates its material reality into the fiction of the imagination.
— *Rosand (2002: 1)*

The situation here is precisely that of overhearing some people conversing in a language that one does not know; we might deduce some details of what is occurring (including the fact that they are conversing at all), but are still a long way from being involved in communication. Alternatively—the same situation but for many less frequently encountered—a group may be conversing using a sign language for the deaf; in this case, depending on familiarity with the nature of sign languages, one might have considerably more difficulty in appropriately classifying the situation at all. Such considerations are important for us because they reveal more of the preconditions that need to be fulfilled before we can talk of being in such a situation ourselves. Adding in those conditions lets us define just what has to be present for a full communicative situation to hold. Returning to our pebble, although we may with considerable justification believe that there has been an intention to communicate, *we have no way of knowing just what that communication is about, nor how it was effected*.

The radicality of the extent of our lack of knowledge here is often underestimated. And yet this is crucial for picking out just what *has* to be in a communicative situation for it to be a communicative situation; this is the 'ontological' angle we are taking as mentioned above. Whenever shortcuts are taken at this stage, any frameworks resulting already have weaknesses and gaps built into them—particularly when we start drawing in more of the multimodality of a situation. As multimodal analysts we must never forget this point because it may, in general, not be at all clear what, precisely, in a complex multimodal ensemble is serving as the actual carrier of a message and what not. In short, since our aim is to explore artefacts and performances for which it is *not* already well known how they go about making their meanings, assuming that we already know is about the worst mistake that one can make!

Radical indeterminacy

Thus, strictly speaking, and we will need to be strict when we proceed into new multimodal territories, we do not yet even know for our pebble what the relevant *sign* involved is. It *might* just be the cross, but it could just as well be the colour or shape of this particular pebble, the orientation of the cross (with respect to the pebble, or the beach, or the moon), or even the shape of the beach (i.e., the particular beach that this particularly placed pebble picks out), or the distance of this pebble from some other landmark, or the depth that the cross is scratched into the pebble and so on. In each case, the cross may be doing quite different 'semiotic work', and so is actually operating as quite a different kind of sign. It may, in fact, just be a pointer to where to look for the 'real' sign. This may sound far-fetched but it is, in fact, quite central because it forces us to ask just where the knowledge necessary for fixing the sign we are talking about comes from. It is only by considering how we move beyond this that we reach communicative situations proper.

Sign vehicle   One of the first steps to be taken to fixing the sign appropriately is to ask more specifically just what physical properties or manifestations of some semiotic activity are to be taken as part of the communication. Unless we know which of these *material regularities* that appear to be indexical of a communicative situation are those actually intended to 'carry' the sign, i.e., to be the 'sign vehicle', we can say little more. And, again, those material regularities can, in theory, be anything. For our pebble, they may reside in the pebble itself or in the specific location or orientation of that pebble in its environment, and so on. We need to *know* which of these to attend to, otherwise again there is no basis for a real communicative situation.

?   "Compare a momentary electrocardiogram with a Hokusai drawing of Mt. Fujiyama. The black wiggly lines on white backgrounds may be exactly the same in the two cases. Yet the one is a diagram and the other a picture. What makes the difference?"

(Goodman 1969: 229)

Expressing this more finely, we need to know what *distinctions* in the material situation are those that are being drawn upon to form the sign vehicle: is it just the presence of the cross rather than its absence (on all the other pebbles on the beach perhaps)? Or is it after all the shape of the pebble?—this is the case mentioned above where the function of the cross is quite different: it frames the unit that is to be treated as a sign rather than being the sign itself.

As a further illustration of this radical indeterminacy let us briefly consider one well discussed case of some of its potential consequences for design. The Pioneer 10 spacecraft was the first probe intended to leave the Solar System (after flying past several planets to photograph them). The idea then arose that perhaps it would be good to include some kind of 'message' with the probe in case it would, at some point on its distant travels, be discovered by extraterrestrial beings (Sagan et al. 1972). This turned the science fiction notion (and occasionally philosophical question) of how to communicate with completely unknown intelligences into a very practical design problem.

As might be expected, the resulting design, put together as is often the case under rather tight time constraints, was subsequently subjected to considerable critique—critique most often concerned with the (political, ideological) appropriateness of its depicted messages. In particular, the message was held to be hopelessly Western-culture specific (the plaque contained outline engravings of probably Western male and female figures) and gender repressive (the male figure is standing more in the foreground raising a hand in greeting; the woman stands 'in the background', allegedly doing nothing). When we consider the actual context of the plaque, however, these issues begin to be overshadowed by the entire issue of radical semiotic indeterminacy.

The plaque also attempted to communicate a range of information, some quite technical—for example, the position of the probe's origin in the galaxy is shown with the aid of abstract depictions of planets (engraved circles), an engraving of the outline of the probe (a 2D projection from the side), a long arrow showing the planet of departure and its trajectory within the Solar System, and an even more abstract representation of the position of the Solar System with respect to some galactic 'landmarks'. Here, more semiotically inclined detractors already began to question the assumption that all species would recognise the depiction of an arrow for directions and paths. If, the argument goes, a culture had never developed the bow and arrow (let us imagine a culture of space-travelling dolphins perhaps), then even the 'direction arrow' might prove quite opaque to them. Gombrich (1982*b*: 150–151) provides further discussion of this kind.

Our characterisation of communicative situations make us push this even further. For example, the plaque is made up of narrow grooves cut into a robust base material. How, then, can its designers assume that such grooves will be decoded as 'visual images' at all, let alone supporting the complex mapping necessary between the spatial regions so demarcated and the intended objects, movements, and trajectories? This would surely depend on the actual embodied perceptual capabilities of the extraterrestrials who find the plaque. This is, therefore, a rather more complex version of our situation of the pebble with a cross on the beach: without knowing more, we only have an indication that some sign producer(s) wanted to communicate (something to someone) but not of what or how. We will leave the question of how one might go about ascertaining just what regularities are depicting what as an exercise for the reader; and to help, Figure 3.1 shows the plaque, although not in the way that it is usually shown.

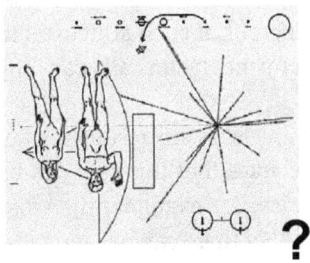

Fig. 3.1. The plaque on the Pioneer space probe shown upside down – although how would the extraterrestrials finding the plaque know this?

Some level of *distinguishable material regularities* is therefore a precondition for a full and effective communicative situation to be able to exist. Until we know about these regularities, we are onlookers who can only guess what may have been going on.

Material regularities

3.2.4 Communicative situations achieved

Comm-
unities

The fact that we need to have certain knowledge about what material regularities are relevant in order to begin talking of 'communicative signs' at all leads us on immediately to a further component that is critical for actually having a communicative situation. The precondition is that, unless we are 'communicating' just to ourselves–for example, we may have marked just this pebble with just such a cross to tell ourselves where the treasure is buried—this knowledge of the distinguishable material regularities is *shared across some collection of individuals*. That is, a community of receivers knows how to interpret the material regularities found. Only when this condition obtains, can we begin to talk of genuinely communicative situations.

> ...communication depends on some 'interpretative community' having decided that some aspect of the world has been articulated in order to be interpreted.
> — *Kress and van Leeuwen (2001: 8)*

Social
status

Moreover, as always the case with 'social knowledge(s)', that knowledge may well be distributed unequally among the community's members. Some may know how to 'read' the material regularities, some may also know how to 'produce' them—the two skills do not necessarily go together, although they may. This naturally correlates with differential access to 'technology': some may know how to burn sticks so as to be able to make a mark, other might not. And the same holds for other means of making such marks, etching, painting, drawing, etc.; as Kress (2010) emphasises, this naturally has a strong connection to the distribution of *power* in societies.

The 'marks' may also be sufficiently complex, or difficult to produce, that *entire groups* are involved in their creation (including production and distribution). At no point in our modelling, therefore, can we make the assumption that we are dealing with 'individual' acts. Although introductory texts often talk of 'speakers' or 'hearers' and so on as if singular, the relevant 'agents' involved may well be groups, again possibly with internally differentiated roles. The actual filling out of such structures in communicative situations will depend on a whole host of further factors and can only be ascertained in each specific case.

Picking out the material regularities that are relevant is by no means always straightforward—although we will return to some of the ways that this step can be assisted below. Although it may seem simple to recognise, for example, the cross marked on our pebble, even for this case we noted above that there can be, and most likely will be, considerable additional information that could, in principle, require interpretation. The pebble may, for example, have seaweed or marks of erosion that have nothing to do with the *intended* material regularities. Whenever we have real material artefacts, such issues will arise—as in cases such as scratches on vinyl records, dust on old style slides used for projecting photographic images, cracked screens on smartphones, and so on.

We need then to be cautious when dealing with artefacts and performances where we might not be sure about just what material regularities are being used intentionally. This is often relevant for conducting analyses because we must decide whether some *particular* material properties of something being analysed are communicative or not. This might well not always be clear. That is, it may be clear that we are dealing with some object that bears some communicative marks—but it might not be clear that *all* the distinguishable marks are functioning in this way. For example, whereas the scratches and clicks on an old vinyl music record might not have been intended and are a result of material limitations in methods of production and reproduction, it might also be the case that identically sounding scratches and clicks introduced on a purely digitally produced composition could well be intended, perhaps as a sign of simulated age or origin. These are issues where, again, more background knowledge about the artefacts' or performances' production is important. One needs to be open to these possibilities during analysis but not to overdo the 'interpretative' impulse – without good reason (and we will see some of these reasons as we proceed), such traces can generally be put aside, at least in the early stages of analysis.

Further 'additional' sources of information may then always be gleaned from a mark itself—in particular from its *manner of production* or *performance*. Thus, the cross on our pebble may have been produced roughly or with considerable precision, with a firm hand or as a rushed gesture. All of these can offer further information but are not necessarily communicative. This means that they can be *explained* since they are generally indexes (→ §2.5.2 [p. 59]) of physical events. But that does not turn them into communication of the kind we are foregrounding here. The additional information does not constitute the communicated content. It is, however, *only the knowledge of the community* concerning which variations in the material regularities are to be considered significant that allows this distinction.

There will, moreover, always be considerable variation possible even within the categories that have been declared significant for communication. And this variation may then be indexical of a host of further information concerning the sign's production independently of what was intended. The distinction between 'content' and additional information about that message's style of production remains and there are typically many 'symptoms' of this kind. This applies to all kinds of communicative situations and their employed materials. Talking fast and loud, for example, may be a symptom of excitement or of drug consumption—this information is not (usually) intendedly communicated but nevertheless comes along 'for free' by virtue of indexicality. We will return to this component of possible communicative situations in a little more detail when we consider the *performance* of communicative situations in more detail. For the present, we just note that paying attention to such variation is also going to be necessary during multimodal analysis—but more to delineate which kinds of material regularities receive what kinds of description (e.g., interpretation *vs.* explanation).

Various kinds of knowledge must therefore be present for a communicative situation to come about. Knowledge that allows distinctive material regularities to be identified is one of these. In addition, however, the knowledge of the community of potential receivers in the communicative situation must also include the practical knowledge of how to turn the material distinctions available into an *interpretation*—an interpretation that appropriately contextualises the material distinctions found so as to recover not only an intention to communicate but a *meaning* for that intention—in other words, that which was intended. We will return to ways of describing such meanings in particular cases of multimodality below and in subsequent chapters; for communicative situations as such, however, this possibility is necessarily presupposed.

Communicative situations: summary of necessary conditions

The three conditions we have now introduced give us our first characterisations of the *necessary* features of a communication situation:

1. we must know (or assume) some particular range of material regularities that are to be considered to be carrying semiotic activity;
2. this knowledge must be shared (possibly unequally) among a community of users, and
3. a scheme for deriving interpretations from the material regularities identified must also be shared.

Only when these conditions are met can communicative situations occur.

3.2.5 The story so far

Let us now summarise the starting position we have achieved, showing diagrammatically the simplest communicative situation of our beach pebble. We will use this, and the graphical conventions adopted, in subsequent sections to build up some rather more complicated situations as we proceed.

sign consumer   canvas   sign producer

Fig. 3.2. The straightforward beach pebble (almost) communicative situation

Canvas   Figure 3.2 depicts the situation where we have a (in this case unknown but necessarily existent) sign producer on the right; the finder of the pebble on the beach who tries to interpret what has been found on the left; and the pebble with its cross mediating between them in the middle. We call the potential interpreter the sign consumer in order to generalise across 'reader', 'viewer', 'listener' and whatever other terms might be used in diverse communicative situations. Similarly, we also generalise across all possible bearers of meaningful regularities with the term 'canvas'. This is to be understood as anything where we can inscribe material regularities that may then be perceived and taken up in interpretation, regardless of whether actual, virtual (digital), simply produced, performed physically in time, or the result of a complex technological process. This places minimal demands on the materialities of their adopted sign vehicles—almost anything that can carry some material regularities that suggest intentionality and order might serve. There are, however, many more interesting kinds of materiality that can be employed as we shall see as we proceed.

The graphical icons of the figure are then to be read as follows. The 'eye' stands for any act of perception with respect to some interpretable artefact or performance, regardless of sensory channel. The 'hand' similarly stands for any process of production. The pebble with the cross on it then stands in for any produced material regularities (the cross) inscribed in any canvas (the pebble). In the present case, then, the (almost) communicative situation is as we described it above: the sign consumer can do little more than attempt to interpret what is shown on and with the pebble, and the sign producer's role is finished and done. This is an unchanging, static physical object; It cannot be changed or questioned, refined or withdrawn.

The three-way split of the traditional model of the communication system echoed in Figure 3.2 reoccurs in many simple and extended forms. Rose (2012a: 19–20) in the context of *visual studies* talks of three sites at which the meaning of images is made: the site of *production* (the hand in our figure), the site of the *image* (the cross) or *object* (the pebble) itself and the site of *audiencing* (the eye). Rose discusses all three in terms of the three 'modalities' (used in a quite different sense to that pursued here) of technology, composition and the social. These already point to the more appropriate visualisation of communication situations as embedded in communities that we suggest in Figure 3.3. Nevertheless, it is useful here to think about other styles of making meanings beyond the visual and the linguistic: to what extent can we identify these three 'moments' in the construction of communicative situations employing quite different forms of expression and their combinations?

?

The materialities capable of supporting the situation just given are diverse, precisely because so few demands are made—changing materialities within these confines can be done freely without changing the type of communicative situation. This emphasises that it is really the material *regularities* that are at issue: not just the material itself. And it is for this reason that knowing how to read is a skill transferable across paper, papyrus, the blackboard or PowerPoint projection. Our cross might appear in various places and still function as a cross because it is only the patterns of regularities that are at issue.

Before proceeding, we also need to recall what was said above about the necessity of having a community of sign users. The situation as shown in Figure 3.2 corresponds more to the old postal model of communication—the pebble and its mark is then the exchanged message sent from sender to receiver. This is what keeps this situation at the level of a guessing game; the necessary relations between receiver and sender, between producer and consumer, are not represented. As we argued above, this is not adequate for a communicative situation and in fact turns understanding much of how communication works so flexibly into an impossibly difficult problem.

What we actually have is then more the situation shown in Figure 3.3. Here we can see that the consumer and producer in the previous figure are just shorthand for *individual roles* that happen to be being played at some particular time, within some particular communicative situation, drawn from an entire community of prospective sign producers and consumers. It is relatively common that writers will stress some notion of 'context' wrapped around the postal model; an emphasis on a community within which signs are exchanged is also found in approaches ranging from semiotics (e.g., Lotman 1990), to social theory (e.g., Halliday 1978; Latour 2005), to human-computer interaction and distributed cognition (e.g., Benyon 2014: 514). It is therefore always to be included. More specifically here, however, we consider context structured in a very particular way, constituted by configurations of social relationships and the roles, including communicative roles, that those relationships provide. *It is this anchoring in a community that turns that task of understanding what the canvas is carrying from a piece of enlightened guesswork to communication.*

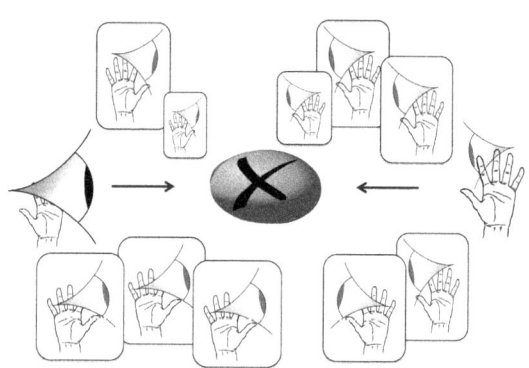

Fig. 3.3. Communicators are always roles taken up by members of a communication community: the *actual* sign producer and consumer are drawn from this *potential* background

Potential/
actual

This background should be understood implicitly for all of the diagrams that follow, although we will often omit the 'cloud' of other potential members in order to focus on what is being added in each case. But we should never forget that those that are shown are roles that may be distributed among community members in various ways. The different communicative situations we depict will build on various constellations of role assignments. Sometimes a community member will play the role primarily of consumer, sometimes of producer, sometimes both. This necessarily differentiates between what is *possible* and what is actually happening within a concrete specific communicative situation/event. Our diagrams now will pick out specific situations, always drawn against the implicit background of the entire community of sign users that *could* have been involved but were not.

Now, with this foundation out of the way, let us now start expanding the communicative situation to reflect both more complex situations and more complex materialities.

3.2.6 Functionality, materiality and communication

The three conditions on communicative situations set out in earlier subsections ($\rightarrow$ §3.2.4 [p. 86]) would be where many traditional accounts of semiotics and signs would stop—we have something that might be a sign (the 'signifier') and some potential meanings for the sign (the 'signified'). Some accounts might even have stopped beforehand by taking the community of sign users more for granted: something which we should never do because they are a foundational component for any communicative situation. This applies even more strongly to multimodality research because here there tends to be far more diversification among the communities of sign users in these contexts than appears to be the case, at least superficially, when we are considering language or pictures. With disjoint multimodal skill sets being more the norm than the exception, diversification among sign users takes on a new significance.

We now need to go several steps further. What we have achieved up until now is a slightly more filled out version of our original very abstract statement about what happens when signs are interpreted—even though we have not yet said very much about actual processes of interpretation and the methods we can use for their study. We will again approach these aspects very cautiously and, to do this, we will refocus attention on the 'component' in the middle of our communicative situation diagrams—in particular, on the 'pebble' or, more abstractly, the *canvas* that carries material regularities.

This time, however, we will dig deeper into the relationship between the nature of any canvas we may be concerned with and the *kinds of meanings that can be made with that canvas*. This is the issue of materiality. We mentioned above several aspects concerned with the materiality of canvases and how this may be relevant for multimodality. We also learned in the previous chapter that it is precisely the material aspect of meaning making that is now playing an ever more important role in pushing disciplines beyond their original boundaries. As we saw, it has become far more broadly accepted these days in many forms of semiotics as well as in cultural studies, art history, linguistics and many more, that materiality cannot be neglected any further. Some approaches have even begun attempting to make it assume a central place. Thus, even for our pebble, we have to be aware that it is not just the fact that some cross is abstractly present, it is a placed on some particular physical object with particular properties, through a specific performance, and these physical aspects may also be 'communicating' to us a host of further information.

Many authors have subsequently focused on materiality as an intrinsic component of drawing meanings from the world. For example, van Leeuwen (1999) picks out two aspects of materiality in particular that influence our engagement with entities:

'provenance', i.e., indications of where something comes from, and 'experiential' properties, of how one can engage materially and physically with some entity. In the context of our current discussion, we must also assign materiality an important role since it is constitutive for any communication that occurs. In particular, and to advance our exploration of communicative situations, we will be concerned with just *how different materialities may support different kinds of communicative situations*.

We will see that this can be of considerable help when we turn to practical analysis. Finding out just what *could* be done with any materiality will constrain what it is that *is being done* in any specific case of use. Thus, considering this aspect more closely will provide several further guidelines useful for classifying and engaging analytically with communicative situations as they become more complex. To get us going, therefore, let us pick out some last details of our trusty beach pebble—its sheer materiality may still help us in some ways rather than others and, crucially, this will also go beyond the kinds of 'traces' and 'symptoms' already mentioned above.

Affordances   For example, the more the pebble supports recognition of the cross, e.g., by having a smooth surface, showing enough contrast between the cross and the background, etc., the easier our task of interpreting communication becomes. In many discussions nowadays it has become common to discuss such 'fitness' for a purpose in terms of a notion drawn from the field of 'ecological psychology' originally developed by Gibson (1977), that of *affordances*. Affordances were originally proposed in order to be able to characterise perception as a far more active and functionally-oriented engagement with the world than simply seeing (or hearing, or smelling, or touching, etc.) what is 'there'. Instead, perception is considered in terms of the *activities that an environment or objects in an environment support for an agent*: hence the term 'ecological' psychology. More concretely, then, the shape and position of a door handle 'affords' turning and opening a door (unless it is very poorly designed), a book affords the turning of pages, a mouse affords clicking, and so on.

When seen in terms of affordances, an object is directly perceived in terms of the possibilities for action that it opens up for an agent in an environment. More recent studies of the parts of the brain that 'light up' when perceiving highly functional objects like door handles generally offer good support for Gibson's proposal. Seeing particular functionally-loaded shapes appears indeed to cause relevant motor routines (for turning, for holding, etc.) to be activated.

Design   Returning then to our pebble, it might have been selected or deliberately shaped with certain properties in mind rather than others, thereby bringing particular affordances into play. Imagine, for example, that we have a more flattish pebble, with two sides—thus doubling the 'storage capacity'! The affordance that the pebble then exhibits of having two sides might well be made use of for communication. This moves us into another area that has grown substantially in recent times when considering all manner of multimodal communication, that of *design*. Design—and, in particular, *functional* design (there are other kinds)—plays a central role for many communicative situations precisely because it can be seen as deliberately creating or increasing affordances for intended use and may, under the right conditions, also serve to *communicate* those intended uses.

The functional quality of designed objects lies in their being meant to be used in a given way, and this use is part of what it means to be that thing in the first place.

— *Forsey (2013: 31)*

This notion will be expanded on considerably as we proceed but, for the present, it will suffice to note that all communicative artefacts or performances can also be seen in terms of their affordances—i.e., how they support their own use—and those affordances may be deliberately shaped by appropriate design. This does not mean, of course, that all 'designed artefacts' are communicative—but some certainly are and considering these in terms of their 'fitness for purpose' provides a natural link to issues of 'usability'. That is, we can ask *how well designed artefacts meet their purpose*, for whom and under what conditions. This is to consider to what extent some multimodal object or performance supports its communicative functions *materially* and by design.

3.2.7 From beach pebbles to the holodeck in three easy steps

In this final subsection concerned with communicative situations, we will raise the bar considerably, building on the constructs that we have so far identified as being constitutive for such situations. At no point have we limited the forms of communications that may be involved and so we are still set up to let this apply across the board to any instances of multimodality that we want to address. We started with a naturally found object, our beach pebble, as a potential material object which may be inscribed with semiotically charged indications of meaning—i.e., our scratched out cross. It will now be instructive to start considering more closely the consequences of employing very different kinds of materialities for the canvases in our communicative situations so as to explore how multimodal meaning-making employing them plays out.

The basic idea here is that considering what a material can do is a good methodological clue towards interpreting what it is that a material might be being used to do. Conversely, considering carefully just what distinct kinds of communicative situations there are can aid in revealing just what it is that materialities in communicative situations need to achieve (for example, by design). So we have a two-way street here running between communicative situations and materialities. Not all materialities can support all kinds of communicative situations.

To begin, let us extend the materiality involved in our simplest communicative situation in several ways which do not qualitatively change the nature of that communicative situation. We might, for example, add the ability to show moving sign vehicles: then we have not a pebble but something like a film, TV or other video. This remains similar to the pebble in as far that the 'marks', once made, cannot be changed or withdrawn (although there is, of course, rather more potential here for supporting other, more complex communicative practices as well). Below we will call all such materials and the 'texts' that can be inscribed on them *immutable* to indicate their unchanging nature—once written, filmed, painted, etc., they remain as created. This has consequences for any communicative forms that are developed employing canvases of this type.

Not all materials are like this, of course. If we write with chalk on a blackboard, for example, the marks can be removed, changed, replaced quite easily. This already changes the communicative situation and, consequently, what can be done in such situations. Moreover, such changes or developments cannot happen on their own: there needs to be some actual communicating agents present—which moves us away from the situation on the beach, in the cinema, watching TV, etc. In fact, we need here already to draw on the whole community of potential sign consumers and producers that we emphasised above because they now start playing an active role.

We saw that the community of users provides individuals, or groups of individuals that can fill the structural roles in our communicative situation diagram. In principle, these fillers are relatively interchangeable. When we have a canvas that allows the marks it is carrying to be changed, then some members of the community of sign users (power distributions allowing) may take up both roles of consumer and producer—and this is even before we address some of the modern discussions of such supposedly 'new' constructs as new media 'prosumers', who both consume and produce media content.

To see this let us take an entire group of potential sign makers and, instead of a pebble, we take a piece of paper and a pencil. This changes many critical features of the situation, even if we make the perhaps unlikely simplifying assumption that all participants in the situation are entirely focused on the marks being made on the piece of paper and are not interacting in any other way. This could, for example, occur in a board or card game situation perhaps; or with a group of people only remotely present interacting via some visual display but without voice contact. Using the graphical conventions that we have introduced, the kind of communicative situation at issue here can be depicted as shown in Figure 3.4.

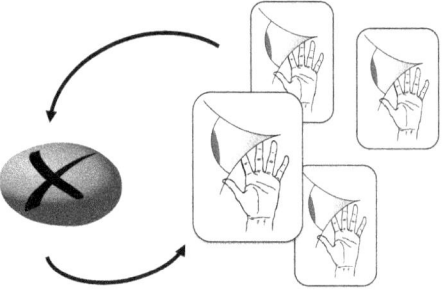

Fig. 3.4. Many sign producers/consumers interacting with respect to a shared external material canvas

Now we have the ability to interpret (signified by our 'eye' icons) *and* to produce (signified by our 'hand' icons) combined within single agents—always drawn from the larger pool constituted by the community of course. A group of these agents then may *together and collectively* produce material regularities on the pebble, paper, blackboard or virtual display that are, in turn, interpreted by the group. Again, we can have variations in the materiality and the precise interaction thus supported—the canvas may be 'write only' (e.g., a piece of paper and pencil with no capability of erasing marks after all, an online chat system without the option of deleting previous contributions, and so on), or allow more flexible changes as the interaction proceeds. Any of the regularities that the materiality of the canvas supports may be taken up for communicative effect and purpose.

There is also no requirement that the group of interactants are equal in their cap- **Roles** abilities with respect to interpreting or producing signs. Imagine with the paper and pencil that only one of those sitting around the piece of paper has a pencil and the others must just look on jealously. As we emphasised above, this is a very general attribute whenever we have groups of individuals involved—the roles that the members of the groups can take on collectively, in smaller subgroupings, and individually are always quite variable. We illustrated this with our pebble: even when we find that pebble on the beach and know what the cross might mean, this does not entail or necessitate that we are also in a position to make such crosses, nor that all of the members of the originating group were able to make these signs. Advances in technology are often claimed to distribute relations between sign producers and consumers more equally. At its most egalitarian, everyone around the table will have pencils and can make their contributions.

It is by no means difficult to find actual cases of this kind of situation. One highly developed example occurs when architects are designing buildings, where there is much visual and graphic communication mediated by the architectural sketches being made. In such situations, particular forms may be given particular, interaction-specific meanings, which then hold for the duration of the interaction or until they are changed. This is then a dynamic development of a very specific kind: the text expressive resources *themselves* change over the duration of an interaction. Moreover, individuals may also change their minds, withdraw or take back suggestions of form or shape, and so on. Forms which are unclear, with respect perhaps to their relative position, might be clarified by connection. This can also happen in many other forms of semiotic activity, however: imagine, for example, working with a draft document that contains suggestions for alternative paragraphs, alternative phrasing and so on. Here also the sign producers directly influence the form of the text that is taken to 'hold'. Another, more restricted situation of a similar kind can be seen in Cohn's (2013c) discussion of so-called Western Australian 'sand narratives', where tribal members tell stories graphically by making marks in sand.

The precise details of any such form of communication need to be examined on their own merits as part of empirical multimodal research. Particular properties of the situation can, however, be strongly determinative of the options taken up—and this is precisely the property that we will use extensively in our use cases in the last part of the book. For example, if all the participants can see (hear, interact with) the canvas and furthermore *know* that the others are in the same position, then this can be built on in their design of their communication. If this is not the case, then other possibilities follow. All of these aspects need to be borne in mind when considering analysis.

Time   A further source of difference is the temporal duration of the sign-making activity that has now been added in. If we can see the other sign producers making their contributions, then we have clearly added a temporal duration to the marks being made and this can itself be an important resource for changing the manner of regularities that are perceptible. Whether a line is made quickly or slowly is just as much a change in potential material regularities as having lines that intersect or not. This extends beyond any single 'mark' of course and adds a history of development: the signs, whatever they may be, are made in a particular order, and this information is also available for manipulation. We can therefore distinguish carriers of semiotic distinctions according to whether they change over time (as part of their semiotic activity) or are static and unchanging. This turns out to be a very basic and fundamental dimension of difference that is very useful for characterising some of the things that signs can and cannot do.

From here on in, the situations become interestingly more complex. First, let us consider what is no doubt the most primordial communicative situation of all, particularly as far as spoken verbal language is concerned—face-to-face dialogic interaction. For such situations, we must generalise the sign vehicle and the canvas to be not an external material object at all, but rather the *bodies* of the interacting sign producers/consumers themselves. This immediately adds a considerable flexibility to the kinds of material distinctions that can be used for communication: since we have still made no restrictions of sensory channels, the capabilities of this materiality stretch over spoken language, gesture, body posture, distances between interactants and much more. These will generally unfold in time but, in contradistinction to the scratches on our pebble, are *transient*: that is, once something has been said, or gestured, it is generally gone and cannot be inspected again. This is again a fundamental property of materialities that is almost certain to have consequences for the kinds of communicative situations (and corresponding texts) that may develop.

We show this situation graphically in Figure 3.5. Here all of the roles of sign producers, sign consumers and canvas are combined and we have an actual group of communicating interactants, or performers, drawn from the pool of potential communicators. There are many communicative situations of this kind, or which emulate this situation as far as they can, precisely because it is so foundational for human communication as such. In general, this will depend on the capabilities of the provided materiality, whether this is fully embodied or some restricted version of that (for example, on the telephone without video support).

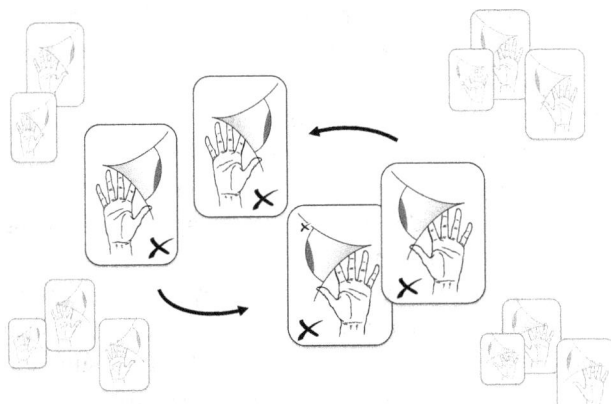

Fig. 3.5. Face-to-face interactive communication: the group of actual interactants are selected from a potential speech community, whose members may take on various additional roles, such as overhearers, listeners and so on

Organisational systems well known from such interactions will generally reoccur, playing a role for other communicative situations as well—for example, one of the most well known and long studied properties of face-to-face interaction is 'turn-taking' (cf. Sacks et al. 1974) by means of which the options for participants making contributions are regulated (or rather: regulate themselves). And, as will be the case whenever there are groups involved, the rights to contribute may be restricted by role allocations of various kinds.

Note again how we are still proceeding in small steps: if we had jumped from the found pebble to spoken, face-to-face interaction in general, we would have missed much that it is necessary to be able to distinguish on the way. Our accounts of different communicative situations would as a consequence have been intrinsically *fragmented*, just as could be observed from the variety of approaches we saw in Chapter 2. In contrast, we derive the properties of communicative situations here from the distributions of roles and the canvases involved. The type of canvas involved in face-to-face communication, for example, is what we can term a *symmetric (mutable and transient), dynamic communicative situation*. Roles may be more restricted (for example, only one pencil), or distributed in more interesting ways, ranging on the one hand through to strictly asymmetric situations, where only one person may draw (speak, sign, etc.) and, on the other, through to more or less non-interactive situations, where the right to make contributions is constrained. These role configurations must also be examined on their own merits—first, however, a material must be able to support multiple contributions of this kind.

The engagement of sign consumers with the canvases invoked can vary, therefore, according to whether the communicative situation involves an externalised, distant re-presentation of some content, or individuals may be *in* the communicative situation themselves, playing roles as interactants. When interpreters are themselves part of the situation, we can equally well talk of *performers*. Moreover, here, just as with the

Performance

manner of scratching used for our cross above, there is information that may come along within the material distinctions being drawn but separate from those intendedly shaped regularities imposed on the material.

When considering language performances, a traditional use made of this distinction, one which we will have to modify substantially however, is to divide verbal and non-verbal communication: verbal is what is said and meant, non-verbal, e.g., gestures and tone of voice, gives further unintended information. It may now be clear just why this is inadequate: we cannot distinguish such categories until we have mapped out the material distinctions that are applicable. There will always be such information and it will often be a powerful component of what is communicated (cf. Choi et al. 2006)—what does not work, particularly in the area of multimodality, is to assume that this aligns safely with some area of 'non-verbal signals', although often treated as such (Burgoon et al. 2011); we will see this in more detail when we return to corresponding use cases.

In summary, this type of communicative situation in general includes events occurring in time: the 'canvas' that meaning is inscribed on in that case is the complete physical situation in which the interaction takes place, constituted not only by the setting, objects, environment and so on but also the verbal utterances, facial expressions, postures and distances maintained by the interactants. Very similar canvases might also be *distanced*, however, as when watching a play.

The distinctions shown so far still correspond to communicative situations often discussed in studies of interaction and signs. The next canvas variation moves us into new territory and combines canvas, producer and consumer in quite a different way. In the move from our first, pebble-on-the-beach situation, the sign consumer is clearly quite separate and distinct from the canvas offered by the pebble: that is, the consumer is external to the canvas. Let us now reverse the perspectives and increase the size of the pebble and its markings so that it *encompasses* the sign consumer. That is, rather than the consumer being external to the canvas, let us assume that the consumer is placed *within* the canvas. This is suggested graphically in Figure 3.6, where we see the pebble much increased in size and the consumer placed inside.

Fig. 3.6. An embedded sign consumer

The kinds of situation characterised by this configuration are those of, for example, first-person computer games or embodied virtual reality setups. These are not acted with, or interpreted, from the outside: the interpreter is placed within their respective gameworlds or depictions. Nevertheless, these situations can again always be modified by the materialities involved: if we are playing a computer game on a traditional 2D computer screen, then we are still 'embedded' even though there is not a full, naturalistic 3D environment. Many further kinds of abstractions can also be applied.

Our graphic depiction here tries to remind us that the canvas and its signs are still created by a sign producer (or producers, we make no assumptions about whether these are group or individual activities). This is an essential component for this to qualify as a communicative situation and differentiates the situation from straightforward, natural embodiment, where one is just in the world perceiving it as usual. Although there are approaches to multimodality that attempt to apply multimodal analyses to the 'natural environment', we will again be more cautious. And, as before, it will be useful to start with some borderline cases and cases that do not quite 'make the grade'. There are now also cases that nicely transcend the boundary between natural situations and designed situations—as, for example, in all cases of *augmented reality*, where designed information is placed 'over' the natural environment by some intervening interface (which can, again, be either 2D or immersive). Such situations do not turn the natural environment into a communicative artefact, but they do open up a dialogue between the designed components and what those components are placed in relation to (cf. Krotz 2012).

Note, however, that it may not always be straightforward to decide whether we are in a 'communicative situation' or not: what, for example, about *designed* environments. An interesting case is raised by Kress and van Leeuwen's consideration of a particular child's bedroom 'as a text'—i.e., something that we may consider for interpretation (→ §4.3.2 [p. 131]). They suggest this on the following grounds:

> "This child's bedroom is clearly a pedagogic tool, a medium for communicating to the child, in the language of interior design, the qualities ..., the pleasures ..., the duties ..., and the kind of future her parents desire for her. This destiny, moreover, is communicated to her in a language that is to be *lived*, lived as an individual identity-building and identity-confirming experience in that individual bedroom." (Kress and van Leeuwen 2001: 15)

Kress and van Leeuwen present this as an illustrative example at the very beginning of their introductory discussion, despite the challenging philosophical, theoretical and practical issues of analysis that such a case raises. Let us unpack in slightly more detail the issues involved using the characterisations of communicative situations that we have presented so far.

Clearly, we have a canvas that has been deliberately constructed in a particular way and in order to have, at least indirectly, a particular kind of effect, or range of effects, on its principally intended 'audience': the child. At least some of our necessary conditions for communicative situations are thereby met. It is in this case, as with the video games mentioned above, far more difficult, however, to state what kind of 'content' is involved. This therefore serves as a good way of problematising the notion of content so that it can be appropriately broadened. In the case of the game, or the room, the design decisions made are not intended to lead to simple interpretations on the part of the 'receiver', but instead are specifically intended to make *certain manners and courses of behaviour more likely, or easier, than others*.

There are many communicative situations of this type. They even include such everyday artefacts as narrative films: these are intended not only to 'show' a story but to move their audience along particular, largely predictable, trajectories of emotional response (cf. Grodal 2009). In fact, this is one good candidate for a potential definition of film genres: a comedy is going to make the audience laugh, a horror film is going to scare them, and so on. Note that this is still perfectly compatible with the extremely broad conception of possible 'interpretants' for sign vehicles within Peirce's model of semiotics. Any limitation to consider only simpler 'messages' is thus unwarranted both for Peirce and the kind of multimodal foundation that we are setting out here.

Then, just as more traditional texts, as we shall set out in more detail below, 'guide' their readers or hearers along particular likely trajectories of interpretation, so constructed environments and audiovisual artefacts similarly 'guide' the emotional responses and likely reactions of their recipients. Indeed, there is in this respect rather little grounds to make distinctions: all of these forms of sign-making operate in rather similar ways to achieve similar effects. A written narrative attempts to be just as 'gripping' and 'immersive' as a film; it simply employs rather different resources to do so.

Thus, to summarise: in the film situation, the film makers construct an artefact that when interacted with produces a particular intended range of physical and cognitive responses. The only difference with the situation in the child's bedroom then appears to be that the behaviours to be carried out in the child's bedroom are far more 'generic': they are actions that might need to be carried out in any case, but which are pushed in certain directions rather than others. The film more tightly directs and binds in time the reactions of its receivers, but this is simply a matter of degree. Somewhere in between the two cases are computer gaming situations, where there is less direct control but still considerable freedom in some (designed) environment.

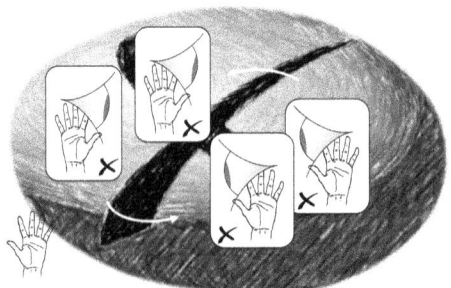

Fig. 3.7. Embodied face-to-face interaction in a virtual environment: the 'holodeck' situation

This then leads us to the logical next step in our development of communicative situations, depicted graphically in Figure 3.7. Here we take the group of interacting agents, as given above for face-to-face interaction, and place the entire ensemble *within* a created environment: that is, we place our group on Star Trek's holodeck. This is a complex situation with its own fairly challenging demands placed on the materiality and, even more, on the external producer of the environment using that materiality. We will return to some of the current debates and problems concerning such situations when we discuss modern computer games as an example use case for multimodal analysis in Part III of the book. The more the interactants here are allowed to influence and alter their embedding gameworld, the more challenging the situation becomes—and, again, just as explained above, there will always be variable roles and associated powers for those roles. This kind of situation has in fact already been picked out for considerable discussion, particularly concerning its relationship to the possibilities for *narrative* (Murray 1997). For the purposes of this chapter, however, we will remain focused on identifying the basic properties of the communicative situations that go along with such interactions, since these are largely determinative of what can be done with the 'medium' at all.

We have so far considered most of the theoretically possible configurations and re-configurations of combinations of sign producers, sign consumers and employed canvases. Each of these places constraints on the affordances required of supporting materialities. There is, however, one final configuration to consider, and that is when the canvas and the text it carries is combined with the sign producer. We suggest this possibility graphically in Figure 3.8.

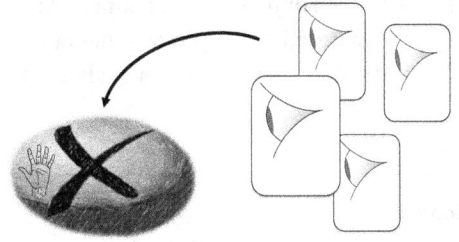

Fig. 3.8. Self-producing texts: cybertext and complex canvases

We take the simplest such case for ease of discussion—that is, a group of sign consumers, with no power or rights of access themselves, interpret a text that is able to change or generate itself. This refers to what Aarseth (1997: 75) calls 'cybertexts', texts which are produced by computation of some kind. Such texts may unfold entirely independently of their consumers, as in some cases of more experimental hypertext fiction or animated text, where the text that is to be read moves or changes with time. Alternatively, there may be more or less minimal interaction allowed for the consumers, such as clicking on links to navigate a hypertext. In this case, our graphic should include (probably rather small) hand icons within each of the observers, showing their role in the construction of whatever is then taken as a text. This is actually the externalised version of the embedding computer game situation of the holodeck variety. The externalised nature of the interaction then usually calls for additional ways of signalling action, usually constituted by some limited repertoire of operations permitted the sign consumers, such as clicking or entering various commands ('jump', 'run', 'shoot', 'build a house', etc.).

Generalised
transience

It is possible to make a further bifurcation at this point according to how the 'computation' is employed—and, in fact, we do not always need computation as changes may also be brought about by physical processes. On the one hand, then, we have actual changes in the text—what cybertext is aiming at—and, on the other hand, we have changes in how the text is accessed or made available. In particular, the canvas may support a whole range of different kinds of transience. For example, a simple linear verbal text may scroll horizontally across a screen while progressively disappearing so that there is only a limited time to read the text as it passes—many film credits use such a presentation strategy. A similar effect might also be achieved chemically, for example with magic ink that might disappear after being written.

With computational support arbitrarily complex forms of transience can be produced and we still have very little information about the effects that this might have. Consider, as a possible illustration, text which progressively fades after being read, only to have the key words reappear on screen (perhaps even over the top of later text) at regular or irregular intervals. We know very little about the effects on reception that this might have; it might even offer aesthetic and literary possibilities that could be beneficially put to work—perhaps someone has even already tried it!

What we can see, therefore, is that there is a very different kind of engagement possible in such communicative situations where roles of producer/consumer can be exchanged or shared. And to the extent that the material employed as a medium of interaction allows it, considerable fluidity and possibilities for fine-tuning shared interpretations may be opened up.

3.2.8 Summary of communicative situations

We have now set out some basic conditions that need to be fulfilled for a communicative situation to exist. These conditions lead directly to some further relevant distinctions. In particular, let us consider the relationship between the 'interpreter', i.e., the person who, as member of the relevant community of sign users, is recognising material traces and contextualising their use to produce interpretations or reactions (we will have more to say on the mechanisms of this production later), and the situation. It is rather traditional to consider such relationships in terms of communicated 'messages'—this is very much the view given in linguistics and in some of the more simple models of communication found in communication studies (cf. e.g. Cobley and

Schulz 2013). But for many multimodal artefacts and performances we will want to say more.

Canvases are the locus of semiotic activity: they present the interface that a medium provides for the interpreters of the 'messages' that the medium carries. Canvases may be complex and articulated in subcanvases of various kinds. One of the most complex is that formed by the world and face-to-face conversational participants. The bodies, sounds and postures of those involved in the interaction and situated in space each constitute a subcanvas for the others and for the situation as a whole. This shows how important it is to break down larger complex communicative situations into smaller, component parts. Each of these parts may be described in terms of a medium/canvas with its own materiality and affordances for communication, which may then in turn involve different forms of expression (semiotic modes: Chapter 4).

Summary: Canvases may vary in terms of their spatial dimensionality (2D, 3D) and whether they allow those communicating to be 'in' them, as embodied performers, or are distanced, as in playing a computer game. Or, performance is not supported, and asymmetric roles lead more to viewers and audience, as in a play or a film. They may even be static (as in a painting) or unfolding in time.

The canvas may be distanced, or one might be in the interaction oneself, in which case the canvas is extended by the further materiality of one's own body. So we can see that there are both 'production' possibilities and 'reception' possibilities to combine and, as emphasised by Drucker (2013), these are always to be considered actively, in terms of the possibility for interaction and interpretation they provide, not simply as physical properties of a situation or artefact. There are several characterisations of distinct kinds of canvases, or media, in the literature. Elleström (2010), for example, suggests: body, external physical, external 'immaterial': sound, projected light, and so on as well as some basic organisational dimensions that we will also draw on below.

Each of these have different kinds of affordances and so may leave consequences for the semiotic modes that they may carry. We will address this more systematically in the chapter following when we have completed the first round of discussion undertaken here. In all cases, however, it will be important to bear in mind that is not the material itself that is often relevant, but rather the *semiotic construction of access* to that material.

Semiotic access

3.3 Medium (and media)

We have mentioned at several points above the importance of the physical (or technological) carrier of 'communication' that may be being employed. It is now time to make this material aspect of any communicative situation do more work for us for the purposes of carrying out analyses. The particular canvases in which the material distinctions carrying communication may be inscribed have a variety of properties.

We sketched several of these in the summary to the previous section. These properties can help us considerably in focusing analyses of any multimodal situations we are confronted with. Knowing what material distinctions a canvas can support tells us much about the kinds of communicative forms, or semiotic modes, that may occur employing that canvas.

Media specificity

Note that this is *not* to say that particular materials can only carry particular kinds of meanings. This line of argumentation, which we also encountered in the previous chapter (→ §2.4.3 [p. 44]), most often misleads since it makes predictions that are easily falsified. It is perfectly possible for a canvas without movement to represent movement, as we see probably most dramatically in comics and other visual narratives; it is also straightforward for space to represent time, for colour to represent sound, or for space to represent touch. Such situations are not particularly special and we will see more of how such meaning-making activities work in the next chapter. What we are focusing attention on here, however, is the fact that while we cannot derive from a particular canvas any statements about *what* can be represented, we can say much about *how* it might be represented. A particular canvas will only make certain material distinctions available and not others, and this is what is available for meaning-making.

This point notwithstanding, we may nevertheless consider some kinds of potential for making meanings as *native* to the canvas. This can be seen informally as 'taking the lazy way out'—that is, if we want to represent colour, and the canvas supports colour variation, then we might just go ahead and use that material distinction for that range of meaning! Note that we do not *have* to take this course: it is always a 'decision' to be made by some community of users. Naturally, however, the more the properties of the canvas are used for distinctions that do not align, or perhaps even go against, the meanings that they are intended to invoke, the less likely the entire configuration is to be functional or effective. There is a long discussion of this issue in visual studies related to potential limits on conventionality (cf. Goodman 1969; Gombrich 1982a).

Considering what is native to canvases will be useful for addressing the many different uses that may be made of those canvases, many of which we will see spelled out in detail in our use cases in the final part of the book. In this section, therefore, we bring the chapter to a close with a systematic classification of canvases according to the material distinctions that they support. More specifically, we reverse the direction of characterisation in order to focus on the distinctions in our classification rather than on specific canvases or materials because it is these that are communicatively, or 'semiotically', relevant. We made this point above with respect to the transferability of certain regularities across material distinctions when discussing how the mark on our pebble could just as well (in certain respects) appear on a piece of paper or projected on a screen. Asking what some canvas 'can do' in some communicative situation will help characterise both what any communicative form using that canvas can make use of and what that form may need to 'work against' in order to cue its intended meanings. Both these aspects contribute to making any analyses that we pursue more focused.

The view of canvases as 'abstract' or 'generalised' materialities that we develop also offers a re-construal of the notion of *medium* that is particularly suited for addressing questions of multimodality—that is, 'medium' can now also become 'multimodality ready'. As we saw in the previous chapter, 'medium' is claimed from several disciplines and traditions and generally does slightly different work in each. This can already be seen in the title of this section: whereas 'medium' (in the singular) is commonly discussed in terms similarly to discussions of 'materiality', i.e, the stuff in which communication is inscribed, or the 'in between' that carries communication, 'media' (in the plural) are more commonly characterised in the context of 'the mass media' or 'communications media' that are inherently linked more strongly to social institutions and the distribution of communicative events. Both facets are important and need to find an appropriate place in an account.

As suggested in the previous chapter, however, extending the notion of 'medium', or 'media', out from media studies or media philosophy to deal with multimodality often runs into a familiar range of difficulties. In short, there is a need to deal with phenomena that stretch the original concept to, and sometimes beyond, breaking point. In this chapter and the next, we will divide the semiotic work to be done more equally since it is not necessary that all necessary theoretical work is done by a single construct. In the remainder of this section, our identification and classification of the canvases of communicative situations will overlap with some of the descriptive work that is commonly attempted with 'medium' when seen from its material side. In the next chapter, we will show in more detail how our semiotic construction of multimodality can then relate canvases to 'media' when seen more from their institutional and social side as well.

Clearing the ground in this way will place our 'multimodality ready' notion of medium within a constellation of constructs that offer a robust and useful category for guiding empirical multimodal research. We make particular use of this in Chapter 7, when we explain how to navigate our selected use cases as examples of practical analyses.

3.3.1 A 'simplest systematics' for the canvases of communicative situations

We saw in the previous section a rather wide range of communicative situations each relying on rather different affordances of the materialities to carry meaningful distinctions. Now we bring all of these possibilities together within a single multidimensional classification that we can use to position any communicative situation that we are likely to encounter. This will give us a classification device that we can use to index methods and approaches for analysis, depending on just which kinds of communicative situations we want to analyse. To articulate this classification, we will again work from the simplest situations we characterised to the more complex.

We also want to ensure, however, that we can identify continuities and similarities whenever possible. Certain standardly made distinctions—such as, for example, digital *vs.* analogue, or 'new media' *vs.* 'old'—are less useful in this respect and so will not be employed. We will see how they are nevertheless subsumed by our classification as we proceed.

We structure the discussion along several dimensions that directly feed into the classification scheme. The dimensions themselves are grouped according to broad 'facets' related to different kinds of material properties of the canvases involved—although again we emphasise that the materiality must always be considered relative to the demands made of it semiotically rather than simply considering physical properties; these semiotic demands will be taken up in their own right in the next chapter.

The first set of relatively straightforward dimensions we can identify are the following:

- We have suggested that materialities may support either static or dynamic (re-)-presentations. If such depictions are inherently dynamic, i.e., what is depicted changes with time independently of any action of a reader, viewer, etc., then they are **dynamic**; if such change is precluded, as with the cross on our beach pebble, then they are **static**.

- We have also seen that we require a dimension that covers the spatial dimensionality of the material and the (re-)presentations it affords: some canvases are flat and two-dimensional (**2D**), others are extended in depth, providing three dimensions (**3D**). Note that for current purposes this always refers to the properties of the material, and not to what is depicted within that material. Thus, a flat screen showing a natural landscape is still only two-dimensional (and so will need to employ additional conventions and projections if seeking to depict depth).

- Also related directly to canvas properties are issues of **transience**. Any of the texts produced according to the possibilities we are giving here may vary in their transience properties—that is: marks may disappear immediately, as with spoken language, or range over all kinds of intermediate forms of transience up to leaving 'permanent' marks, as on paper, indentations in clay tablets, or engravings in glass. Properties of transience are generally given by the material or medium employed and so are properties of the canvas and can certainly influence the kinds of meaning-making activities that are possible because the loads placed on memory and attention are quite different.

- There is also the broad distinction that we drew above between whether a sign consumer is external to the depictions or texts presented or *within* those depictions or texts. This is characterised in terms of **participant** *vs.* **observer**. Participants are necessarily *part of* the world being depicted, for example, when actors in a gameworld are on the holodeck. For participants, the perspective adopted is necessarily first-person. For observers, the relationship between the sign interpreter and the material interpreted is one of 'distance' or separateness.

It is important to note that these three dimensions can interact freely; that is why we describe them as 'dimensions'. Thus, the participant/observer distinction applies equally to 3D depictions: an observer may move around 'within' a virtual 3D technical diagram, but is not thereby 'part of' the diagram—i.e., there is (usually) no possibility of a first-person perspective *with respect to an object in what is depicted*. Again, this difference leads to quite different conventions and communicative systems being employed, which can then be usefully held distinct during multimodal analysis. Similarly the three-dimensional external depiction that is being moved within may itself be either static or dynamic; both options are possible.

The second set of dimensions emerging from the discussion above includes some rather more complex categories. We characterise these in terms of the work that 'readers' may themselves need to put into the *creation* of the 'text' they are engaging with in an artefact or performance. Clearly, with the mark on our pebble, the reader does not have to do anything with the mark apart from perceive it in some way—remember that our use of 'reader' here explicitly does not commit to any particular sensory channels: the mark on the pebble may be seen, felt, smelled or any combination that works! This situation represents the minimal level of work required of the interpreter of a sign.

There are several kinds of communicative situations, however, where the 'reader', 'consumer', 'player' has to invest significantly more effort to get at what is to be interpreted. Usefully for our discussion here, this particular kind of interaction between reader/user/viewer and what is read, used, viewed has been taken up quite extensively in the field of 'cybertext' studies. 'Cyber' in this context refers simply to the notion of 'cybernetics', originally proposed by Norbert Wiener in the 1940s, in which the idea of information carrying 'feedback loops' was seen as an important way of describing more complex computational systems as well as combinations of machine and biological agents within single systems. We will not be concerned with this development at all here: the only relevance it has is to explain why cybertexts were called cybertexts—these are also then 'texts', a term we use here as a shorthand that we will clarify theoretically in the next chapter (→ §4.3.2 [p. 131]), where some kind of 'feedback' cycle between the text and the text's consumers is assumed.

Literary studies of exploratory works of hyperfiction, for example, took the notion of interaction between reader and text as an important new direction for literary and artistic creation. As mentioned briefly above, the principal impulse that established this direction of research was that of Espen Aarseth (1997). We now build on some of the distinctions that Aarseth proposes for the purposes of our classification also, although Aarseth's concern was a very different one to ours. This means that many of the questions or distinctions that he probed will not be relevant for our current aims, which will allow us to cut the cloth somewhat differently.

For present purposes, then, we will only be concerned when the kind of activity or engagement required of the reader, viewer, player, etc. 'extends' the medium and so impinges on the kinds of communicative situations possible. In particular, we will take the term *ergodic* from Aarseth (1997) as one of the dimensions we need to incorporate.

Cybertext

Ergodic

Aarseth's motivation for selecting 'ergodic' as a term can be seen in the etymology he gives for the word: i.e., it combines the Greek word *ergon*, for 'work', and *hodos*, for 'path'. So in 'ergodic literature', the reader has to do work to select or make the path that they follow and it is this activity that 'creates' the text they encounter. This is then clearly suggested for something like hypertext, one of Aarseth's major concerns, where it is the reader's selection of links that drives the text forward.

Aarseth then defines what he specifically characterises as 'ergodic literature' as follows:

> "In ergodic literature, nontrivial effort is required to allow the reader to traverse the text. If ergodic literature is to make sense as a concept, there must also be nonergodic literature, where the effort to traverse the text is trivial, with no extranoematic [i.e., not just in the head] responsibilities placed on the reader except (for example) eye movement and the periodic or arbitrary turning of pages." (Aarseth 1997: 1; plus our own annotation)

Aarseth is essentially thinking here about linear verbal text as his model for what kinds of physical entities carry texts and so we will have to differ from Aarseth's characterisation in several respects. As we will see in Chapter 5 when we describe the empirical methods involving eye-tracking studies, for example, it is well known that there is very little that is 'trivial' about how our eyes obtain information about what we are perceiving—but this was not at that point of interest to Aarseth. When we move to full multimodality, any assumption of linear verbal text as a starting point makes less sense. As we saw in the range of communicative situations set out in the previous section, this would not be able to take us very far.

What we will maintain and now build on, however, is Aarseth's notion of the user/reader/viewer/hearer *having to participate* in order to co-construct the 'text' that is emerging within some communicative situation. Clearly this can happen to a greater or lesser degree, and in different ways, and it is this space of variation opened up by Aarseth that is useful for our current task—even though we end up with a rather more general characterisation of communicative situations than that originally envisaged by Aarseth.

Since we are now taking several terms from work on ergodic literatures and gaming, we should also clarify where some of the terms we are defining differ from their usage in those literatures so as to avoid confusion or misunderstanding. For example, some research uses the distinction static/dynamic to refer not to the presence of time and change within the medial representations as we did above but to whether the *text itself* (regardless of whether involving movement or not) changes. We see this as potentially confusing because work in most other fields uses static/dynamic in the way we have done here. For this reason, we will use the terms 'mutable'/'immutable' to refer to the texts themselves changing. As seen in ergodics and gaming, this is indeed a separate dimension to that of temporality or not within what is depicted in the text and so has also to be included even within a minimal classification such as that we are developing here.

Some simple examples will clarify this: a comic strip where the *contents* of panels may be changed by the 'reader' is static but mutable; whereas a film where scenes may be replaced for others by the 'viewer' is dynamic and mutable. Traditional films and comics are not of this kind—they are dynamic and static respectively, but both are immutable. A piece of 'hyper-cinema', where the viewer can explore a pre-given but perhaps not always accessible network of scenes, is also dynamic and immutable, because the text itself (the network of film fragments) does not change—even though the viewer may not find all of that text. We position this alternative more precisely below.

Communicative situations can thus be set out on a dimension based on the degree and extent of these 'ergodic' requirements and capabilities. We suggest four distinguished levels of required work that usefully distinguish quite different classes of possibility, extending the hierarchy of levels for types of 'texts' given by Aarseth (1997: 64). In addition, we begin our transition here from talking about types of texts to classifying the kinds of *canvases* necessary to support texts, or communicative situations, of the kinds we have identified.

1. First, the least complex pole of this dimension takes in traditional, linear and unchangeable texts that just need to be read, preferably ignoring the fact that that may happen to be printed on a page or shown on a screen that needs to be scrolled. This is the form of presentation that Twyman (1987) describes as 'linear interrupted'—the text in this case tries its hardest just to appear as a linear string of items. This is naturally limited to quite specific *kinds* of texts and actually occurs rather rarely. Again, our beach pebble with its single cross would fall into this category. We term canvases that only support marks of this kind **linear**. Linearity

2. Second, this becomes more complex when linearity is relaxed and the text or depiction makes use of, at least, the two-dimensionality of its supporting materiality—what Bateman (2008: 175) defines as 'page-flow'. With this development, possibilities of layout and other diagrammatic depictions emerge in which the order of consumption of any depictions is no longer fixed to the material presentation itself. That is: readers/viewers have themselves to determine how the information presented is to be brought together. This intrinsic 'nonlinearity' in the depiction leads to significant semiotic possibilities that are not available in linearly presented materials. If our pebble were big enough, we might use the space of its surface in this way—again emphasising the important role of the *use* that is made of material rather than just the material itself. Page-flow

Although this area of potential was excluded in Aarseth's discussion on the grounds that the work of the reader/viewer is accomplished 'internally', by perception and interpretation, in the context of multimodality we cannot ignore work of this kind. Accordingly we characterise such depictions and materialities as **micro-ergodic**. By this, we mean to capture that there is still substantial work to be done by the sign interpreter, even though this may require eye-tracking or brain studies to be revealed. Again, a further range of meaning-making possib- Micro-ergodic ilities are supported when the canvas is capable of such combinations of signs. We characterise the kind of work that the reader/viewer must perform here as *composition*.

Note that it is only with the kinds of materiality required here that we enter the entire realm of pictorial and other visual depictions at all. Although these are often stated to stand in contrast to, for example, 'textual' renderings by virtue of being 'all at once' depictions rather than the linear characterisation of the verbal, this is not accurate. There is always a selective inspection of pictorial material as has been revealed by eye-tracking and other kinds of experimentation (→ §5.4 [p. 159]). For this reason, largely following Bucher (2011), we consider the designation non-linear as appropriate for such canvases as well.

Immutable ergodic

3. Third, we cross the border into the territory that Aarseth describes as properly ergodic, where the reader/viewer has to do substantial work to engage with the text they encounter. In contrast to Aarseth, however, we also here explicitly allow the full range of both 2D and 3D micro-ergodic materials to participate in such depictions. The additional kind of work that the reader/viewer must perform here is characterised, again following Aarseth, as *exploration*. That is, there may be a web of 'textual' nodes to explore linked in a hyper-structure, but the content and organisation of that web does not itself change: it can only be explored, mapped out, etc. This web (and the text it constitutes) is thus **immutable** as distinguished above.

Mutable ergodic

4. Fourth and finally, we reach the most complex category that we will need to address: **mutable ergodic**. In these cases, the reader/viewer/user/player may themselves alter the organisation or content of the text. Although this is not really considered by those concerned with cybertexts, it is interesting that this characterisation is also the classification that must be adopted for completely traditional, face-to-face spoken interaction. This also serves to demonstrate just how complex the 'primordial' speech situation is! Conversations are constructed dynamically by their participants and so are necessarily mutable. They may also make free use of any materials that lie to hand, and so can draw on the full range of other kinds of 'texts' that are presented in our classification.

It might not yet be quite clear to you why drawing these kinds of distinctions, and drawing them in this way, will be useful. In fact, a classification of this form offers some fairly radical ways of recasting a host of existing problems cross-cutting a diverse body of literature scattered across several disciplines. In order to fully unfold this potential, however, we need to add the constructs set out in the next chapter. It may then be useful to return to this chapter as needed when working through the chapters which follow and, particularly, when we build on the classification in our multimodal 'navigator' in Chapter 7.

3.3.2 Summary of the classification

For the time being, this completes our general classification of the kinds of canvases that may be drawn upon by communicative situations. We summarise the distinctions in this section graphically in Figure 3.9.

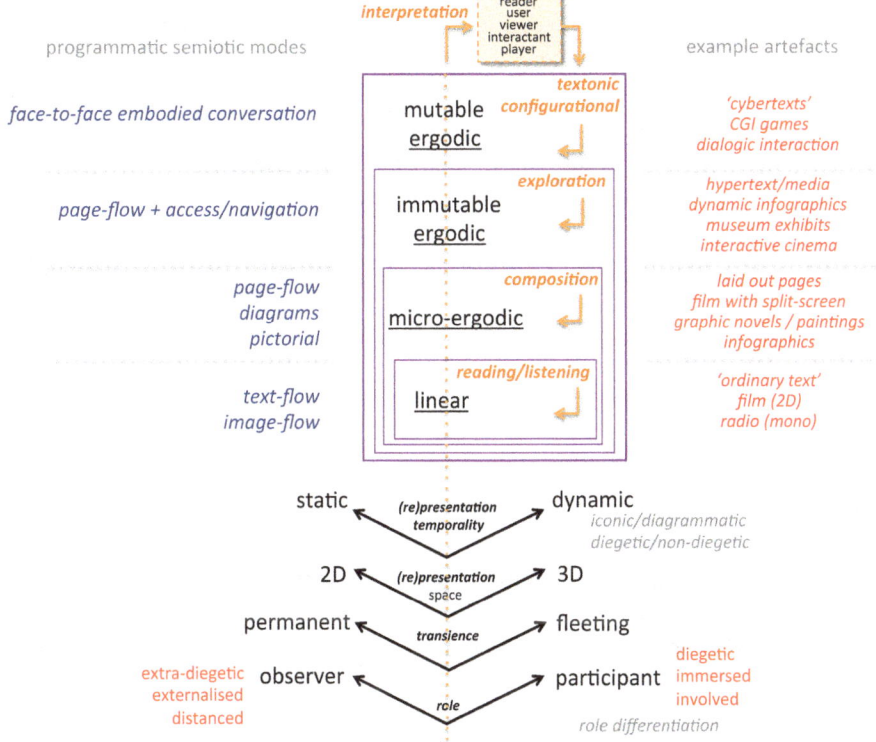

Fig. 3.9. A simplest systematics of communicative media according to the affordances of their involved canvases

The lowermost four distinctions repeat the basic 'material' properties of our first set of dimensions introduced above. The central boxes then characterise the increasing degree of 'work' that a reader, interpreter, viewer, player, etc. must perform in order to get to, or to create, the object of interpretation. The boxes are embedded one within the other to capture the principle that the more 'powerful' canvases can generally include less powerful canvases in their scope: that is, just because a canvas supports a user changing its inscribed 'text' does not *require* that this happens. On the right-hand side of these boxes are example artefacts that illustrate the kind of media that employ each of the types of canvas described. These are by no means intended to be exhaustive and some may be further differentiated internally: more precise characterisations will be offered for some of them in our use cases in Part III of the book. The examples suggested here are purely illustrative but should nevertheless give some idea concerning just what kinds of artefacts are intended.

The dashed line and arrow running upward through the nested boxes marks a path of interpretation for the designated 'sign consumers'; the corresponding downward arrows pick out strategies that those sign consumers need to employ to engage with the respective canvases. These latter then extend the 'exploration' and 'configurational' strategies, or 'user functions', suggested by Aarseth (1997: 64–65) for cybertexts by including more directly perceptually anchored notions of 'composing' blocks of information into signifying wholes in the case micro-ergodic canvases and straightforward reading or listening for linear media and their canvases.

Finally, the items on the left-hand side of the boxes list potential 'semiotic modes' that employ the possibilities of their associated canvases. The nature of semiotic modes will be a major topic of the next chapter—we list them primarily without further comment at this point.

3.4 What this chapter was about: the 'take-home message'

We have now provided a rather extensive consideration of just what communicative situations are independently of any particular commitments to sensory modalities, to verbal language, to pictorial representations, or any other specific form of communication. We used this to draw out distinctive properties of the 'canvases' that can appear in such situations for supporting different communicative practices. Finally, we organised these distinctive properties into a classification scheme for media and their canvases in general.

It should now be possible to position any communicative situation and the materiality it involves according to this classification system. Doing this allows the multimodal analyst to focus attention on just those semiotic systems that support communication within the bounds defined. In subsequent chapters we will show this classification at work as the first step both in organising any material under analysis and in selecting the prospective analysis methods that can be employed to characterise the material under analysis multimodally.

Achieving clarity concerning just what constitutes some case of multimodal communication is critical for pursuing analyses in a productive fashion so as to achieve incisive results that go beyond simply describing what is being analysed. As communicative situations become more complex, perhaps drawing on new technological capabilities and combinations of meaning-making strategies, being able to pick apart the constitutive contributions of material and what is done with that material will prove crucial. As multimodal analysis proceeds, we will see that we can gradually fill in the 'programmatic' semiotic modes listed in our diagram above with empirically motivated characterisation that in turn can be used to explain and predict communicative behaviour using them.

Table 3.1. Four basic orienting questions to be asked before multimodal analysis of any communicative situation

	Who are the 'sign makers'?—this can be any collection of agents, which may also exhibit diversified roles with differing rights and obligations. For one of the most complex such groupings, consider the hundreds of names that now regularly appear at the end of major films.
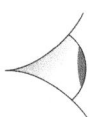	Who are the intended 'sign consumers'?—this can again be any collection of agents, also possibly exhibiting diversified reception roles. Examples might include medical professionals interpreting an X-ray image together with a patient, art critiques and laypersons in a gallery, textbooks for teachers and pupils, and so on.
	What is the canvas in which, or on which, the sign makers are working? —this can be absolutely anything extending in space and time and where material distinctions can be drawn, including the bodies of the sign makers and their environment.
	What is the *time profile* of the entire constellation? – are the signs made once and for all and then left, are they produced synchronised with the actions of some of the producers and consumers, do they exist independently of their producers at all? And do they last?

Finally, then, let us summarise the practical consequences for analysis that the discussion in this chapter leaves us with. *Before* beginning any analyses, you need to answer the questions set out in Table 3.1 in order to be explicit about the boundary conditions that apply for your objects of investigation. These questions orient themselves to our diagrammatic summaries of the distinct kinds of communicative situations possible given throughout Section 3.2 above. Remember also, therefore, that the various roles in the communicative situation may be combined into very many different configurations. The canvas may be combined with the sign maker in the case of automated webpage production, the sign consumer may be 'within' the sign in the case of immersive video and so on.

Consequences for analysis

The answers you give to the questions may consequently be more or less complex. When complicated, this does not mean that you need to address every last subtlety revealed, but it *does* mean that you will be more aware of just what is being left out and what could potentially have further implications for your study. This then offers a set of strong organising principles for engaging flexibly with any particular piece of practical multimodal analysis that you undertake.

4 What is multimodality?
Semiotic modes and a new 'textuality'

Orientation

In this chapter we provide the second and final batch of principal definitions for key multimodal terms: first, the central notion of 'modality' will be clarified and, second, the primary use of semiotic modes in the service of creating 'texts' will be explained. The relationships between modalities, texts and the communicative situations and media we introduced in the previous chapter will also be illustrated. The result offers a robust and very broad conceptual framework for dealing with all kinds of multimodality, wherever and however it occurs.

The previous chapter gave us the tools necessary to distinguish fine-grained details of various media—or, as we defined them there, of 'canvases' within which, or on which, signifying practices can be inscribed. In this chapter, we address the different forms these signifying practices can take—in particular, we set out a detailed account of just what the 'semiotic modes' are that we have been mentioning throughout the book. Semiotic modes actually form the foundation for any real account of 'multimodality' and so they need to be characterised in a way that is robust enough for driving empirical research while still being open to theoretical and methodological developments as they occur. Definitions should in fact be evaluated on the basis of how much they allow 'predictions' concerning modes that do not even exist yet—otherwise work on multimodality will always be trailing behind what has been done. We therefore begin with a precise description and definition of semiotic mode.

In a second part of the chapter, we place this characterisation of the modes of multimodality against a background that gives important central positions to the notions of 'discourse' and 'text'. Both terms are highly contested and slippery in use, packing entire histories of debate and disagreement across disciplines where they may be either employed or deliberately avoided. Both are, however, essential for picking out how semiotic modes are used in concrete contexts of application. Discourse, or rather the particular way in which we focus on phenomena of discourse here, is a basic mechanism that we assume to operate in all semiotic modes and, at the same time, supports our use of 'text' to talk about the results of using semiotic modes for communication.

4.1 Semiotic mode

In the preceding chapters we have mentioned several of the ways 'modes' have been treated across some quite diverse approaches and disciplines. In Chapter 1 we gave several quotations illustrating such definitions and pointed to some of the reoccurring problems that arise when definitions remain vague or programmatic. Equally, we saw problems of identifying semiotic modes too rigidly with individual biological sensory channels. Our conclusion was that there were always certain difficulties that remained unsurmounted—particularly concerning, on the one hand, distinctions between modes and the kinds of overlaps that different materials seem to require and, on the other, the relations that need to be drawn between modes and specific communities of users.

...the question of whether X is a mode or not is a question specific to a particular community. As laypersons we may regard visual image to be a mode, while a professional photographer will say that photography has rules and practices, elements and materiality quite different from that of painting and that the two are distinct modes.

— *Kress et al. (2000: 43)*

Now, with the background that we have constructed in the previous chapter, it will be possible to take the next step of providing a working definition that can be put to use in further rounds of analysis. This will then feed into our characterisation of how to go about organising your multimodal research in the chapter following and the many use cases presented in the final part of the book.

4.1.1 The necessary components of a semiotic mode

All accounts of semiotic modes that have been put forward in the literature tend to agree on one point: on the one hand, modes appear to have a material dimension, relating back to the sensory channels that are used to perceive them but, on the other hand, they also exhibit a semiotic dimension, i.e., the material used is given some kind of significance by its users. We saw this from the perspective of the various kinds of communicative situations that can develop in the previous chapter.

Opinions then begin to diverge concerning the question of how the material and the semiotic are related. Some are prepared to talk of modes in terms of perception— and hence of the materiality, others are prepared to countenance semiotic modes that 'free' themselves of materiality, supporting the idea that a given mode may have different material manifestations. The common assumption that language is a semiotic mode that may appear in both written (i.e., visual) and spoken (i.e., aural) form is one consequence of this. Many have discussed this distinction before, as it relates broadly to Saussure's distinction between signifier and signified (→ §2.5.1 [p. 53]), to Hjelmslev's widely cited distinction between *expression* and *content* (Hjelmslev 1961 [1943]), and so on. The distinction as such also corresponds to the traditional view of semiotic codes, which also have two sides—largely following Saussure.

Material vs. semiotics

For our present purposes of securing a better foundation for multimodality research, however, we now take a different course. We will anchor *both* facets, the material and the semiotic dimension, into our working definition. That is, we take it to be a constitutive feature of a semiotic mode that it has both a material side and a semiotic side. The material side determines—as we saw in the previous chapter—the kinds of manipulations any semiotic use of that material may draw on: this is captured in the notion of 'canvas' that we employed (→ §3.2.5 [p. 87]). The semiotic side determines which kinds of distinctions in that material are actually 'meaningful' in and for that mode. We then push a step further and make a further division 'within' the semiotic side. Rather than simply talking about the 'signified' or 'content' for some expression, we separate out two very different levels of organisation of that content.

Content: form

Our first division within content requires us to characterise the particular forms that can be made from any selected material—that is, any given material might be used by a semiotic mode in specific ways that may well differ from how the 'same' material might be used by other semiotic modes. We thus do not make the assumption that all 'visual' canvases are used in the same way; similarly, it is not enough to talk simply of 'sound' being the material, since one semiotic mode (e.g., 'music' perhaps) does very different things with that material than another does (e.g., spoken language).

The lack of this important degree of detail is one of the main reasons why empirical analysis of multimodal artefacts and performances has proved itself to be a slow business. Talking of the 'visual', for example, draws too loose a net to reveal differences in use, differences that are essential for characterising semiotic modes. Written language as a visual entity, for example, is used completely differently than are pictures, another class of visual entities. In the terms we have introduced concerning (virtual) canvases, these semiotic modes are imposing different organisations on their material which we now need to gain practical and theoretical control over.

Materiality always brings *its own constraints* to bear, of course. For example, any broadly spatial material is going, in all likelihood, to show things to be more or less closely positioned to one another. This is what our perceptual system makes directly available. That is, we cannot see a spatial field without also having our brain tell us that certain things are closer together than others. We cannot 'decide' not to notice this—our perceptual system is wired to include this information (→ §2.2.1 [p. 27]). Similarly with temporal materials, up until a certain granularity, we will also perceive certain things as occurring *before* or *after* others. Again, this is not something that we can choose to ignore. But, on top of this foundation, semiotic considerations will necessarily intervene and provide access to these distinctions in the service of their own, quite diverse kinds of signification. One might see spatial proximity as indicating that some things are 'about the same topic', others might see spatial proximity as indicating some things as 'belonging to the same word' or 'having the same value', yet another might take it as indicating that two things occurred close together *in time*. We simply will not know until we have specified what semiotic mode is at work and performed the prequisite empirical studies to find out.

McGurk effect

The fine-grained interaction of semiotic modes and (the perception of) materials can also be complex. One well known illustration of this is offered by the McGurk effect (McGurk and MacDonald 1976). The McGurk effect demonstrates that how we will *hear* certain speech sounds depends on the shape of the mouth that we *see*. This perhaps surprising result is extremely strong; we cannot decide 'not' to perceive in this way.

There are several videos available online which let you do the experiment yourself (and on yourself!). Thus not only are distinctions built into the possibilities provided by certain materials—which, it should be emphasised again here, may cross 'sensory' divisions—but the kind of perceptual access that we receive to those materials is also mediated by the semiotic modes that apply. Just *what* a semiotic mode does with the material possibilities on offer is a matter for each individual semiotic mode.

Going further, a semiotic mode may also define particular ways of combining its signs or sign parts so that specific meanings may be recovered or evoked. Obviously the most developed semiotic modes in this respect are any involving language. Consequently it is here that notions of 'grammar' are strongest, although certain principles of a broadly similar nature have been suggested to appear in other modalities as well. It is also possible, however, that a semiotic mode works more with simpler organisations more reminiscent of a 'lexicon'. Whereas a 'grammar' combines signs productively, a lexicon lists the possible forms directly. Such *lexically*-organised semiotic resources consist of collections of signs with little additional organisation, simply associating forms and meanings. In contrast to this, *grammatically*-organised semiotic resources place their distinguishable signs within a productive system of meaning potential. This is one important way of providing the power to compose simpler signs into complex signs—i.e., by employing structural mechanisms analogous (but usually different in detail) to those of grammar in language.

David Machin's introduction to multimodality characterises these alternatives in terms of the following contrasting schemes (Machin 2007: 5):

> Lexis vs. grammar

- *lexical approach*: Simple sign → meaning
- *grammatical approach*: Complex sign → lexicogrammar → meaning

The schemes are typically considered to form a continuum. Any particular semiotic mode may then lie somewhere between the two poles—either being more or less lexical, or more or less grammatical, i.e., making use of more complex kinds of combinations.

The necessary link we emphasised above between semiotic modes and communities of users also plays an important role here. Just as power and knowledge are distributed within society, so is the ability to deploy particular semiotic modes. Thus, expert users of a semiotic resource (such as a professional photographer) might use those resources in a way that composes distinct material effects to create meanings that are simply not available to the novice. Similarly, the learner of a new language might only have available a collection of lexical items, while the conditions under which these items might have a desired effect and the ability to compose these signs into more complex utterances only emerges later. In other words, the use of the language (as a resource for meaning) moves from one that is lexically organised to one that is grammatically organised. This gives distributions across both members of a society and within members of a society across time.

> No semiotic resource is by 'nature' either 'lexically' or 'grammatically' organised. ... it is possible that a mode is 'grammatical' to some of its users and 'lexical' to others, especially where there is a gap between producers and consumers, and where the producer's knowledge is kept more or less secret.
> — *Kress and van Leeuwen (2001: 113)*

The account of semiotic modes we have given so far is then still broadly similar to several of those offered in existing introductions to multimodality. And, in contrast to traditional semiotics, all of these explicitly put more emphasis on the possibility of specifying what happens when signs, or parts of signs, *combine*. We see here the strong influence of linguistics and, in particular, approaches to grammar that were not available when the basic tenets of semiotics were set out by Saussure, Peirce and others (→ §2.5 [p. 51]).

This is precisely what we need when we turn away from the rather straightforward examples discussed in much of the semiotics literature. Real signs typically come in complex constellations and this raises particular methodological challenges of its own, going far beyond the basic notion of a sign 'standing in' for something else—such as we might see in an account of a traffic light of the form 'red means stop'. The colour red used in a painting together with the image of a knife, for example, definitely does not mean stop in this particular combination and we need to describe what is happening in order to explain the difference in meaning. This is part of the value of talking about an organisation like grammar: we begin to have a tool that allows us to address complex 'signs' with their own rich and detailed internal structure—whether that be a painting, a verbal text, a piece of music, or face-to-face interactions between speakers using a collection of electronic devices (as in the first example we discussed in Chapter 1).

Content: discourse

Our second division within content moves beyond such approaches and specifically addresses the issues of how we *make sense of* any selection of organisations of form in material. Adopting this as an explicit requirement for a semiotic mode takes us into new territory, although it does have correlates in a variety of other positions— particularly with those working broadly in terms of 'pragmatics'. Here we orient to a different tradition, however, and will talk instead of the **discourse semantics** of a semiotic mode. In Chapter 2 we suggested that we would be seeing discourse as playing a significant role for interpretation in any semiotic mode (→ §2.5.2 [p. 63]). We now fill out more precisely what we mean by and, in the sections following, show its implications.

Discourse semantics

As a first introduction to the notion, however, we see the task of discourse semantics within any semiotic mode as *relating particular deployments of 'semiotically-charged' material to the contextualised communicative purposes they can take up*. The discourse semantics of a semiotic mode thus provides the *interpretative mechanisms* necessary (i) for relating any particular forms distinguished in a semiotic mode to their contexts of use and (ii) for demarcating the *intended range* of interpretations of those forms. In short: all semiotic modes are taken to include a discourse semantics, otherwise we would not have a place for capturing how to relate particular uses of these modes to their interpretations in context.

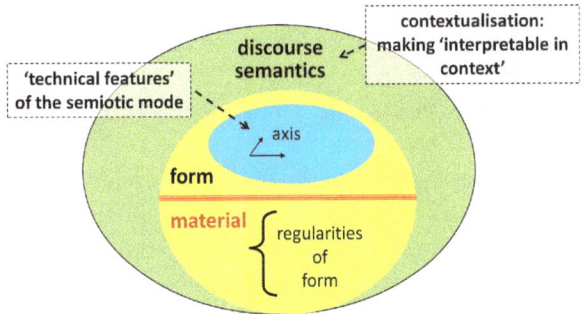

Fig. 4.1. Abstract definition of a semiotic mode. All semiotic modes combine three semiotic levels or 'strata': the material substrate or dimension, the technical features organised along several axes of descriptions (abbreviated as 'form') and the level of discourse semantics.

For ease of reference and as a visual reminder, the overall three-way division across materiality, form, and discourse semantics that we have introduced so far in this chapter is shown in Figure 4.1; more detailed descriptions of this definition of semiotic mode can be found in Bateman (2011, 2016). This remains very abstract of course, but it is good to have this part of the definition behind us so that we can proceed to its implications for practical analysis.

Going back and forth between the concrete examples and the abstract definition will be helpful for understanding both what is intended and its applications. We will see many cases being played out in the analyses discussed in our use case chapters in the final part of the book. There too, it will be worthwhile returning regularly to the present abstract description in order to fully familiarise yourself both with what is said and with its consequences for multimodal description and explanation. The ramifications of the definitions, both of multimodality and of semiotic modes, will concern us throughout the rest of the book. Once we have the notion of a 'discourse semantics' firmly in place as part of our notion of mode, certain questions and problems of multimodal analysis become far easier to deal with, as we shall now see.

4.1.2 The role of discourse semantics for semiotic modes

As a first example of making the notion of discourse semantics do some work for us during analysis, consider Kress's (2003) account of the difference between 'writing' and 'image' that he discusses when talking about literacy and which we mentioned in Section 2.4.4 in Chapter 2. Kress talks about the 'affordances' of the two 'modes' (cf.

Kress 2014: 55–58; Jewitt 2014a: 24–25), and we made use of this above in our discussion of media canvases. Kress presents 'writing' and, even more, 'speech' as necessarily organised by the logic of time and sequence, and 'image' by the logic of space and simultaneity. This is important and always needs to be considered when discussing and distinguishing modes—i.e., what are the affordances of those modes? But we can now also be more discriminating to tease apart rather different contributions, each of which may have different consequences not only for literacy but for analyses in general. And this teasing apart will turn out to be essential for developing more incisive methodologies for analysis.

Kress suggests that if, in a biology class, a pupil has to describe the relationships between the nucleus of a cell and the cell as a whole, this will fall out differently depending on whether the information is expressed in language or in an image—"*the world told* is a different world to *the world shown*" (Kress 2003: 1; original emphasis). In language (specifically here in English but in many other languages as well) one has to use a relating verb, in this case one of 'possession', 'the cell *has* a nucleus'. Naming this relation is a commitment that is required by the use of language. In contrast, when drawing the cell, the pupil has to put the nucleus somewhere with a specific location in the boundaries drawn for the cell—again, this is a *commitment* forced by using the mode of drawing. Where this needs to be modified, however, is by recognising that the actual commitments made are not a property purely of the canvas used, i.e., of having a visual canvas. The actual commitments made can only be ascertained when the *discourse semantics* of the mode being used are also considered. That is, although the nucleus may be placed at some particular measurable point within the boundary of the cell, *this does not mean that the drawer is committing to cells always being in that particular specific measurable position.*

The degree of precision that is intended in fact depends on a far finer-grained decision concerning the mode in use—in a scientific technical diagram (including architecture plans) for example, there may well be the kind of commitment to a (scaled) representation of space and position; in a drawn sketch of the cell there is generally no such commitment. The commitment made is actually far weaker: commonly of the form that the nucleus is 'somewhere in the cell like this'. There is therefore a significant buffer between what is 'actually' measurable from some physical qualities and *what is being done with the physical qualities*. And for this latter, we need the discourse semantics in order to tell us just what the distinctions are being used for.

For this reason, we consider it is better to restrict the use of 'affordance'—appealing though the term is—more to its original senses of being related quite closely to perception and not broaden it to take in what can be done with modes generally. Spatial and visual organisation quite correctly *affords* the expression of relative positions, containment, touching and so on. But what a particular semiotic mode then *does* with those affordances, i.e., the affordances of the canvas or material of any mode, depends on the discourse semantics. This applies equally to language. Although there is a rather strict temporal sequence in, for example, spoken language, this does not mean that that temporal sequence has the same import when we are working out what was said. That is, just because two sentences follow each other, does not mean that what the second sentence is telling us happened after what the first sentence is telling us: all we need is a sequence of words such as 'But before that…' and things are quite different.

This flexibility of interpretation again shows a discourse semantics (in this case that of spoken or written language) intervening between the material distinctions drawn in the canvas and what we *mean* with those distinctions. To aid discrimination, we label what can be done with a semiotic mode with another term that Kress suggests, **reach**—i.e., different modalities may have different 'reaches' in terms of what can be done with them and what not. The reach of a semiotic mode will usually be a refinement and extension of what its material carrier affords.

Characterising semiotic modes clearly with respect to their commitment to particular discourse semantics is definitely important for literacy, since those discourse semantics have to be learned; but it is also crucial for describing semiotic modes at all, since without this level of information, we have few grounds for making sensible distinctions. Each semiotic mode has its own discourse semantics, and this can then be used to explain why even some quite similar looking representations may have rather different properties. The discourse semantics is the necessary first step in making sure that we not only say what modes are being used, but can also respond to this appropriately in our actual descriptions and explanations.

This semantics is important because it helps us to avoid making over-general statements. Accounts in the Kress and van Leeuwen tradition often state, for example, that relative spatial position (on a page or screen) has an *intrinsic* meaning in its own right: e.g., the top of the page is more 'ideal', 'how things should be', whereas the bottom of the page is more 'real', 'how things are'. This meaning is then assumed across all uses of pages or screens without further empirical testing or evaluation. We need instead to separate out the phenomena involved more carefully. There are affordances of the canvas that clearly make available notions of relative vertical positioning; and these, according to some interesting neuropsychological studies, even appear to have some differences in 'meaning' that are most likely linked with our embodied perception and being in the world (cf., e.g., Coventry et al. 2008). But what these properties of the canvas are actually used for in a given semiotic mode is a quite separate question: some may be correlated with Kress and van Leeuwen's ideal/real distinction, others may not. What is then important is the actual empirical work necessary to know the answer. As with much earlier work in multimodality, the proposals made need to be couched as suggestive hypotheses concerning what might be going on which need then to be investigated empirically. They should *not* be taken as ready-formed principles of multimodal organisation in general.

Finally, it needs to be noted that it is very unlikely that a semiotic mode will 'go against' the affordances of its material. That is, to take the example of the drawing of the cell again, placing the nucleus spatially within the drawn boundaries of the

> Semiotic Reach

cell is hardly going to mean, in any semiotic mode, that the nucleus is to be understood as *not* containing the nucleus! Similarly, if something is to be expressed as being hot, it is going to be unlikely that some semiotic mode evolves that represents
this through a dull blue colour, or that a jagged, piercing sound (note already the spatial metaphors) be represented in some visuospatial mode as a smooth curved line
(cf. Ramachandran and Hubbard 2001). We mentioned this material contribution to
meaning for both sound and vision in Chapter 2. In the account that we are proposing
here, this is another reason that we do not allow the 'material' and the 'semiotic' to
come apart. Both dimensions always need to be considered together in order to avoid
making our responses to signifying material mysterious.

This actually picks up again on some semiotics fundamentals already present in
Peirce's work (cf. Krois 2011: 207). Since all our dealings with 'signs' are via embodied
perception, we will always have full bodily responses to the marks and traces we encounter. That is, we do not just see a drawn 'line' as a visual element; we also perceive
aspects of the force and manner in which the line was created—a principle elevated
into an art form in its own right in Japanese traditional calligraphy, for example. Correlations of this kind have been described further within a semiotic model by Theo
van Leeuwen (1999) in terms of 'experiential meanings'. We must always note, however, that these are *correlations* and not a shortcut to the meanings being made by a
semiotic mode. More extensive empirical work is necessary here as elsewhere.

Discourse semantics do add significantly more to the *power* of semiotic modes,
however—that is, you can do more with a semiotic mode because of the presence of
its discourse semantics. Let us look at just one very simple example; further examples
follow in later chapters. It is often said that 'pictures can't say no', that is, they cannot
express 'negation' because they can only show what is there, not what is *not* there (e.g.
Worth 1982). This is largely true *until we add the possibilities of discourse semantics.*
Consider the warning sign shown on the left-hand side of Figure 4.2 and ask yourself what it might mean. One of the problems with this sign is that it attempts to show
'everything at once', in a single sign. Certain information is communicated by resemblance as usual in such representations: we have, presumably, someone on a boat on
the water, and someone in the water, and the entire situation is indicated to be one to
avoid by the conventionalised visual 'lexical item' of the broad red line crossing out
the situation depicted. But what is the arrow? Arrows have a variety of interpretations
and what is intended here is not particularly clear, although one may be able to guess.
It is not considered good design for such artefacts that their intended users must guess
what they mean, however.

A range of such signs and the problems that interpreters have with them have been
studied systematically by Adams et al. (2010) in order to find ways of improving on
the situation. Adams et al. (2010) argue, and then demonstrate in a series of empirical
tests, that a much clearer rendition of the intended warning can be given if the warning
is divided into two parts—their redesign for the current sign is shown on the right-
hand side of Figure 4.2. We suggest here that the reason why this redesign functions

Fig. 4.2. Left: A warning sign: but against what? Right: The warning sign as redesigned by Adams et al. (2010); used by kind permission of the authors

so much better can be found in the discourse semantics that applies. In contrast to the left-hand sign, where we only had one combined source of information, here we have the explicit construction of a *contrast* relationship—one of the most commonly occurring discourse relations found across many semiotic modes and so one of the interpretive mechanisms that discourse semantics provide.

The contrast relation states that we should be able to identify two situations that differ in recognisable ways. This means that we no longer see each situation by itself, but that we are asked as interpreters to consider each one in relationship to the other. Note that this is not 'present in the material', since we only have two frames in proximity to one another and this can mean very many different things—as we can quickly confirm by looking at sequences of panels in a comic, for example (cf. Chapter 12). In the present case, the fact that we have two situations, and one is marked with a green tick and the other with a red cross, gives strong clues that we would be well served by assuming contrast. Once we have this contrast as a hypothesis—because that is what all discourse interpretations are, hypotheses—then we also need to find the *point* of that contrast, which we can do by seeing what is different across the two images.

In one, marked with the green tick, there is nothing in the water; in the other, marked with the red cross, there is. The first image of the pair then clearly means that it is acceptable to jump into the water only *when there is nothing in the way*. Note that this is also a useful and intended generalisation beyond the far too specific situation depicted in the sign on the left of the figure, which only suggests that one should look out for swimmers in the water! The redesigned sign places its depictions in an appropriate discourse context and so allows application of corresponding discourse semantics. As a consequence, the images are readily able to express the negation that there is nothing in the water. This demonstrates rather clearly that the power of the visual pictorial semiotic mode has been increased.

4.1.3 Mode families

Applying the definitions given above has the consequence that certain communicative resources are separated into distinct semiotic modes in ways that might not always match your expectations. For example, since spoken language and written language obviously have very different materialities, then simply by virtue of our definition these two just have to be considered separate semiotic modes.

Writing *vs.* speaking

While this might at first glance appear more complex than it need be, actually the reverse is true, precisely because it insists that we take a proper view of just what gives the materialities involved signifying form *without* presupposing that we are just dealing with language plus some extras. For spoken language, we saw in Chapter 2 and will see more in Chapter 8 how complex the material can become, including not only the sound signals (and then even different facets of the sound signals!) but gesture, proximity, gaze and so on. These are actively used by the semiotic mode of spoken language to broaden its 'reach' (see above). A very different situation holds for the semiotic mode of written language, which uses the resources of typography, page composition, size, colour and so on in order again to extend its reach, but in rather different directions. Taking these aspects together means that we would do very well to pull the two distinct communicative resources apart. This gives us more methodological support when we consider moving between the semiotic mode of writing and that of speaking, since it becomes clearer just what kinds of meaning are most at risk in the 'translation' involved.

There is still the challenge, however, not to lose the fact that it is of course nevertheless the case that written language and spoken language have a considerable amount of 'semiotic capabilities' in common as well. We will describe semiotic modes that can be related in this way—i.e., as making use of semiotic organisations that appear to be at least partially 'shared'—as belonging to semiotic mode **families**. Thus spoken language and written language are members of the semiotic mode family 'verbal language'. This is different from saying that we have a single mode with various materialities (as many previous accounts have suggested), in that variations can be expected ranging over all the levels involved, from materiality to discourse semantics.

Modes related in this rather intimate way do not 'just happen'. There has to be a (potentially very long) history of interaction between media. For example, written language began in much simpler attempts to provide a non-transient record of *some aspects* of information that was expressed in verbal language. Communities of users had to do considerable work with the written mode in order to make it available for a substantial portion of the spoken mode—and it is naturally still not a complete equivalent. Moreover, other material affordances of visual media have made their own contributions along the way as well. And so enough has now been 'transferred' to warrant considering written language as a semiotic mode in its own right and not simply a 'notational' system for the sounds of spoken language. For established written languages, therefore, the very different mediality of a visually perceptible canvas is used to 'depict'—we return to this notion below—distinctions that allow the partial application of the semiotic modes involved in the medium of spoken language. And, again, over time, the richness of visual perceptible materials naturally encouraged further semiotic modes (typography, layout, etc.) to develop that have contributed to an increasingly strong separation (or 're-factorisation' using the software engineering term) of resources from those of spoken verbal language.

Resemblance across semiotic modes can then be ascertained at various levels of specificity. We reserve the use of semiotic mode families for cases of substantial 'overlap'. Other resemblances, such as the almost universal requirement of building 'units' of some kind—what some approaches have termed *semiotic principles* (e.g., Kress and van Leeuwen 2001; Jewitt et al. 2016)—may arise due to the nature of semiotic systems themselves and so are distinguished here. There may also well be neurocognitive reasons that semiotic modes come to show resemblances due to the general 'architecture' of our neurobiological makeup (e.g., Jackendoff 1987, 1991; Jackendoff and Lerdahl 2006; Cohn 2016). Again, there are more open questions here than there are answers, and so there is plenty to do in empirically testing our knowledge of semiotic modes!

4.2 Modes and media

When examining individual artefacts or performances, any number of communicative modes may be operative and so careful empirical analysis is necessary to distinguish among them. This follows from the fact that materials (both actual and virtual) are able to support a host of simultaneously co-varying dimensions. This does not mean, however, that we must always start from scratch—it is certainly possible to determine likely constraints both on the semiotic modes that may apply and on their precise manner of application. Perhaps the most prominent source of such constraint is the *medium* within which the artefact or performance is couched—this is why the previous chapter carefully set out how communicative situations, and in particular their 'canvases', can be categorised. Medium-specificity as such is then also a necessary component of multimodal analysis—that is: just which modes may be operative and how they are combining may exhibit medium-specific properties and so knowing more about the medium can help guide subsequent empirical investigation.

A medium is best seen as a historically stabilised site for the deployment and distribution of some selection of semiotic modes for the achievement of varied communicative purposes.   **!**

That certain semiotic modalities regularly combine and others not is itself a sociohistorically constructed circumstance, responding to the uses that are made of some medium, the affordances of the materialities being combined in that medium and the capabilities of the semiotic modes involved. For example: books are a medium, traditionally mobilising the semiotic modes of written text, typography, page layout and so on. When we encounter a book, we know that certain semiotic modes will be likely, others less likely, and others will not be possible at all (at least not *directly*—more on this below). This relationship between media and semiotic modes is suggested graphically in Figure 4.3.

We consider a medium a socially and historically situated practice and, in the suggestive phrasing of Winkler (2008: 213), a *biotope for semiosis*. Semiotic modes participate in a medium and realise meaning through their material or 'virtual canvas' as well as their semiotic side. It is thus not the medium itself which realises meaning! This is itself a considerably expanded consideration of what, in several functional linguistic approaches, has been called 'mode' or 'channel' of communication (cf. Martin 1985: 17) and also shows connections with Ryan's (2004a) suggestion that medium can be separated into 'transmission' and 'semiotic' components. We mention this further in one of our use cases below (→ §12.2 [p. 314]).

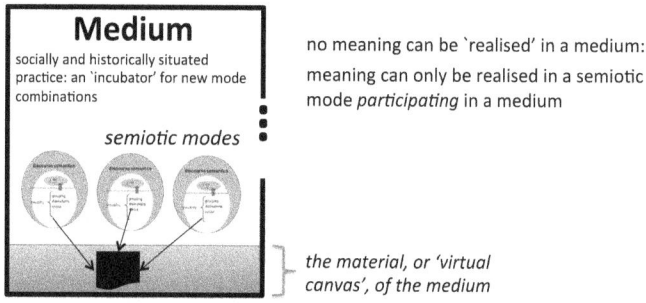

Fig. 4.3. Relation between semiotic modes and media

Media have a further range of quite distinct properties that are very important for helping guide multimodal analyses. Particularly relevant is the basic phenomenon that certain media can 'represent' others. We can characterise this situation quite precisely by mobilising what we have now learned about the relationship between semiotic modes and media. What we have called the 'canvas' of a medium is necessarily some selection of the possibly quite different 'materials' contributed by the participating semiotic modes. Any particular medium does not need to fully utilise the material possibilities that a semiotic mode brings—i.e., any particular media may not necessarily provide full sensorial access to the options a semiotic mode in principle spans.

This may sound complex but is actually quite common. We find it in situations, for example, when technological developments lead to a new medium that in certain respects exhibits deficits with respect to an older, pre-existing medium *but those deficits are counterbalanced by new capabilities*. Consider the respective introductions of the printing press, the telephone and the web. Telephone communication, particularly in its early days, was certainly a poor affair compared to natural, face-to-face interaction: but the fact that one no longer had to be in the same place was obviously an enormous gain that outweighed the lack of vision, access to gesture and facial expression and so on. Thus, even though the medium of the telephone was 'carrying' spoken language, only certain facets of the materiality generally employed by face-to-face interaction were being supported. Very similar observations have been made for the printing press, which at the outset was very limited compared to the visual richness of earlier book materials, and the web, whose early webpages with basic HTML similarly looked very poor from the perspective of the typography of the already existing and mature medium of print.

This potential mismatch between the material of the semiotic mode and that provided by some medium gives rise to the phenomenon of what Hartmut Stöckl has termed 'medial variants' (Stöckl 2004). Medial variants describe situations where there appears to be the same or a similar 'semiotic' organisation being used with different materialities—just as was the case with written and spoken language as discussed above. Here we distinguish, however, between the very close relationships of participating in 'mode families' and a far looser degree of connection brought about by media having semiotic modes in common. This distinction is first and foremost a theoretical one that opens up a space of possible relationships that must be filled in by empirical investigation. What we want to avoid is being forced to group different phenomena together prematurely simply because we lack the theoretical terms to describe their differences.

Medial variants

Teasing apart the relations between different signifying practices is, for example, often made more complex by the fact that we can have many semiotic modes contributing to a single medium. Some of these may also be used by other media, some may not. Consider the situation when comparing 'static image' and 'moving image'. Stöckl (2004) also classifies this as a case of 'medial variants', a position found implicitly in many accounts of visual communication. This sees the members of the pair spoken-language/written-language and the members of the pair static-image/moving-image as being related to one another in similar ways. In contrast to this, we suggest that the semiotic potentials of the latter pair differ so substantially that there are almost certainly separate systems in operation. More on the potential course of semiotic development of spoken and written language is suggested in Bateman (2011), while the emergence of distinct semiotic modes for film is taken up in Bateman and Schmidt (2012: 130–144); our use cases for comics (Chapter 12) and films (Chapter 13) below also bring out their substantial differences (as well as some points of overlap).

Describing the distinct semiotic modes that the media of static and dynamic images individually employ then offers a more natural characterisation of what is going on. On the one hand, modes of pictorial representations may be shared; however, on the other hand, modes of shot transitions and structures made possible by shot sequences (e.g., shot/reverse-shot and so on) are clearly not. The extent to which it is possible, or necessary, to 'enlarge' static pictorial representations to also describe systems of dynamic pictorial representation is then turned into an empirical issue. For good methodological reasons, one can investigate both aspects separately until sufficient evidence for a tighter relationship is found. This is another illustration of why it is methodologically helpful to make the distinctions we are suggesting. We will need far more fine-grained and empirically well-founded analyses of how these media are functioning before taking the discussion further.

Depictive media   Another kind of relationship between media that may easily be conflated within the idea of medial variants occurs when some medial form appears to reproduce another. Consider, for example, a 'dance' being shown as a 'photograph' within a 'film' and so on; or, as very common nowadays, phenomena such as reading a digital version of a newspaper in a web browser. We will refer to this particular kind of relationship between media as one of *media depiction*. Again, the possibility of separating out media depictions from other kinds of 'medial variants' discussed in the literature provides strong methodological guidance concerning how such things are to be analysed, as we shall see.

We adopt the term 'depiction' in relation to its more common usage with respect to pictorial representations. For multimodality, it is useful to generalise beyond the pictorial case and allow all possibly available material features to 'depict' (cf. Bateman 2016). Media can then be said to operate as depictions when their (virtual) material offers sufficient foundation for the application of semiotic modes from *other* media—that is, when the materials formed by the combination of the materials of the contributing modes and their technical capabilities make it possible to apply semiotic modes from other artefacts.

Such cases do not require any exact equivalence in material form—all that is necessary is that we can locate some 'sub-slice' of material distinctions sufficient for supporting similar distinctions from the 'depicted' semiotic modes. This then makes it possible to re-use particular techniques or mechanisms across media. Representational pictures may consequently be drawn, painted, sketched with a mouse, and so on. These are all very different media but the material distinctions provided by their canvases generally overlap with the distinctions required by the other semiotic modes and their media and so allow transfer and alignment of semiotic mode use. As another example, the virtual canvas created by current web browser technology is very different to that created by print technology and we are therefore dealing with quite different media. There is, however, still sufficient commonality to offer material support that is in many ways 'overlapping' (although there are still many differences, cf. Bateman et al. 2004).

This framework allows some basic questions raised for multimodal analysis across differed media to be answered quite simply. Analysis involving content-flexible technical devices is often seen as a general challenge and analysts are sometimes unclear how to proceed. Stöckl (2014: 276) suggests differentiating between the medium (e.g., tablet computer) and 'communicative forms' (see also Holly 2011: 155)—e.g., newspapers, radio, e-mail, etc.—in order to capture this phenomenon. However, 'medium', as we are using it here, is often closer to 'communicative form' in intention than physical material. In our sense, therefore, the tablet computer is not a medium, but a virtual canvas that might be employed by a variety of media. Taking this perspective allows us to avoid assuming hybrids or fuzzy boundaries that are not present and which are not necessary for characterising the multimodal meaning at work. Instead, we have a continuum of media, with an increasing number of 'virtual canvases', spanning from any medium that is used to implement another—including perhaps the use of parchment to replace painting on walls at one extreme—right through to an interactive immersive digital environment, such as the holodeck (→ §3.2.7 [p. 99]), on the other.

The blurring of the 'digital'/'non-digital' divide that results is quite deliberate because this distinction has often been overstated. As we discuss in our use case on analysing webpages (Chapter 15), the organisation of virtual digital forms still most often involves recreations of other media. This precisely fits our model of media depiction. Of course, there are some things that such virtual media can do that a sheet of paper, for example, cannot, and these also need to be considered, although not as a complete break in relation to existing media. Relevant discussions here include those of Lev Manovich (2001), who offers several characterisations of what is distinctive about the 'new media', including their reliance on numerical representation (giving rise to modularity and copyability) and programmability (supporting mutable canvases). It can then be investigated to what extent media constructions using such media mobilise these possibilities for their depictions, and subsequent extensions, of existing media.

 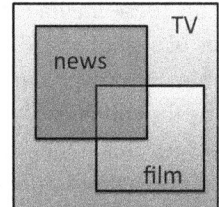

Fig. 4.4. Two common kinds of depiction relations between media: embedding and blending

In general, therefore, media can be related to each other in several ways. There may be relationships brought about by a shared or overlapping use of semiotic modes. There may also be relationships supported by depiction, which can itself play out in several forms. The two principal cases of depiction are suggested graphically in Figure 4.4, where we see the classic case of 'embedding' on the left and a more complex case of combination, or 'blending', on the right. Embedding is the case, for example, for a photograph in a newspaper on a website, whereas combination or blending happens when a TV program uses the combined resources of film and news reports. In all cases, an 'embedding' medium sets the possibilities for representation and communicative form use for any 'embedded' media. Each layer of medial depiction may make available additional media-specific affordances, while also restricting what can be used from the embedded layers. If 'too much' is lost in the depiction—and whether too much is lost or not will depend on the specific communicative purposes of the entire enterprise—then explicit 'adaptation' or repurposing steps may need to be undertaken.

The broad framework relating media and semiotic modes that we have now introduced opens up a host of still unresolved empirical questions for systematic investigation. For example, we do not yet know whether 'film' or 'comics' contribute their own semiotic modes or are better considered depictive media involving semiotic modes drawn from other media. We may well have strong intuitions or preconceptions about these issues, but the actual empirical work remains to be done. There are certainly some good arguments that at least some of the regularities and specific workings of these media offer good candidates for semiotic mode status. The criterion to be applied is that of finding an appropriate discourse semantics that explains how particular 'slices' of patterns through the employed material are to be related and contextualised. However, the task is made more complex by the fact that there will also be other semiotic modes that apply over the 'same' material, drawing on different slices in order to carry their own signifying practices.

Mode com-
binations

'Depiction' provides a methodologically beneficial barrier that prevents treatments of semiotic modes and their combinations unravelling—that is: just because a photograph is shown within a film, or a newspaper on a tablet computer, does not mean that we suddenly stop having the expected and conventionalised properties of photographs or newspapers and how these media employ semiotic modes to make meanings. There is no 'combination' of semiotic modes intrinsically involved in such cases. Although more creative combinations can then be generated (e.g., the increasingly frequent 'trick' in film of having the contents of a photograph move, or allowing zooming and hyperlinking on a tablet), this alters the medium depicted and in so doing brings different semiotic modes together in the service of communication. We will use this boundary function below, and particularly in Chapter 7, when we set out methodological guidelines for multimodal analysis in general. Important bases for this analysis are offered both by of semiotic modes, as introduced above, and by the mechanisms of 'text' and 'genre' that we clarify in the next subsection.

4.3 Genre, text, discourse and multimodality

When analysing artefacts and performances of any kind, it is important to have a good understanding of the conventions within which, and possibly against which, those artefacts and performances are working. If one does not know this in advance, then several distinct kinds of studies are to be recommended. First, and more theoretical, one should familiarise oneself with the literature in the area from any disciplines that have addressed the artefacts or performances in question. Second, rather more practically, one may perform preliminary *corpus-based* investigations of collections of similar objects of analysis to gain a 'data-driven' overview of how they are organised and what kind of phenomena appear (→ §5.3 [p. 152]).

Any relations found between general conventions assumed and individual objects of analysis can normally be usefully characterised in terms of *genres*, *texts* and *discourses*. These terms have been extended substantially beyond their original uses in relation to primarily linguistic or literary objects of study and so we explain here how this can be employed in the context of multimodal research. Nevertheless, when considering these constructs multimodally, it is important not to extend them *too* far! For example, and as we shall see below, some characterisations end up calling almost any object of analysis a 'text' and this loses much of what is useful about the term. We will be concerned, therefore, only with extensions of the scope of the applicability of these terms that also maintain (or better strengthen) their utility for supporting empirical research.

4.3.1 Multimodal genre

Essentially 'genre' is a way of characterising patterns of conventions that some society or culture develops to get particular kinds of 'communicative work' done. Thus, 'narrative' is a genre that gets the work of storytelling done; a 'scientific article' is a genre that gets the work of scientific argument and reporting results done; 'comedy' is a genre that gets the job of making people laugh done; and so on.

The relation between genres, as recognised socially constructed usages of some communicative media, and texts, the individual carriers of communication that may be positioned with respect to genres, is best seen as one of *participation*. Texts participate in genres; they do not 'belong' to genres. This view of characterising genres goes back to an important article by the French philosopher Jacques Derrida (1980). Derrida argues that texts and genres cannot be separated; any text must necessarily be seen as existing with respect to genres. This can also be closely related to the view of genre pursued in sociosemiotic theory (Martin and Rose 2008). The educationalist and social semiotician Jay Lemke describes this as follows:

> "We construct genres by construing certain sorts of semantic patterning in what we consider to be distinct texts, and we say that such texts belong to the same genre. Co-generic texts are privileged intertexts for each other's interpretation." (Lemke 1999)

And so, when readers, viewers, players, hearers, etc. allocate certain artefacts or performances to particular classes of communicative event, those classes bring with them certain interpretive frames and expectations. These frames guide readers, viewers, players and hearers, etc. to make sense of what they are seeing, hearing, doing, etc. An awareness of appropriate genres is thus also a cornerstone of any approach to literacy, multimodal literacy included.

The utility of genres as an organising frame applies equally from the perspective of communicative *production*: we can take the intention of creating an artefact or performance as a mobilisation of precisely those constraints that signal that some interpretative frames are to be applied rather than others. The 'decisions' taken during production—regardless of whether these are explicitly taken as a matter of conscious design or implicitly by virtue of tacit knowledge—then rely more or less explicitly on the conventions and practices established for the class of artefacts or performance in which the communication is participating.

> A genre is a class of communicative events that share a recognisable communicative purpose, that exhibit a schematic structure supporting the achievement of that purpose, and which show similarities in form, style, content, structure and intended audience.
>
> — *Swales (1990: 58)*

These very general notions of the contribution of genre as a characterisation of expectations, conventions and comparisons for analysis apply equally in the multimodal context. A more extensive consideration of 'genre' in relation to static page-based artefacts is set out in Bateman (2008), while van Leeuwen (2005*b*) picks up the discussion in relation to design. For our current purposes, we will position genre with respect to the accounts of semiotic modes and media given so far so that the interrelationships are clear. This then also shows how genre forms one of the central methodological pillars for conducting any kind of multimodal research.

Visual genre   Genre has been approached not only in studies focusing on linguistic communicative acts but also in several areas of visual research—including static documents of various kinds (e.g., Kostelnick and Hassett 2003) and artefacts involving movement such as film (e.g., Altman 1999). This is often pursued in an empirical-historical fashion, documenting how some forms have changed over time. Genre is also a prominent term found in studies of 'new media', where the somewhat problematic notion of the *cybergenre* has been proposed (cf. Shepherd and Watters 1998). In all these areas there is active consideration of the interaction between classifications of visual form, design and interpretation.

Approaches to genre are consequently now often explicitly 'transmedial', or 'intermedial', drawing their examples freely from literature, film, web pages, video, computer games, theatre, TV, radio programmes, painting and many more. Such work shows the reoccurring goal of exploring the extent to which common generic properties can be found within and across intrinsically multimodal artefacts (cf. Hallenberger 2002; Frow 2006; Lewis et al. 2007; Wolf and Bernhart 2007; Altman 2008).

Nevertheless, whereas the informal use of 'genre' as a type of communicative artefact or situation is relatively straightforward, its use in scientific investigation still faces considerable problems of definition. It is increasingly assumed that what is picked out by use of the term 'genre' across diverse disciplines and approaches should actually have something more in common than rough family resemblance or analogy. Particularly as we move into *endemic* multimodality of the kind seen throughout this book, the seductive path has been to relax disciplinary boundaries and to apply genre across the board as a theoretical construct covering very different kinds of communicative artefacts and situations.

For the multimodal case, styles of presentation must include any possibilities that the semiotic modes provided in a medium offer. This opens a path that offers characterisations of cases of 'transmedial' communication in terms of genres that are performable with respect to various media. The genres define communicative purposes and it is then the discourse semantics of the modes provided by targeted media that must attempt to achieve those purposes. This then sees genres as bundles of strategies for achieving particular communicative aims in particular ways, including selecting particular media. They thus necessarily bind notions of social action, particularly communicative action, and styles of presentation, on the one hand, and semiotic modes as ways of achieving those actions, on the other.

4.3.2 Multimodal text and textuality

Similar to the notion of 'genre', the term *text* has also been generalised substantially over the years (→ §2.5 [p. 52]). There are many presuppositions or prejudices concerning 'text' that we need to explicitly counter as they appropriately align neither with our usage in this book nor with a suitably strong empirical framework for conducting multimodal research. In fact, we need to 'reclaim' the notion of text for the purposes of multimodality and give it, on the one hand, a rather broader meaning than many traditional accounts assume and, on the other hand, a more focused usage that improves its potential to guide and support analyses.

> Reclaiming 'text'

As we discussed briefly in Chapter 2, the domains of application of the term 'text' expanded steadily over the latter half of the twentieth century (→ §2.5 [p. 51]). The word took on such a generalised meaning that almost anything subjected to semiotic analysis was automatically to be considered a 'text'. This has been problematic because 'text' is also strongly associated with many verbal, linguistic characterisations. Subsequent analyses did not unravel this contradictory situation whereby 'text' was any object of semiotic analysis and simultaneously something with largely linguistic properties.

As we set out in Chapter 2, we consider this conflation to be misplaced. Visuals (and other materials) afford very different ways of making meanings that cannot be simply reconciled with the ways of making meanings that we find in language. This means that we agree entirely with the many critiques that can be found in, for example, visual studies of labelling almost all semiotic artefacts and performances 'text' in a linguistically-oriented sense. Consider the following quotations from two prominent theorists of the visual, James Elkins and W.J.T. Mitchell:

"I hope that we all hope that images are not language, and pictures are not writing. Visual studies has a lot invested in the notion that the visual world and the verbal world really are different, even beyond the many arguments that persuasively demonstrate that we think we read images all the time." (Elkins 2003: 128)

"The pictorial turn is ... rather a postlinguistic, postsemiotic rediscovery of the picture as a complex interplay between visuality, apparatus, institutions, discourse, bodies, and figurality. It is the realization that spectatorship (the look, the gaze, the glance, the practices of observation, surveillance, and visual pleasure) may be as deep a problem as various forms of reading (decipherment, decoding, interpretation, etc.) and that visual experience or 'visual literacy' might not be fully explicable on the model of textuality." (Mitchell 1994: 16)

Although these critiques are in many way completely justified, they are also predicated on the assumption that 'text' is irreconcilably the kind of entity that we find in discussions of language—and it is this contradiction that needs to be resolved.

We see the way forward here to be rather different, therefore. It is necessary to revise 'models of textuality' so that the notion of 'text' is *not* limited to the ways of meaning making we find in language, while still offering useful guidelines for conducting analyses. Such a model follows naturally from our definitions above. The intrinsic properties of a text are necessarily dependent on the semiotic modes that produce it—it *may* be verbal, but it may equally be a visual depiction, a piece of music, or a physical movement. The reason for maintaining, or rather re-instating, the term 'text' for this diversity is to draw attention to a very specific property shared by all of these communicative or signifying artefacts or actions: i.e., the property that *they are structured in order to be interpreted*. Distinctions in materials are deployed in order to trigger or support particular lines of discourse interpretation.

! A text can be defined as a unit that is produced as a result of deploying any semiotic modes that a medium might provide in order to produce a particular and intended structuring of the material of the medium's canvas so as to support interpretation by appeal to those semiotic modes. That is, expressed more simply, a text is what you get whenever you actually use the semiotic modes of a medium to mean something.

This supports a further important distinction, that between *text* and *discourse*. These terms are often used interchangeably, even in the context of multimodality, although traditions may vary in this regard. On the one hand, *text* is often taken as a concrete artefact which can be analysed with discourse analytical methods and with regard to its embedding in a certain communicative situation and context. At the same time, individual occurrences are often described as discourses, in particular in situations in which a concrete media discourse, such as *filmic discourse*, for example, is discussed. For current purposes, we take the materiality of the text as a decisive factor for distinguishing *text* and *discourse*. This in turn grants us analytic access to text-internal criteria of structure and coherence, and the qualities of *texture* and *textuality*.

A central intrinsic property of texts then follows by the role placed by the combined discourse semantics of any semiotic modes employed. The material distinctions mobilised in the text will function as 'instructions' for deriving interpretations. These instructions typically operate by allowing interpreters to attempt to 'maximise coherence', following the principles that the schemes for interpretation of the semiotic modes provide. Thus, whenever we have a text, we necessarily have a communicative situation fulfilling the conditions defined in the previous chapter. And whenever we have a communicative situation, there may be texts produced in that situation.

The situation is the context in which the text is embedded and which defines and specifies the various knowledge and information sources, social and cultural structures, as well as the social activity that influences the interpretation of the text. This is the *discourse* and discursive context in which the material *text* is embedded. As a consequence, we can distinguish *text* and *discourse* from each other by referring to the materiality of the *text* and the higher level of abstractness for *discourse*. The relationship between text and discourse is one of 'realisation' and *text* and *textuality* refer to the realisation of the entire set of semiotic modes in some act of communication as well as the semiotic relations involved.

4.3.3 Multimodal discourse

'Discourse' as a term appears in many disciplines and is often used rather differently depending on community. Two rather broad senses are the following. First, 'discourse' may be used in a way that is more or less synonymous with a 'way of thinking about' or 'a general style of meaning-making'. Thus, we might hear of the 'discourse of capitalism', the 'discourse of gender', and so on. We will refer to this, somewhat informally, as 'discourse' with a 'big D', i.e., *Discourse*, or 'discourse-in-the-large'. Second, 'discourse' may be used for referring to phenomena at a much more local scale, on the level of the text, as we said above. Many approaches in linguistics, for example, consider discourse analysis to be concerned with individual interactions, exploring how some particular text is organised. This kind of discourse (with a 'small d', or discourse-in-the-small) is thus about fine-grained analyses of particular instances of communication and the mechanisms by which utterances, turns, sentences, etc. can be flexibly joined together to make 'larger' texts and, eventually, even Discourses (with a 'big D').

'Discourse' with a 'big D' began to take on an important role in many theories of society and culture as an outgrowth of the so-called 'linguistic turn' that occurred in many fields in the middle of the 20th century (→ §18.1 [p. 380]). The basic idea here revolved around seeing our individual and cultural understandings of ourselves, our societies and our world as essentially constructed through acts of communication. Many previously accepted tenets of the structure and organisation of reality, particularly social reality, were recast at that time as the results of ways of speaking about them or as 'social constructions' (cf. Berger and Luckmann 1966; Searle 1995). We saw

Discourse vs. discourse

this particularly in the developments concerning media and their role in society discussed in Chapter 2 (→ §2.6 [p. 64]).

It is important not to see these two views—i.e., of discourse-in-the-large and discourse-in-the-small—as being in any way in opposition to one another. Both perspectives are crucial, even though we focus for current purposes primarily on the smaller-scale use. This focus is motivated by the fact that this is what we need when we begin close empirical analysis of multimodal communication. Nevertheless, many of the *reasons* why particular fine-grained discourses are used is precisely to be in the service of large-scale Discourses. We argue, therefore, that the fine-grained view can serve as an important bridge linking smaller, more 'micro' acts of communication with the broader, 'macro' acts that constitute society and culture (→ §2.6.2 [p. 69]). We need a considerable range of tools and experiences to be able to reveal the fine-grained detail of multimodal meaning-making activities, and discourse-in-the-small provides many of these as we shall see throughout our use case discussions.

Weather vs. climate

The two views can also be drawn together by using an analogy proposed by the linguist and social semiotician Michael A.K. Halliday (Halliday 1978, 1985). Halliday suggests viewing complex dynamic systems such as language in the same kind of terms that we employ for the relation between climate and the weather. Most people will have some idea about the climate of, for example, London or the Amazon jungle; geographers and climate researchers, on the other hand, will have quite precise classifications for this climate. And yet, each day in these and any other location the weather varies: it might one day be warmer, another colder. The climate of a location is then really nothing more than an 'average' over individual weather situations. If the weather gradually becomes warmer over time, then eventually the climate will need to be considered to have changed as well.

This shows that 'weather' and 'climate' are *not two separate phenomena*. The difference between the two is again precisely that of the difference between the level of Discourse (with a 'big D') and occurrences of discourse (with a 'small D') and offers a straightforward way of conceptualising the relation between the two. The micro-view will give us accounts of the weather, thus the discourse at the micro-scale; the macro-view looks more like a description in terms of climate, thus in terms of Discourse at the broadest scale of society and culture.

This way of relating the phenomena also addresses one of the most difficult issues that have been raised with respect to complex dynamic systems such as language at all: that of the relation between the individual occurrence, e.g., the individual sentence that appears to follow a general set of rules or principles, and the system-as-a-whole, which appears to gradually change over time, as when one 'language' becomes what is described as another with the passage of time. Old English does not look very much like modern-day English, for example, so there has clearly been a change. Halliday's proposal is that each individual occurrence both responds to the general expectations established by the 'system' that is in force and itself modifies those expectations. Over time, a succession of modifications may be sufficient to measure as a change in the system of expectations—that is: the climate will have changed.

Discourse in this sense is then both *inherently dynamic* and *local*. It offers the relationships that hold together sequences of turns in a conversation, or the sequences of sentences in a text so that these sequences become coherent communicative acts in their own right, achieving particular communicative purposes rather than others.

Our principal use of 'discourse' here will, then, orient to the second smaller-scale sense precisely because of our designated aim of orienting to close multimodal empirical analysis. The moment-by-moment, act-by-act unfolding of signifying activities is placed (temporarily) in the foreground. For multimodality, however, we generalise further and so use 'discourse' for all kinds of such local meaning-making mechanisms, regardless of whether these are between utterances, between utterances that point using gestures to things in the world, to pictures and text placed together on a page, or to pop-up menus and depictions in the environment in an augmented reality situation, and so on. Discourse is what happens when we make sense of these contributions to an unfolding communicative situation.

This view of discourse already lets us start putting some much needed organisation on top of the diversity of approaches seen in previous chapters. We need to ask in each case what kinds of activities are being pursued, with what kinds of goals, and with what kinds of expressive forms and materials. This will be taken up in depth when we move towards discussing use cases of actual analyses according to the methodological framework we set out for all use cases in Chapter 7.

4.4 What this chapter was about: the 'take-home message'

In this chapter we have set out a detailed view of just what the elusive 'modes' of multimodality can be. We have defined them as an extension beyond most traditional models deriving from semiotics, extending them to necessarily include material and discoursal facets. This was then placed in relation to media and genre. The framework as set out has many methodological consequences for how analysis can be pursued—and this is the primary reason that we have needed to go into this level of foundational detail at this point. One result of this discussion was a framework in which we can also look at relations between media at various levels. Three distinct kinds of intermedial or transmedial relationships were identified; these are summarised graphically in Figure 4.5 together with their position with respect to genres, texts and discourses, all considered multimodally.

Although it might appear easier to 'just analyse' whatever object of analysis has been selected for study, this would be mistaken. The more multimodality an object of study mobilises, the more crucial a finely articulated analytic framework becomes to

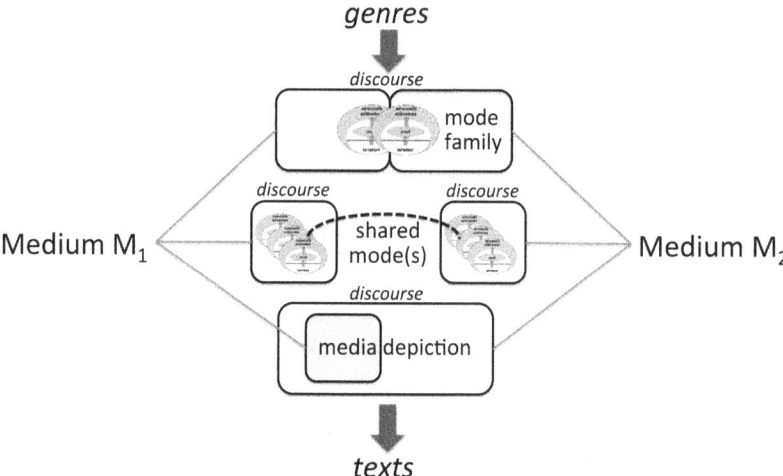

Fig. 4.5. Overview of possible relations between media and their positions with respect to genre and text. Media may have developed from common origins (top option: mode family), may overlap with respect to the semiotic modes that they mobilise (middle option), or may depict other media (bottom option). Discourse semantics provides a means for relating semiotic modes throughout.

avoid being swamped by the detail. The definition of communicative situations from the previous chapter provides a first cut through the complexity to isolate just what material possibilities are on offer for making meanings. We turn this into a concrete methodological procedure in Chapter 7. The approach to semiotic modes set out in the present chapter then provides a detailed method for factoring in the ways of making meanings actually articulated and performed by communities of users so as to make them accessible to empirical investigation.

One by-product of this characterisation of semiotic modes is that they generally turn out to be much 'smaller' than modes or codes commonly discussed in the previous multimodality literature—as in the cases of, for example, the 'visual mode' or 'language'. Semiotic modes may be developed for quite specific purposes, such as particular forms of diagrams, forms of layout, forms of relating pictorial information and spoken or written language, and so on. Again: it is an *empirical* issue just which modes there are. We still have very fragmented knowledge of just what semiotic modes exist and where they are used: in this sense as well, the field of multimodality is very much at its beginnings. Exciting further developments can be expected almost regardless of what objects of analysis are selected—provided that sufficient methodological constraint is brought to bear to achieve results!

Part II: **Methods and analysis**

In this part of the book we introduce the primary methods that have been developed or applied across various disciplines in a manner useful for performing multimodal analyses. We then show how to characterise any object or situation so that it may be approached for detailed multimodal analysis.

5 The scope and diversity of empirical research methods for multimodality

Orientation

In this chapter we set out some of the principal methods and methodological considerations to be borne in mind when conducting, or better, already when planning, some multimodal research. We also describe approaches that are generally applicable to broad ranges of multimodal investigations, as these can always be considered for your own research questions and practice. As in other places in the book, we will be suggesting the value of broad-disciplinarity—and this will extend to considering combinations of methods. Methods need to be selected for their fitness to a task, not because of disciplinary habit.

5.1 What are methods? What methods are there?

Methods are ways of guiding your investigation of a research question. They set out particular techniques that support this process from beginning to end—ranging from when you are thinking of what data to analyse right through to presenting your results and planning your *next* round of research. Most methods can be used for a variety of different kinds of research questions and so the selection of methods may be made rather flexibly. It is often the case that several distinct methods can be usefully applied to a single research question, and so particular approaches can in fact benefit by considering their objects of analyses from several different methodological perspectives.

As we set out in our note of warning at the beginning of Chapter 2, however, some disciplines see this slightly differently, strongly prescribing that certain methods should be used rather than others (→ §2.1 [p. 23]). We will remain fairly neutral on this point as it is really a central part of how any particular discipline defines itself and its subject matter. You will need to make your own decisions concerning just what methods are going to help you better understand the multimodal phenomena you are interested in. Our own position on this matter is clear enough: methods should be seen as tools for addressing particular kinds of research questions and therefore be deployed as such, that is, in the service of research questions and not because of a particular disciplinary preference.

This does *not* give license for confused or inappropriate applications of methods. Any methods selected should be applied properly and in the manner for which they have been designed. Moreover, certain kinds of methods may just be incompatible with the orientations and presuppositions of some specific discipline or question; this also needs to be respected. Our call for employing multiple methods is thus quite challenging, as it is sometimes difficult enough for anyone to become fully competent in the methods of a single discipline. Nevertheless, this is the kind of skill set that is most suitable for multimodal research—which is actually another reason why such research is often done better in teams.

Qualitative/
quantitative
The broadest division in methods—and one that is itself the subject of considerable, not always amicable, debate—is that between so-called *qualitative* methods and so-called *quantitative* methods. Qualitative methods generally involve accounts with interpretative categories of some kind; quantitative methods involve somehow measurable properties of some data that then can be counted and subjected to statistical analyses. Qualitative analysis is more broadly applied in the humanities; quantitative analysis in the natural and social sciences, although some disciplines—such as linguistics—sit uncomfortably between this division.

Discussions of the relative merits of these orientations, the so-called "paradigm debate" within methods research, are often overtly ideological and evaluative. Quantitative research may be described as 'mechanical' in contrast to the 'creative' role played by induction of patterns from more abstract data seen in qualitative research, or alternatively, the admission of interpretative categories in qualitative research can be viewed with suspicion by those seeking 'stronger' and more predictive results that may be supported independently of 'untrustworthy' personal evaluations. Questions here revolve around definitions of terms such as 'reliability' rather than 'accuracy' and a focus on various forms of validity apart from those of the scientific method.

Such diverse orientations are themselves often linked back to basic assumptions made about the nature of the (social) world and our (possible) knowledge of that world. Is the world largely a product of the constructive practices of its members and so requires adding how individuals and groups see that world into the equation? Or is that world more objectively given, making it susceptible to probing in a less interpretative fashion? Differing answers to these questions generally align with preferences for different groups of methods (cf. Rossman and Rallis 2012: 42–45).

Mixed
methods
Here we will take the position that many kinds of methods are required in any complete cycle of research—at some point categories will need to be formed and this process will draw on a variety of interpretative schemes and, at some other point, it will be useful to consider just how well such schemes are matching data that can be collected and measured. This aligns with what nowadays has increasingly been described as *mixed methods research* (cf. Teddlie and Tashakkori 2011), a style of pursuing research that explicitly combines qualitative and quantitative approaches. This does not then require that *all* research eventually lead to quantifiable categories, but it equally does not rule out attempts to achieve such categories on the grounds they they are necessarily reductive and unworthy of intellectual attention.

Particularly with recent advances in physiological measures and brain studies, it becomes possible to look for correlates of more interpretative categories than ever before. A new dimension is thereby added to *reception studies*—the study of how materials, including communicative acts, performances, art works and so on, are received by those who perceive them. Former behavioural measures of an external kind, such as heart rate, skin resistance, facial expression and so on, can be augmented by brain studies. This is certain to remain a strong source of additional 'triangulation' across results obtained by different methods in the future. Similarly, with the radical increase in the sheer quantities of data that can be examined, however that data is classified, there are new ways of exploring how well interpretative categories and patterns made from those categories characterise broader populations.

Our definition of methodological eclecticism goes beyond simply combining qualitative (QUAL) and quantitative (QUAN) methods ... For us *methodological eclecticism* involves *selecting and then synergistically integrating the most appropriate techniques from a myriad of QUAL, QUAN, and mixed methods* in order to more thoroughly investigate a phenomenon of interest.

— *Teddlie and Tashakkori (2011: 286; original emphasis)*

The move to include more varied methods in multimodal research is still relatively new. Far broader studies and 'triangulations' of results appealing to a variety of methods and levels of abstraction should now be considered a high priority. Qualitative semiotic descriptions that may have been offered—such as those suggested by Kress and van Leeuwen (→ §2.4.5 [p. 48]), for example—all need to be considered as *hypotheses to be tested* with more quantitative methods. Bell (2001) suggests approaching this via content analysis, while Holsanova et al. (2006) performs studies employing eye-tracking methods and Knox (2007) presents results from corpus-based studies of particular genres and media-types—all methods that we introduce in this and the next chapter.

Hypotheses

Until far more data has been analysed, collated and reviewed the precise limits and strength of hypotheses will remain open. The methods described in this chapter are thus all to be seen in this light: they are ways of helping us move beyond the programmatic observations still common in the field. In some areas, such as the multimodality of face-to-face interaction, such methods—particularly corpus-based methods—are already considered central; in others, the step to performing evaluative cross-checks is still to be taken.

5.2 Getting data?

Because in this book we are promoting a broadly empirical approach to multimodal issues of all kinds, a very early step to be considered when undertaking an investigation is just where the data is going to come from. For the purposes of analysis, 'data' is going to be any body of material that you collect or elicit with the hope that the data exhibits a sufficient quantity and range of the phenomena that you are interested in to provide grist for the analytic mill. After doing the analysis of that data, one should know more about how the phenomena studied operate than one did before. Methods, and methodology, are there to raise the chances of success of the enterprise.

5.2.1 How to choose data

Sampling

The first question that arises is just how to go about choosing what you want to analyse. Most of the specific methods, or types of methods, that follow may be applied to data collected in different ways, and so we briefly set out the basic concepts of selecting here before going on to the methods that follow. Choosing data for empirical studies is often called *sampling* because one almost always is working with some 'sample' of all the data that could theoretically be relevant, not with an exhaustive collection. Typically one wants the sample to be 'representative' for the data collection as a whole so that if you find out something about the sample it is then also relevant for the collection as a whole; we return to this topic in more detail in the following chapter on basic statistics.

In order to get a representative sample, selection of data on some kind of random basis is usually the way to go; only then can one apply methods supporting the expansion of sample results to results in general. There are three common data selection methods of this type. The first is 'random sampling', where instances are selected equally, i.e., without bias, from the entire set of potential candidates. The second is 'systematic sampling', where one adopts some simple scheme for the selection of data that is applied until one has enough (again, for questions of what is 'enough', see the next chapter). Examples of simple schemes would be 'every 12th document', or 'every webpage visited on a Monday', and so on—i.e., the criteria for selection are separate to the data itself. And the third is 'stratified sampling', which also picks randomly but only from members of the potential data pool that have certain pre-specified features—i.e., the criteria for selection are inherent to the data. Of course, it may not be absolutely clear just which features condition the data and which not, but that would be an outcome of further study!

The last method of stratified sampling is primarily useful for focusing analytic attention so that subsequent research is more likely to reveal interesting patterns. In general, if some phenomena that one is interested in is rare, then one will need more data exhibiting that phenomena in order to come to any conclusions—stratification is then a way of increasing the likelihood that enough *relevant* data will be available. As an example, if we were interested in analysing the use of camera angle in photographs taken by people aged between 30 and 40, then it would be rather pointless to use random sampling where we first select photographs at random and then hope that among those photographs there will also be some taken by people of the desired age. There may turn out to be so few of these that we do not have enough data to really address our question.

In addition, when selecting features, one also needs to make sure that one does not rule out just what was intended to be studied. Analysing some kind of difference in gaming activities according to gender is not going to work well if we employ a stratified approach that only picks out those with the feature: 'identifying themselves as male'. The point of stratified sampling is to remove biases, not to add them! Nevertheless, stratified sampling is often the best alternative; one must simply be clear in interpreting and reporting results that the data has been 'pre-sorted' to a degree. Any patterns found can only be generalised to the kinds of individuals or phenomena possessing the stratification features.

If one has no idea in advance just how rare or frequent a phenomenon of interest might be, then the best option may be to make a broader random selection and attempt to characterise the relative frequencies of various phenomena in a trial or preparatory round of experimentation. This might then allow more focused selections of stratification criteria for subsequent studies. The quantity of data that one needs for such an exploratory study can be calculated on a statistical basis—predictably, the less data that one takes the less confident one can be that the distribution of features in that sample will reflect the dataset as a whole.

However, since most measures of this kind are employed in situations of very large testing—as in health, voting surveys, and similar—the extent to which they are practically applicable to multimodal research may vary. For datasets up to around 75, the size of a representative sample comes out as barely less than the dataset itself; from 75 to 400, one would need from two thirds to half the size of the dataset; it is only for much larger datasets that the needed sample size begins to stabilise at a number that is 'low' (i.e., still around 350) when compared to the size of the dataset as a whole (cf. Krejcie and Morgan 1970). In the area of multimodal research, these are already large numbers. However, if the likelihood of particular phenomena occurring is already known to be quite high, or can be assumed to be high for some reason, then much smaller sample sizes will already provide sufficient data to show effects (Krippendorff 2004: 122); some of the methods for estimating effect sizes are discussed in Chapter 6.

Sample sizes

Krippendorff (2004: 124) also describes the useful 'split-half technique' for assessing the adequacy of a sample. This works by randomly splitting a sample in half and comparing the results obtained in analysis across the two halves. If the same results are achieved, then this is supporting evidence for the size of the sample having been sufficient. If different results follow, then it may well be that the original sample did not contain enough data to be representative. This can be repeated with selection of different 'halves' in order to see how robust, i.e., remaining unchanged, the assessment is. This allows the data itself to guide the determination of size, which can certainly be beneficial if one does not know in advance how phenomena in the data are distributed.

Split-half technique

Some other sampling techniques that are used include: just asking who you happen to find ('convenience sampling'), and selecting from some pre-selection that you have good grounds to believe are representative in some way ('judgement sampling'). These may then be combined with the random methods. Clearly, all of these deviations from a purely random procedure have drawbacks and need to be considered carefully beforehand. Krippendorff (2004: 111–124, Chapter 6) goes into considerably more detail about these and other sampling techniques.

5.2.2 Ethnographic methods, protocols, interviews and questionnaires

Increasingly relevant as a means of obtaining information about multimodality, and particularly about multimodal practices, are methods drawing on ethnography. Ethnography involves viewing what is going on with some community of practice *from the inside*—that is, an ethnographer would typically sit in and observe ongoing practices in contexts that are natural for the community under study, in the most extended cases actually 'living in' those communities as a member as far as possible. In multimodal research, this is weakened somewhat and so the settings studied might include, for example, the production of a TV broadcast, or the planning and design of a film, or a group of architects discussing and deciding on building designs, and so on. The data collected is then based on close observations of what is actually done as well as interactions, interviews and so on with the participants.

Ethnographic methods may be relevant whenever the interest of research is focused on actual interactions, always multimodal, that bring about some result. Here ethnography overlaps with the goals and methods employed by many studying interaction such as, for example, Conversation Analysis (Goodwin 2001). In cases where the focus of research is more on the *product* of such interactions, the connection is more indirect. We see, for example, ethnographic methods discussed by Burn (2014) and Gilje (2011, 2015) in relation to video and film-making. This tells us a considerable amount about interaction during the design process; we may also require, however, multimodal analyses of the products of that design as well. Addressing and combining these different methodological perspectives is addressed more explicitly in Chapter 7 under the topic of fixing the 'focus' of research. On the one hand, a different focus can make different methods relevant; and, on the other hand, a single task may itself demand a variety of interlinked foci.

Multimodality is also increasingly a methodological issue for ethnographic researchers themselves because the data being collected is itself increasingly multimodal. Whereas previously research notes may have been predominantly textual, it is now common practice to include at least video recordings and sound recordings. The resulting information is complex and so needs to be considered an object of multimodal analysis in its own right (cf. Goodwin 2001; Collier 2001; Mondada 2016).

Protocols   When turning the records of interactions into data that can be analysed, a variety of issues are raised. Typically, the result will be some protocol of what occurred. This may already employ certain categories or types of information that are considered potentially useful and may be structured to a greater or lesser degree. This begins to move field notes towards more organised 'transcriptions', a process we return to below in Section 5.2.3. Whereas direct recordings of actions in context provide one type of data—a type particularly favoured by those working in sociological interactional traditions and their descendants—more structured materials can be obtained from interviews and questionnaires.

If one uses interviews or questionnaires in order to make the most of the limited time available during any practical period of study, then they have to be very carefully designed. Avoiding problems with good design can make all the difference between actually gathering some usable data and failing. Since questionnaires are examples of social interaction, they run all the dangers that commonly occur in interaction of all kinds. One classic scientific study showing the consequences of potentially misleading questions is reported by Loftus (1975) and her investigation of the reliability of eyewitness reports. The most general result in this study was that the answers given depend on the phrasing of the questions—which is already enough to make it clear that any such data collection needs to be extremely cautious in order not to invalidate the very data with which one wants to work.

As a consequence, writing good questionnaires is an art in its own right and there are various sources that offer guidelines: these apply to all kinds of surveys and are not specific to multimodality. If you are going to collect information by giving questionnaires, it is certainly advisable to familiarise yourself with these issues beforehand, and to run 'pre-tests' on your questionnaire to see if any unforeseen problems in answering the questions arise. In short, however, questions in questionnaires should obey at least the following precepts: *Question-naires*

- Questions need to avoid 'leading' the answerer in some directions of interpretation rather than others—asking whether someone saw the dog, or noticed the sound, are leading questions because they suggest that there was something to be seen or noticed; such questions are well known to sway answerers in the direction of saying that they did see or notice what is asked about; these need to be expressed more neutrally therefore.
- Questions need to be 'clear' in the sense that they do not combine different aspects of what is being asked about—typically this might occur in a question containing an 'or'; for example, if one asks 'was the question clear or was it redundant?', then this brings together two possibly separate facets that the answerer then has to juggle with: they may have found a question to be both clear and redundant, etc. Questions should therefore avoid these kinds of 'doubled up' queries.

Questionnaires may also be problematic because the information that you are looking for might not even be available to any individual questioned. For example, people are not aware of the paths their eyes take when reading or interpreting and, moreover, *cannot become aware* of this because it all happens too fast and is not subject to conscious control. Thus it makes little sense to ask about this directly. Recipient studies, i.e., the study of how the consumers of some materials respond to that material, needs to be very aware of these methodological issues.

A better strategy is generally to concentrate on probing *consequences* of any behaviour under study rather than asking about it directly. In general, people will be very bad at reporting their own behaviour and so indirect methods are essential. One might find out about some particular reading paths, for example, not by asking where people thought they were looking but instead probing things that they should know, or have seen, had they looked in one direction rather than another. Working out how to probe behaviour in this way needs to be thought about very carefully and overlaps with issues of experimental design, to which we return below.

5.2.3 Transcriptions and coding data

A reoccurring theme across various disciplines concerned with analysing data of various kinds has been how to make that data more accessible to research questions. If we just take written language, for example, and we want to find all uses of the noun 'can', then we immediately face difficulties when searching for such uses because it is difficult to weed out other uses of that string of letters, such as the 'can' of 'Mary can surf very well' or the rather different 'can' of 'let's can the fish'.

When working with images, the situation is even more difficult: if we have a collection of scanned images of some kind, searching to see how some visual motif such as 'the sun' is used is extremely difficult—moving from an image to some more abstract characterisation of that image and its content is not straightforward. And again, if working with pieces of music, selecting all use of some particular tone sequence or melody or rhythm presents challenges at the very limit of what can currently be done technologically. In all these cases, however, being able to search for and find particular patterns of interest to a research question is a prerequisite for carrying out that research at all.

The most widespread technique for addressing this problem is to 'enrich' the data by adding particular categories that have been determined to be potentially relevant for the studies being performed. Once such information has been provided, research can move along far more effectively. That is: while it might take considerable effort to find relevant cases of particular phenomena in the 'raw data', if that data has been *coded*, or *annotated* in a way that makes those phenomena accessible, then one can search for the coding or annotation categories instead.

Annotation   Coding and annotation of this kind thus form an integral method for conducting research in many contexts, and for multimodal research is absolutely essential since it is, in many cases, difficult to operate *directly* with the raw data in any case. In work on language, this has been a substantial issue ever since researchers began exploring interaction and spoken-language data. Since it is generally still difficult to manipulate the sound or video recordings themselves, the actual data was replaced with a representation, or *transcription* that was expressed in a form that could be easily manipulated—such as text. This was an important enabling tool for research.

Before it became straightforward to record and manipulate audio files, for example, research on spoken language would commonly work with written transcriptions of what was spoken instead. And this is still the most common way that results are reported in journals and other traditional forms of publication. Finding other levels of abstraction to access data is then even more essential for more complex multimodal data. Although a growing number of journals now encourage submission of supporting media that might contain more complex data, these are most often cited in some textual or static visual form in the articles themselves.

Such transcriptions were at first, and sometimes are still, seen as 'standing in' for the original data. Their primary motivation was to 'fix' transient phenomena in an inspectable form. For more complex time-based media, such as film or TV programmes, a common strategy is to use tables of information where the rows may indicate some selected analytic unit and the columns contain different facets of the information conveyed (cf., e.g., Baldry and Thibault 2006). Very much the same construct has been used in film studies for a long time; many detailed analyses are set out in this form (cf., e.g., Bellour 2000 [1976]). In the case of film, the tables are referred to as 'film protocols' or 'shot analyses'. There it is common to present segments of film under discussion setting out the prominent features of the data on a shot-by-shot, scene-by-scene or, more rarely, frame-by-frame basis. In fact, it is probably in film analysis that such tables have been pushed the furthest. This can readily lead the table to become overloaded with information, however, as the range of *potentially relevant* features in film is immense (cf. Table 13.1). Overloading is particularly a problem whenever there is any attempt to consider transcription as *replacements* for the original data because then there is really an attempt to capture 'everything', which is impossible.

Consequently, such representations have also received particular critique (Crawford 1985), both for their potential over-complexity (are all the details necessary for an analysis, for example?) and the often very time-demanding nature of their production. Many branches of film studies no longer set out film protocols at all. Viewed multimodally, the remediation from a dynamic audiovisual canvas to a static visuo-spatial canvas certainly places considerable strain on any adopted notations and it is by no means clear that the resulting information is helpful when attempting to deal with larger bodies of data. Such tables need then to be transferred to 'database tables' for online storage, retrieval and exploration, but this is still too rarely done. Our descriptions both of corpus-based and computational methods later in this chapter provide approaches that encourage such work further.

Describing data in terms of some collection of categories can also be done in order to 'reduce' the sheer bulk of data that needs to be considered to something more manageable. For example, if a research task is concerned with the kinds of programmes being shown on various TV channels, assigning the programmes of the data to various categories in a pre-processing step—such as 'advertisements', 'news', 'drama' and so on—makes it straightforward to get a good overview of the range of offerings. This kind of approach, which can become quite elaborate, is generally called *content analysis* since the categories are usually something related to the content of the data under study (cf. Neuendorf 2002; Rose 2012*b*; Schreier 2012).

Deciding on appropriate coding schemes or transcriptions for your multimodal data is of considerable help when carrying out research. When working with audi-

Transcription

Content analysis

ovisual data, one does not want to search through hours of data each time a question arises: it is then typical to pick out particularly salient occurrences, marked with their time of occurrence in the original data recordings. The categories can be selected for a variety of research questions and it is important to be aware that such categories are never 'neutral'—by selecting some categories rather than others one is automatically making decisions concerning just what aspects of the data are going to be focused on and which not.

One should therefore always make these decisions carefully and be aware that there may be aspects of the data being excluded that could turn out subsequently to have been important. If a theory of dialogic interaction focuses on complete or incomplete syntactic structures as its theoretical units, then a transcription that renders this information and this information alone may be an appropriate choice. If a theory of dialogic interaction in addition claims hesitational and filler phenomena (i.e., 'ahs' and 'umms') to be significant, then a transcription for this research needs to include such information. This is, of course, inherently 'risky'. The limitation of attention that ensues may turn out to omit just the details of the phenomena that would lead to explanatory accounts. For this reason, the style of fine-grained transcription typical of conversation analytic accounts is clearly to be preferred to a regularised focus on syntactic forms when research questions of interactional phenomena are at issue.

A natural consequence of the above is that one should always be cautious about losing information—access to the original raw data collected should be preserved both to enable further checking of any classifications made and to allow extensions or changes in the aspects of the data subjected to study. The theoretical implications of any particular coding choice are emphasised well in the seminal paper by Ochs (1979); this should be required reading for anyone who is considering transcribing data. Note that it would be fundamentally mistaken, therefore, to take any fine-grained transcription as capturing 'all' aspects of the data in a theoretically unmediated form.

Transcoding   Extending the discussion of such transcriptions to consider them as further forms of multimodal representation in their own right is a beneficial step. From this perspective, any transcription of data that a research undertakes is a case of 'transmodal translation' or *transcoding*—for example, we might convert a film into a combination of layout (e.g., a table), text (descriptions of shots) and images (static screenshots). All such recodings then raise issues at the core of multimodal research: what kinds of meanings can the transcription make and which not? The transcription that the researcher works with is rarely a simple transmodal reconstruction of the originating data but represents instead a first step up on a level of analytic abstraction that is intended to support positively the search for meaning-bearing patterns and to hold otherwise inherently transient phenomena in a fixed form. Explicitly considering a transcription as a multimodal transcoding of other data provides a more theoretically-founded conceptualisation of the entire process of transcription.

Several quite distinct notation schemes have been suggested for multimodal data collection. These vary according both to the particular traditions that they are descended from and to the specific research aims. Seeing them all as multimodal transcodings is one way of permitting comparison and evaluation of their respective merits or problems.

One detailed set of conventions is set out in Norris (2002) in order to explore the unfolding of face-to-face interaction in natural contexts, where any parts of those contexts—furniture, clothes, positions, windows and so on—may come to play a role in the discourse. Norris' scheme consists of visual key frames of video recordings of natural interactions (rather than rows in a table) and adds aspects important for her analysis, such as the spoken utterances and their intonation, into the visual images directly. Speakers are indicated by placing their spoken text close to them in the image and intonation is shown typographically by curving the horizontal line of the text to suggest its intonation contour.

A rather different set of conventions is followed by those working in multimodal analyses in the tradition of Conversation Analysis (→ §8.1 [p. 240]). Early CA transcriptions of spoken interaction attempted to present presentational features of languages, such as hesitations, filler sounds (hmm, erh), corrections, overlaps of speakers and so on by fairly direct changes to spelling and typographical indicators such as dashes, commas and full stops. In current multimodal research on interaction, these conventions are maintained. However, in addition to the transcription of the language, textual descriptions of activities carried out, gaze directions, gestures and so on are included with their times of occurrence marked explicitly by arrows in the transcription. An example taken from Heath et al.'s (2010) description of how to use video for qualitative research is reproduced in Figure 5.1.

We can see in this transcription many of the standard problems that are raised by the need to deal with multimodal data. For example, any transcription needs to indicate the synchronisation of events—i.e., how can it be indicated who performs some action, such as 'creasing pages', and how long that action takes? Varying solutions are explored in the annotation schemes set out in Flewitt et al. (2014), Hepburn and Bolden (2013), Bezemer and Mavers (2011) and Bezemer (2014); Jewitt et al. (2016) presents further examples. Most of these kinds of transcriptions have the positive feature that they can be made as detailed as required and may even include pictures, photographs, drawings, etc. of the context in which an interaction is occurring if this is considered useful for understanding what is going on. But they also have the rather more negative feature that they do not support generalisation or empirical studies of larger bodies of data. If we wished to ascertain the frequency of 'glances' we would need to read all the transcripts and take additional notes.

In our section below on multimodal corpora (→ §5.3.2.1 [p. 156]), we will see that the communities employing computational tools for corpus-based multimodal methods not only have a host of quite well worked out and standardised transcription frameworks but also computational tools for supporting their use. These possibilities need to be considered far more than is currently the case in multimodal research, where *ad hoc* coding schemes are still prevalent, with little technical support for man-

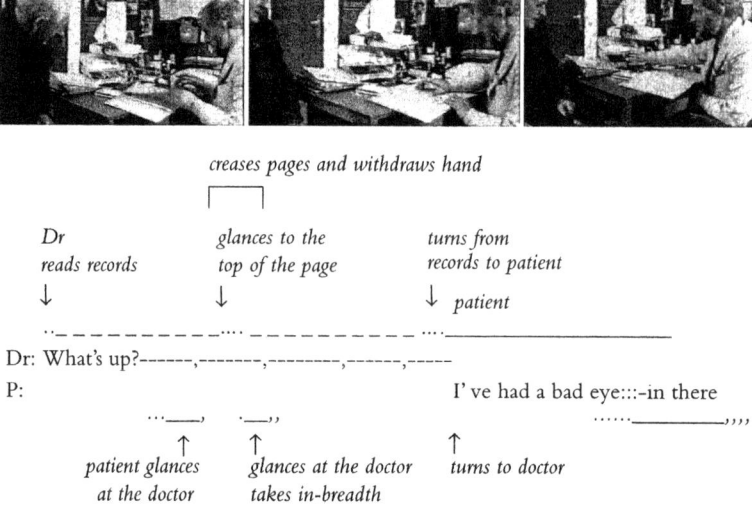

creases pages and withdraws hand

| |

Dr　　　　　　　　*glances to the*　　　　　*turns from*
reads records　　　*top of the page*　　　*records to patient*
↓　　　　　　　　　↓　　　　　　　　↓　*patient*

Dr: What's up?------,-------,--------,------,-----
P:　　　　　　　　　　　　　　　　　　I' ve had a bad eye:::-in there

↑　　　↑　　　　　　　　↑
patient glances　*glances at the doctor*　*turns to doctor*
at the doctor　　*takes in-breadth*

Fig. 5.1. Example of multimodal conversation analytic transcription of a doctor-patient interaction; ©2010 SAGE Publications. Reproduced by permission from Heath et al. (2010, Transcription 3)

aging such basic tasks as temporal alignment with videos, multiple types of simultaneous transcription and so on. All of these functionalities are relatively well understood and supported with modern tools.

Enriching
data

Enriching data with additional information is therefore a natural progression from more informal notes concerning linguistic phenomena that any researcher of language might use, as for example might be produced in field notes, towards more systematic data transcription and the many attempts to regularise linguistic notations, to the explicitly defined annotation schemes and practices of modern corpora that we will discuss below. The kinds of pre-preparation that are attempted depend on the material that has been selected, the research questions being raised of that material and the technological possibilities for enriching the data with additional information that support analytic investigation.

Articulating the rather different natures, functions and requirements of these ways of enriching data helps clarify the issues that need to be addressed when we turn to the corpus-based study of multimodal phenomena. We conclude, therefore, by distinguishing four broad categories of data enrichment: notes, transcription, coding and analysis. Any and all of these may play useful roles in multimodal research:

- *Notes* are in principle the least restricted: any information that might be thought to be useful concerning the artefact or activity under analysis might be added in any form that the analyst chooses. If the form of such notes starts to be regularised in any way, then they begin to transition towards transcription, coding and analysis.

- *Transcription* is then any more systematic attempt to 'transcode' the data: for example, to represent an audio event in a written form that allows the original audio event to be re-created in more or less detail. Phonetic transcriptions are the most commonly found.
- *Coding* is a further step away from the data and towards analysis. The coding of data is generally carried out with respect to categories defined by a coding scheme. The coding scheme identifies criteria for segmentation and for the allocation of segments to categories of the coding scheme. In contrast to transcription it is not generally possible to go from a coding back to the original data in any sense: the coding scheme serves the function of grouping together segments that fall under the same classification; detail below the level of the coding scheme does not play a role beyond providing evidence for particular allocations to coding scheme categories rather than others.
- *Analysis* describes units identified in the data at higher levels of abstraction which typically capture how combinations of elements from the data are combined to form more complex structural configurations. For example, a corpus of spoken language could contain syntactic analyses of the units identified in the corpus. In such a case, transcription might provide a rendition of the spoken speech signal in terms of phonetic categories and lexical items; coding might attribute the lexical items to parts-of-speech; and analysis might run a parser over the coded lexical items to produce syntactic trees. Indeed, any of these stages may be additionally supported by automatic processing techniques as well (→ §5.5 [p. 162]).

Regardless of whether the information is made up of notes, transcriptions, codings or analyses, it must be linked in some way to the specific data which it is characterising. Moreover, multiple descriptions may be added at each level of abstraction, usually reflecting different aspects of the phenomena being studied. These may then even draw on differing theoretical bases as required for the aspects considered—for example, a level of analysis concerning visual composition may be drawing on a completely different set of theoretical constructs to one concerning any characters depicted. Powerful techniques for managing these kinds of multiple description tasks are provided in multimodal corpus studies, as we set out in Section 5.3 below.

5.2.4 Behavioural data and experiments

One final form of obtaining data that we will describe here is to gather information from observing and also measuring what people do in more or less controlled situations. Controlling the situation is important so that what is being measured can be reliably isolated from potential distracting information. Although such research methods are commonly criticised from those who desire *ecological validity*, i.e., naturally occurring behaviour, controlling situations can be a very effective way of pinpointing relations between effects and consequences. For this to work, however, it helps considerably when the hypotheses being investigated are sufficiently sharp and well defined that one knows just what to vary.

This is then an example of combining very different methods for different stages of a research program. In order to formulate hypotheses, one may well need to draw on more qualitative interpretations or models of what might be going on. These are then operationalised so that particular behavioural responses—e.g., heart rate, eye-tracking, choice of some options in a multiple answer sheet, facial expressions, re-action times, etc.—can be gathered and correlated with the experimental conditions varied.

Reliability   Finding out, or establishing, whether we have a *reliable* correlation between some controlled experimental condition and some kind of behavioural response is then the task of statistical methods. We give a basic introduction sufficient for carrying out simple statistical tests of this kind in Chapter 6.

5.3 Corpus-based methods to multimodality

One of the most established approaches in many areas of empirical research nowadays is that of corpus-based studies. Corpus-based studies rely on collecting naturally oc-curring examples of the kind of data that is to be subjected to analysis. Such collec-tions need to be sufficiently broad as to promise 'covering' a substantial amount of the variation relevant for the research question being posed. This is typically addressed simply by increasing the number of examples in the hope that that variation will be present. Thus, for example, if one wants to know about the design choices made in newspaper front pages, it is wise to consider a whole collection of such pages rather than just one or two that one happens to find.

Corpora/   Selection of particular 'corpora' can then be made more systematic in order to
subcorpora   address further questions. For example, one might not just collect newspaper front pages at random, but instead deliberately select them according to their production dates (perhaps grouped into decades or years), or according to their intended readers or their countries of origin. A larger collection is thus organised into smaller 'subcor-pora', each characterised along some dimensions of classification in their own right.

The adoption and subsequent development of corpus-based methods more gener-ally has been made possible by two advances: first, large quantities of recorded 'data' have become readily available and, second, effective tools for searching those data for patterns have been developed. Both aspects need to be present for corpus analysis to be effective: insufficient quantities of data raise the danger that the patterns found are accidental and do not generalise, while a lack of facilities for searching larger scale bodies of data means that researchers cannot get at the patterns, even if they are there—several million words of natural text helps a researcher little if actually reading the texts is the only mode of access to the data!

Clearly the extent to which one can apply corpus methods in the multimodal context then relies on the extent to which sufficiently large samples of data can be made accessible in this way. And so, when the data to be investigated involves multimodality these issues remain and raise significant challenges of their own. For traditional text corpora it has often been possible to take relatively simple approaches to viewing the results of corpus searches by presenting them 'directly' in the form of lists of cases retrieved according to certain criteria. Approaches of this kind quickly show their limitations when working with multimodal data. Methods need to be found both for gathering the necessary quantities of data and for making those sets of data, even if present, *accessible* so that they can be interrogated effectively.

These challenges need to be embraced, however, because it is just as important to base multimodal research on broader collections of actually occurring cases of multimodal meaning-making as has been demonstrated for language research. Fortunately, the considerable experience now gained in data-driven and corpus-based approaches to language research contains much that can now be applied in the context of multimodality research as well. This has been taken the furthest in work on face-to-face spoken interaction where multimodality of many kinds naturally arises. There is still a considerable amount to do to make corpus-based methods available for other areas of multimodality, however—we describe some of the steps taken in this direction below. Broad overviews are available in, for example, Allwood (2008) and Bateman (2014c).

5.3.1 Making data usable for multimodal corpus research

There will always be a distance between the multimodal data being interrogated and the original phenomenon that is the actual goal of study. Restrictions in corpus design will often need to be made on technical grounds and these must be carefully weighed with respect to their potential for revealing or hiding meaningful patterns.

As an example, we can consider face-to-face interaction. Whereas face-to-face conversational interaction between two participants is (ontologically) an embodied, spatially-situated three-dimensional temporally unfolding set of coordinated behaviours (→ §8 [p. 239]), data selected to make up a corpus of such interactions might consist of very different kinds of information: this is the issue of multimodal transcoding mentioned above (→ §5.2.3 [p. 148]). There may be a written transcription of what was said and by whom including written indications of intonation, or an audio recording of what was said, or a video recording with one fixed camera of what occurred, or two close-up video recordings, one for each participant, or a computer-regenerated three-dimensional representation of the activity with full motion-capture of the participants, and so on.

Each kind of representation can be used for a multimodal corpus and might reveal significant properties of the activity under study. They are all nevertheless distinct from the original behaviour in a way that can often be overlooked with monomodal textual corpora. For multimodal corpora, in contrast, it will be common for a variety of levels of description to intervene between original behaviour or artefact and researchers interrogating that behaviour or artefact.

In some areas of multimodality more is known about the probable workings of the modes at issue than in others. When linguistic utterances are present, for example, assumptions concerning their phonological, lexical, morphological, syntactic and semantic organisation are well motivated and so naturally offer appropriate levels of description of *certain aspects* of the phenomenon under investigation. In other areas, such as gesture, there are also well developed proposals for characterising both the forms and functions of any gestures that might occur. In other areas still, however, such as document layout (taken further in our analyses in Chapter 10), accounts are in need of considerably more empirical research.

What will become increasingly important in the future, therefore, is the ability of corpora to combine information with distinct theoretical statuses. Patterns can then be sought and hypotheses verified not primarily against data, but against *other levels of descriptions of data*; we characterised this above in terms of different kinds and levels of transcription (→§5.2.3 [p. 146]). This requires a rather different mindset to that common in more traditional studies, where single researchers might have attempted to characterise many levels of abstraction within a single theory—as in the multimodal conversation analysis transcript shown in Figure 5.1 above. Nowadays, and especially in corpus-work, it is more beneficial to allow descriptions from many theories and even disciplines to co-exist. The resulting 'multi-level' annotation schemes are then productive meeting points for different theoretical approaches: one level may be linguistic (of a particular kind), another may be psychological (driven by experimental data), another still may refer to production aspects, and so on.

Dealing with diverse multimodal materials of this complexity in principled ways that are supportive of empirical research raises substantial challenges of its own and has been a rapidly growing area of research and application since the early 2000s (cf. Granström et al. 2002; Kipp et al. 2009). Nowadays, we are also seeing exciting cross-fertilisation of research methods: as in, for example, the use of motion-capture as practised in the film industry to acquire rich data sets of *natural* interaction including speakers' and listeners' movements, gestures, body postures and so on (e.g., Brenger and Mittelberg 2015). There are also many efforts that are attempting to provide agreed classification or annotation schemes for particular areas that can then be used by different research groups for a variety of research questions (cf. Kranstedt et al. 2002; Allwood, Cerrato, Jokinen, Navarretta and Paggio 2007; Kipp et al. 2007). This marks an important development away from one-off, single studies to larger-scale research in which results from differing perspectives can be combined and used to leverage off one another. The use of *ad hoc* annotation practices is therefore to be avoided as far as possible.

5.3.2 Types of multimodal data and corresponding tools

Different kinds of multimodal data raise different kinds of problems and so it is useful to characterise them with respect to the specific challenges they raise, both for corpus-based approaches and corresponding support tools. Basic dimensions for organising the discussion can be distinguished on the basis of properties of the data to be covered—this is necessary because without addressing these distinct properties, the pre-preparation and, hence, accessibility of the data is compromised and corpus-based work is not then possible.

The broadest distinction relevant for discussions of multimodal corpora is that between *linear* and *non-linear* data. Linear data include material which is essentially organised to unfold along a single dimension of actualisation. This dimension may either be in space, as in traditional written-text data, or in time, as in recordings of spoken language. Non-linear data include material where there is no single organising dimension that can be used for providing access. This generally involves *spatially distributed* information and visual representations, such as those found on pages of documents, printed advertisements, paintings and so on, but also includes other kinds of *logical organisation*, as in websites, for example.

Linear/
non-linear
data

We discussed some of the impact of the distinction between linear and non-linear data in Chapter 2 above. The continuing force of the distinction can be seen in the fact that many accounts of multimodal corpora and tools in fact *only* consider linear data, paying little if any attention to non-linear data. Visual representations, such as photographs or images of paintings, are still more commonly *archived* than maintained in corpora, and a variety of classifications exist for the purposes of search and organisation. And again, a reoccurring common problem in multimodal research as we have seen, quite different communities are typically involved in the two cases — speech and interaction research on the one hand and press photography, art history, medical imaging and others on the other (cf. Chapter 2).

The relatively natural ordering of data that is possible within time-based media is one of the main reasons why speech corpora, using *time-stamped* data, constitute by far the most developed type of multimodal corpora available today. Corpora for non-linear data are less developed and few tools specifically for corpus work have been developed with this area in mind. Unfortunately, tools built for linear data are often unusable for cases of strong non-linearity, although some developments are now being attempted to add spatial information to time-based data. Explorations of the kinds of meanings involved in non-linear communicative artefacts, such as page-based media (including the web), as a consequence remain limited with respect to their empirical foundation—a situation in urgent need of change.

5.3.2.1 Working with linear data

Two of the most widely used and freely available tools for multimodal corpus analysis of linear data compatible with the above considerations are *ELAN*,[1] developed
and maintained by the Max-Planck Institute for Psycholinguistics in Nijmegen (Wittenburg et al. 2006), and *ANVIL*,[2] developed by Michael Kipp while at the DFKI in
Saarbrücken (Kipp 2012). Both are based on the current international standards for
information exchange and so allow data to be freely reused and incorporated in other
projects and tools.

Annotation
tools

As with most tools for temporally-organised multimodal data, annotation of data
within these tools adopts the musical stave metaphor already seen in Eisenstein's
characterisations of film (see Figure 2.1 in Chapter 2 above). Arbitrary numbers of *tiers*,
or *tracks*, can be defined for maintaining different kinds of information that is automatically aligned with the temporal medium being annotated. All information is then
displayed spatially aligned by virtue of the common time frame. Although substantial
projects have been carried out using these tools for analyses of face-to-face interaction, gesture, intonation, body posture and more, the tools themselves are agnostic
concerning both what kinds of information may be added and what kinds of visualisations or analysis reports may be generated. This is in contradistinction to many
more dedicated tools that already commit to particular styles or kinds of analysis.

An illustrative screen shot showing work with ELAN for the analysis of film is given
in Figure 5.2. In the top-half of the figure we find several information windows plus a
video player, while running across the lower-half of the figure we see the particular
tracks defined by the annotator for analysis and the state of work in progress. The current position in the video is marked by a vertical line intersecting the displayed tracks.
The scale of the tracks can be freely changed: in the figure a smaller scale has been
selected in order to see more of the video at once. The tracks here are some of those
typically used in film analysis: the division into shots, some indication of accompanying music, the lines spoken by each of the actors—all synchronised with the film and
each other. Manual annotation is typically done by clicking within the graphically
represented timeline of a specified track as the video is playing and either selecting a
particular classification category defined by the annotation scheme specified for that
track or by typing a free-text description.

The general appearance of ANVIL and most other tools of this kind is similar. They
all share typical usage scenarios as well. An audiovisual artefact is segmented along
several dimensions and the resulting segments are then categorised, often by hand,
i.e., by groups of human annotators. Annotators straightforwardly define categories
for their own purposes or use annotation schemes that have already been defined by
others. The annotated information can then be used for looking for patterns or com-

---

1 http://www.lat-mpi.eu/tools/elan/
2 http://www.anvil-software.de/

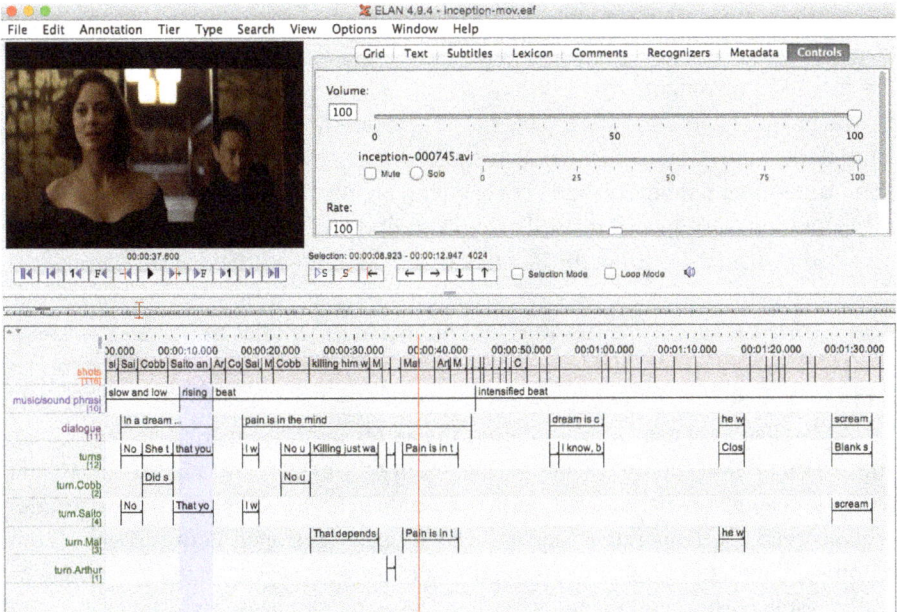

Fig. 5.2. Screenshot of the ELAN tool showing annotation of a fragment from Christopher Nolan's film *Inception* (2010).

piling statistics of use over the corpus, and also exported for use in other tools. When deciding to use any tool of this kind, therefore, it is important to consider the extent to which the software allows you to both *import* and *export* transcribed data.

Importing data transcriptions means that results from other projects and tools can be built upon; exporting data transcriptions means that other tools can be employed. Typically, for example, a transcription tool will not itself allow sophisticated statistical or visualisation techniques to be used—exporting data therefore makes it possible to do much more with your transcribed data as well as making it easier to pass on your analysis to other research groups and projects, which might use quite different tools to your own. Mature tools with extensive user-bases such as ANVIL and ELAN and which support such functionalities have much to offer, therefore, both for the annotation of data and for supporting subsequent searches for meaningful patterns and presenting results.

5.3.2.2 Working with non-linear data

The tools mentioned until now and the annotation schemes they support are strongly linear. For artefacts in which time cannot be invoked as an organising dimension, other solutions must be explored. That is, whenever the form of multimodality being analysed involves data that make use, for example, of the two-dimensional area of the page or a three-dimensional volume within which 'communication' of some kind takes place, it is necessary to ensure that corresponding corpora record this aspect of the data in its own right.

Documents and other static page-based artefacts are good examples of objects of multimodal analysis that do not make any convincing reference to temporal organisation. Annotating data of this kind consequently requires access to spatial properties of the information depicted instead of, or in addition to, temporal properties. In many cases, what is required is not the identification of properties that may be aligned with time-intervals but instead the identification of 'entities' within images, which may then receive various properties as annotations. These entities may belong either to what is represented or to the form of the images themselves, as in their layout organisation.

GeM model A detailed proposal for annotating the latter is given in the layout structure and area model of the Genre and Multimodality framework introduced in Bateman (2008) and taken further in application in Hiippala (2015*b*). Other kinds of annotation schemes and tools for non-linear data have been developed within specific communities; Bateman (2014*c*) offers examples. There is, however, much to be done for really usable corpus analysis tools for non-linear data to result. Here again there is a considerable need for consolidation of efforts across disciplines and approaches.

5.3.3 Challenges and directions for multimodal corpora

For multimodal research, corpus-based approaches will need to move away from simple notions of a corpus as a 'collection of transcribed data' where the transcriptions more or less stand in for the phenomena to be studied. Instead multimodal corpora will increasingly become repositories of data seen from a variety of perspectives. This transition constitutes a major challenge for corpus-based multimodality research: theoretically-motivated distinctions need to be brought to bear to organise material for analytic inspection but, in many areas, it is not yet known just what the distinctions necessary for explanatory accounts will be. Achieving productive states of balance between theoretical focus and empirical openness represents a considerable methodological challenge.

An important step towards achieving results will be the adoption of standards by which community efforts can be multiplied and more powerful mechanisms for revealing patterns can be deployed. Properly annotated data may then support the search for generalisations by allowing examination of potential *correlations* across the various levels of descriptions that corpora provide. Establishing such dependencies represents a crucial step towards understanding how multimodal meaning-making operates. Moreover, for larger scale corpus work an increasing reliance on automatic methods in appropriate combination with manual approaches will be essential; we return to this in Section 5.5 below.

5.4 Eye-tracking methods for multimodality

Eye-trackers are highly accurate cameras pointed towards the viewer's eyes, which track and record eye movements. These observations are processed by analytical software and broken down into distinct events that make up the perceptual process: for the current discussion, we will refer to this bundle of hardware and software as an eye-tracker. Once calibrated by having the viewer look at a specific location monitored by the eye-tracker, it is able to estimate where the viewer's gaze lands when eyes move.

Because eye movements have been strongly linked to cognitive processing of visual input, knowing where viewers look, for how long and under what circumstances is valuable information for researchers in many fields, such as cognitive science, educational and experimental psychology, human-computer interaction and many more, naturally benefiting multimodal research as well (Holsanova 2014*b*).

> Eye tracking gives us insights into the allocation of visual attention in terms of which elements are attended to, for how long, in what order and how carefully.
> — *Holsanova (2012: 253)*

Fixations and saccades

Insights may be gained by attending to what have been established as key events in visual perception: *fixations*, that is, where the eyes are focused and for how long, and *saccades*, which are rapid eye movements that occur between fixations. The material presented to a viewer under observation is referred to as a *stimulus*. Different stimuli and viewing conditions elicit different patterns of fixations and saccades, which provide cues about how the stimuli are processed by the brain (Rayner 1998).

For additional information, eye-tracking may be triangulated with other methods, such as psychophysical measurements, verbal think-aloud protocols, interviews, tests and questionnaires, which can provide "insights into the rationality behind the behaviour, comprehension and interpretation of the material, attitudes, habits, preferences and problems concerning the interaction with multimodal messages in various media" (Holsanova 2014*b*: 292).

To exemplify: Holsanova (2008) combines eye-tracking measurements with think-aloud protocols in a multimodal scoresheet which represents eye movement data together with the viewer's own explanation of the perceptual process. Müller et al. (2012) augment eye-tracking data with psychophysical reaction measurements to investigate how press photographs are perceived and what kinds of emotions they evoke. And Bucher and Niemann (2012) use questionnaires, retrospective interviews, knowledge tests and think-aloud protocols to reconstruct the process of interpretation when observing PowerPoint presentations. Given the complexity of multimodal phenomena and the interpretative process, this kind of *triangulation of methods* may help to shed light on how contextual factors such as background knowledge and the performed task affect visual perception.

Types of
eye-trackers

Eye-trackers come in various forms and sizes. They range from static mounts that can track eye movements very accurately to mobile trackers, which may be attached under the display of a desktop or laptop computer, thus creating a more realistic environment for running experiments. Eye-tracking glasses allow for even greater mobility, freeing the viewer from the computer screen and making them particularly suitable for studying face-to-face interaction in natural conditions (cf., e.g., Holler and Kendrick 2015). Most recently, eye-trackers have been integrated into head mounted displays for studying the perception of both real and virtual environments. Choosing the appropriate eye-tracker depends very much on what is being investigated and under what kinds of conditions (for a comprehensive guide to eye-tracking methods, see Holmqvist et al. 2011).

Scanpaths
and
heatmaps

The data received from an eye-tracker, that is, the measurements describing eye movements over time, may be studied and represented in various ways. Static and animated scanpaths that trace the trajectory of gaze on the object or scene and heatmaps, which use coloured overlays to indicate the areas that are attended to most frequently, provide rough sketches of the visual perceptual processes. Examples of each are shown in Figure 5.3. The scanpaths in the figure already show a number of interesting features that might not have been predicted: for example, people tend to look where the main character is looking in a way that suggests they are trying to work out what he is seeing. It is not always possible to simply 'read' the interesting points from such visualisations, however. Approaches in experimental psychology and cognitive science therefore go much deeper and make extensive use of statistical methods to search for reliable patterns in the data in order to interrogate the processes underlying visual perception in ever increasing detail.

Fig. 5.3. Left: a scanpath showing a single experimental participant's eye behaviour when looking at a moving image from a film (*Conte d'été*, Éric Rohmer, 1996) discussed in Kluss et al. (2016: 272–273). Right: a post-processed heatmap from another film (*Solaris*, Andrei Tarkovsky, 1972) showing aggregated results over several participants—the colours show how often particular areas were looked at, ranging from red (most fixated) to blues and violets (hardly looked at at all); heatmap supplied by Mehul Bhatt

Here the message is that the wealth of data provided by eye-trackers may be ana-lysed in different ways: whereas heatmaps or scanpaths may provide an initial view into visual perception under some conditions, drawing more extensive conclusions requires working with raw data. Moreover, experiments must be designed carefully to ensure that the collected data is valid and appropriate for answering the research questions asked (→ §6 [p. 169]).

As said, eye-tracking is increasingly applied in multimodality research: a key fig-ure in this process has been the cognitive scientist Jana Holsanova. In addition to writing several accessible overviews of eye-tracking research (Holsanova 2014*a,b*), she has contributed a number of valuable case studies. Holsanova et al. (2006), for instance, use eye-tracking to examine general assumptions about the perception of layout space proposed in social semiotic theories of multimodality (→ §2.4.5 [p. 48]), such as information value zones, framing and salience. Boeriis and Holsanova (2012), in turn, compare social semiotic theories of visual segmentation, such as part-whole structures, with eye-tracking data, examining how objects are located, identified and described both individually and in relation to one another. Finally, Holsanova and Nord (2010) consider different media, the artefacts they are used to realise and their perception, suggesting eye-tracking as a potential means of bridging the gap between production and reception: Holsanova et al. (2008) explore these issues in greater de-tail by analysing information graphics, showing how their structure can guide visual perception (→ §11 [p. 279]).

In multimodal research, visual perception is often discussed in terms of 'reading paths' likely taken by the viewer when engaging with some situation or artefact (van Leeuwen 2005*b*; Hiippala 2012). Here the crucial issue is that hypotheses about reading paths must not be confused with measuring actual sequences of eye movements, which researchers working with visual perception refer to as scanpaths. Scanpath analysis provides multimodal researchers with the means to *validate* hypotheses about pos-sible reading paths.

To sum up, eye-tracking enables validating hypotheses about the perception of multimodal artefacts and situations. There are, however, limitations to eye-tracking methods as well: although tracking eye movements can inform us of the allocation of visual attention, it does not tell us why certain parts of the stimuli are being looked at. The viewer may find the part in question worthy of attention for the performed task, or on the contrary, have trouble processing the stimuli. Consider, for instance, eye movements between semantically related text and images: we cannot know whether the eye movements reflect success or problems in integrating the content (Holsanova 2014*b*: 292). As suggested above, this kind of situation obviously requires additional methods for investigating which is the likelier scenario.

Finally, in the case that you do not have access to an eye-tracker, there are also comprehensive overviews of different research topics studied using eye-tracking whose findings may be related to questions pursued in multimodal research. To draw on some examples, these include overviews on the perception of newspapers (Leckner 2012), advertisements (Simola et al. 2014), websites (Nielsen 2006), films (Smith et al. 2012; Loschky, Larson, Magliano and Smith 2015; Kluss et al. 2016) and much more. Referring to one of these reviews will undoubtedly place your research on a much firmer basis than relying on assumptions about visual perception.

5.5 Computational methods in multimodality research

Computers already have a long history of use in research of all kinds—and not only in areas that would be broadly described as 'quantitative'. Since the 1970s a variety of computational tools have been produced with the explicit aim of supporting qualitative research also. Such tools, generally going under the labels of CAQDAS (Computer-assisted Qualitative Data Analysis Software) or QDAS (Qualitative Data Analysis Software) are widely used in the social sciences and have also been applied in some multimodal research as well.

Analytical software

A good overview of the current state of the art and practice using such tools is given in Davidson and di Gregorio (2011). Modern tools can accompany data analysis at all stages: organising data according to coding categories, exploring the data by annotating it and searching for content, integrating data from various sources, as well as helping interpretations by performing searches for patterns of varying sophistication. Relatively well known software of this kind includes, for example, ATLAS.ti and NVivo. It is interesting that we see here again the kind of separation into communities typical when engaging with diverse forms of data. Whereas the tasks of data collection, organisation, analysis and interpretation overlap considerably with those involved in studying corpora (→ §5.3 [p. 152]), the tools employed are largely disjoint.

This will need to change in the future. On the one hand, the tools for qualitative analysis in the social sciences generally do not support well the kind of richly structured annotations that are standard fare for linguistic approaches and which will become even more important for multimodal analyses. On the other, the tools for corpus studies are not so streamlined for support for the entire research cycle. Selection of different tools for different phases of analysis is therefore to be recommended: the more sophisticated the annotations become, the higher the chances are for finding interesting and meaningful patterns. For the use of different tools, interchangeable formats for importing and exporting data become critical; using application-specific or proprietary formats hinder such developments significantly.

In this section, however, our main focus will turn to an even more recent set of possibilities that will be essential for the future growth of empirical multimodal research. In addition to the established functionalities for easing the research process found in traditional tools, we are now beginning to see a growing orientation to data processing that significantly increases the kinds of information that can be extracted from larger bodies of data. This is of particular importance for multimodal research because the kinds of data most often of interest are themselves complex and richly internally structured. It then becomes difficult for even straightforward coding to be carried out in a reasonable time with a sufficiently high degree of reliability. This latter development is what we focus on in this section.

In general, a computational approach to multimodality research refers to how the data is processed by a computer, that is, by performing calculations. The prerequisite for performing any kind of calculation at all, of course, is that the data under analysis can be represented numerically. While this may initially seem to take us very far from some of the methods presented above, computational methods are highly effective in converting multimodal data such as photographs or entire documents into numerical representations. Even more importantly, algorithms that manipulate these numerical representations are getting better at forming abstractions about them in a manner similar to humans, learning to recognise objects, their shape, colour and texture. This takes us into the territory of artificial intelligence, where the concept of multimodality has also been given consideration.

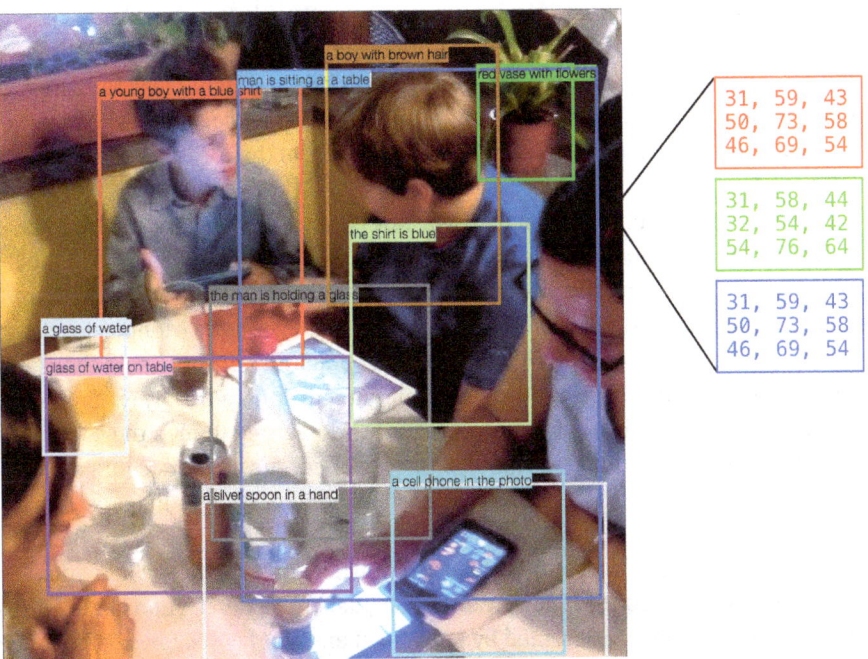

A group of people sitting at a table with food

Fig. 5.4. Objects automatically detected in an image and described in natural language by a neural network (Johnson et al. 2016). Right: the red, green and blue values of a 3 by 3 pixel area. Below: a caption provided by another neural network (Karpathy and Fei-Fei 2015) also fully automatically. Underlying photograph by Oliver Kutz, used with permission

Before diving deeper, let us consider how an image may be represented numerically. Figure 5.4 shows a photograph of the communicative situation that we encountered right at the beginning of the book. Here the three coloured boxes on the right represent the values for the red, green and blue channels, which carry colour information for a small 3 by 3 pixel area in the top-right corner of the photograph. Combining the values for each channel, much like mixing colours from three different buckets of paint, determines the colour of an individual pixel in the image.

For instance, picking the value in the first row and first column of each channel (31+31+31) gives us a very dark colour, which is not surprising as the pixel is located in a rather dark part of the photograph. In short, pixels are the smallest building blocks of an image, which can be easily described using numbers. When grouped together in different directions (including depth), pixels form colours, shapes and textures, which we can perceive visually, while the computer processes them by performing calculations over their numerical values.

Machine learning Below the photograph is a caption, "A group of people sitting at a table with food", generated by the computer using a technique proposed by Karpathy and Fei-Fei (2015). But how can the computer turn numerical representations into natural language? Recent years have witnessed rapid advances in *machine learning*, a subfield of computer science concerned with teaching computers to perform different tasks ranging from the classification of images to automatic gesture detection *without being explicitly programmed to do so*. To put it simply, machine learning is at the core of modern artificial intelligence.

Deep learning One prominent subfield of machine learning is called *deep learning*, which makes use of artificial neural nets whose design mimics the activation of neurons in the human brain upon receiving sensory input. These networks have proven particularly effective for computer vision tasks (LeCun et al. 2015). To generate captions for photographs, Karpathy and Fei-Fei (2015) trained a deep neural network using thousands of photographs, each described by five different human annotators. By examining the human-annotated training data, the neural network learned to map the contents of photographs to their verbal descriptions. Once trained, the network could also generate captions for photographs it had not previously seen, such as the one shown in Figure 5.4.

In addition to captioning entire images, neural networks can detect individual objects in images and describe them in natural language. The descriptions in Figure 5.4 are provided by a network designed by Johnson et al. (2016), which has been trained using Visual Genome, a dataset containing photographs annotated for the objects that they contain and their corresponding locations (Krishna et al. 2016). The neural network searches the image for areas of potential interest, which are then classified and described verbally. As Figure 5.4 shows, the network is able to extract much relevant information from the photograph, although many of the objects overlap each other. Neural networks can also learn to infer even more complex information from images, such as the direction of gaze for human participants (Mukherjee and Robertson 2015).

Image captioning and object detection represent just a few possibilities of leveraging computational approaches in different tasks involving multimodality. Indeed, techniques emerging from this line of work are now beginning to be applied by researchers across the board. For instance, Podlasov and O'Halloran (2013) examined a data set of 2249 street fashion photographs taken in Tokyo, Japan, using computer vision and machine learning to crop the parts containing clothing from each image, split the resulting image into top and bottom halves and then calculate the average red, green and blue colour values of both to capture the colour of the clothes. The resulting six values (three for each half of the image) where then fed to a self-organising map, a particular machine learning algorithm, which organised the photographs according to their features to help identify fashion trends in the data set.

In another study, Hiippala (2015a) applied computer vision techniques to interrogate a particular hypothesis about multilingual written texts, namely that the content presented in different languages tends to be organised into symmetrical layouts. In other words, layout symmetry in bilingual documents implies that their contents are semantically equivalent. Hiippala used a structural similarity (SSIM) algorithm to search for symmetrical designs in a data set of 1373 double-pages collected from in-flight magazines, showing that these kinds of symmetrical designs were primarily used for describing the available services and the airline, and for providing practical information on travelling in two different languages.

While the multimodal studies described above have based their analysis on low-level features of the images, such as pixel values and their placement, another body of research has investigated how computational approaches could be used to assist in the annotation of multimodal corpora (→ §5.3 [p. 152]). Thomas et al. (2010) explore how commercial optical character recognition (OCR) systems could be used to extract the content of page-based multimodal documents and to annotate them for their features. Hiippala (2016c), in turn, proposes an alternative solution for processing documents, built on open source computer vision and machine learning libraries, whose goal is to reduce the amount of time the user has to spend post-processing the annotation generated by the system. To conclude, although computer vision and machine learning techniques are just beginning to be introduced to multimodal research, their contribution is certain to grow over time.

In future, algorithms may be used to complement human annotation, leading to larger and richer databases and corpora (Bateman, Tseng, Seizov, Jacobs, Lüdtke, Müller and Herzog 2016). Combining manual annotation with the results of automatic processing—such as speech processing for spoken language, or shot detection in film and audiovisual data—will become increasingly important. As a consequence, multimodal corpora will need to include levels of abstraction in their annotations ranging from low-level technical features (e.g., for spoken language: acoustic properties; for film: optical flow, colour balance, edge detection, cut detection, etc.), through transcriptions of selected perspectives on the data (e.g., for language: phonetics and intonation; for interaction: gesture, movement and body posture) and the results of

Computer vision

Manual vs. automatic annotation

experimental studies (e.g., for images or film: eye-tracking data), to more abstract analyses (e.g., for interaction: dialogue acts), to hypotheses of category attributions ready for empirical testing (e.g., for interaction: relations between dialogue acts and co-selections of gaze, gesture and intonation; or for documents: rhetorical relations between content elements, and their distance and position relation, types of typographical realisations and eye-tracking predictions).

Supporting access to such combinations of information and the search for meaningful patterns is itself complex and new methods and techniques of visualisation will be crucial (cf. Caldwell and Zappavigna 2011; Manovich 2012; O'Halloran, E and Tan 2014; O'Halloran 2015). In short, within such frameworks, it will become natural to explore corpora by combining mixtures of annotation layers where 'ground' data, i.e., information that is considered reliable with respect to the data, is combined with information that is generated as hypotheses in order to check the actual properties of data against those predicted. This can be applied to all kinds of multimodal data and offers a methodology for driving forward empirical, data-driven approaches to multimodality research.

5.6 Summary and conclusions: selecting tools for the job

In this chapter we have set out a range of methods that can be used for multimodality research. Multimodality and its study is still young enough as a field to make the most of the freedom of applying such diverse methods in the service of the research questions posed. There are already more than enough barriers in existence elsewhere to overcome without taking the methods of multimodality to be already cemented as well! We can summarise the difference intended here by relating approaches and methods to the different 'speech acts' that they perform: approaches deal with 'what'-questions, methods deal with 'how'-questions. And a single 'what'-question might be helped by several complementary answers to the 'how'-question.

Particular ways of doing multimodal research should not, therefore, really be treated as 'approaches' at all—they are ways of supporting research questions. Discussions of methods that link methods too tightly to approaches may then prematurely rule out beneficial ways of framing and answering questions. This applies to all of the methods that we have set out in this chapter, as well as to the statistical methods that we turn to in the chapter following. 'Multimodal transcription', ethnographic studies of multimodal practice, formal modelling of multimodal meaning-making, with experimental methods, recipient studies, reception studies and so on are all valuable ways of organising research. They are all methods that should be considered for their relevance and benefit when considering multimodal research questions and are by no means mutually exclusive. They should never be ruled out *a priori* simply because of approach or disciplinary orientations. This chapter has therefore attempted to provide an orientation to the notion that there are many methods of various kinds that will help you approach your research questions, *regardless of what particular disciplinary orientation you may be coming from*. All are relevant currently for performing good multimodal research.

This does not mean that we are suggesting that all multimodal researchers have to be fully competent with all methods. This is hardly possible because some of them require quite technical knowledge and skills that are not equally available in all disciplines. But there are some basic skills that we would suggest need to be practised by all who wish to engage with multimodality. These are the skills particularly associated with collecting your data, preparing that data so that it is in a form, or forms, that supports further research rather than hindering it, and classifying the phenomena in that data in ways that are reliable and consistent. Whatever research question is being addressed with whatever approach, these issues will be present and it is good to be familiar with the basic skill sets required.

Some other research methods, for example using eye-trackers (→ §5.4 [p. 159]) to find where people are looking when engaging in some activity relevant for multimodality, may require expert knowledge not only in how to use the necessary equipment (although this is getting easier all the time) but also in how to interpret and build on the results. This is not at all straightforward and so if you are not from, or familiar with, a discipline in which these methods are taught, it is probably better to seek suitable collaborations with those that do have this kind of training. As always, multimodality research tends to bloom in an interdisciplinary environment.

It is still important, however, to know something about the various methods that are available, even if one is not going to be the one actually applying them. Several reasons can be mentioned for this. Perhaps one of the most compelling is to avoid making statements that are, at best, weak and, at worst, completely misguided. Again, using eye-tracking as an example, it is important to know the kinds of things that one can find out using eye-tracking and also the kinds of general results that have been, and are being, gained using this method. This awareness makes it far less likely that wrong or inaccurate things are said about 'reading paths' with respect to some multimodal artefact or performance, or concerning relative visual prominence and so on. We, i.e., humans, are generally very poor at making these kinds of properties of our perception explicit and our intuitions may prove misleading or unhelpful. But, because there are such things as eye-trackers, we need not be in doubt about issues of reading paths and visual salience: we can look and see what people actually do.

This is an important feature of scientific research in general. What would previously have been hypotheses or conjectures can gradually be turned over to empirical investigation. All of the methods presented in this chapter are to be seen in this light.

5.7 A word on good scientific practice: quoting multimodal artefacts

A further important feature of scientific research is the correct and complete citation of all kinds of analytical objects we are dealing with in our research—no matter whether we give a presentation in a seminar with PowerPoint or Keynote slides or write a term paper, a Bachelor's or Master's thesis or a scientific article. The rules of good scientific practice include not only the correct use of references to all sources in the literature, but also and in particular credits to the creators (directors, screenwriters, inkers, authors, artists, programmers, etc.) and owners of the various artefacts referred to.

Many formatting and style guides (for example, APA or MLA) today provide clear standards for the accurate citation of newspapers, films, comics or Tweets in a unified way. This will generally include a title of the respective artefact, the name of the producer or production company, the URL (if applicable), and a time stamp saying when the material was last accessed. The use of a DOI (digital object identifier) for the unique identification of electronic documents and other digital objects (such as blogs, for example) and a reference to their access online also makes it easy to properly quote even more dynamic and volatile documents online.

As one example, have a look, for instance, at the APA Style Blog by the American Psychological Association (2016)[3] which provides a list of examples to quote entries in social media, including YouTube comments, blog posts, TED talks and hashtags. If you now have a look at the entry for "American Psychological Association 2016" in our bibliography, you can also find an appropriate reference to this website with its specific URL. It should be a general rule for all your work that you give references to every example, quote or sequence from the data you are using. You should get into the habit of maintaining your own bibliographies, or lists of films, art works, etc. that can then be used to fill in the necessary sections at the end of your written work or presentations. You will see many examples for all these kinds of references both in-text as well as in the bibliography in the use case chapters of this book.

---

3 http://blog.apastyle.org/apastyle/social-media/

6 Are your results saying anything? Some basics

Orientation
In this chapter we introduce the basics of assessing the statistical significance of any quantitative results that your empirical analyses might have produced. This is an essential part of empirical research that has been neglected far too long in the majority of multimodal research. If you come from a discipline where it is already second nature to apply statistical tests for everything you do, or you know how this works for some other reason, then you can safely skip this chapter. But if you do not, then it is **recommended in the strongest terms that you read through this chapter!** At worst, you will then know how to avoid attempting to report or publish results which are invalid or perhaps even wrong; at best, you may even find this area of knowledge far more interesting and much less intimidating than you thought!

One of the most important components of empirical research is being able to ascertain whether the results that you have achieved are actually saying what you thought they say. What we do in this chapter is therefore to show some of the basic techniques that can be used for this purpose and to explain in very simple terms both why they work and why they are necessary. For a few of the less straightforward but still very common techniques that you will find used in the literature, we will also explain how such reports can be understood, at least in broad outline—thus providing a 'reading competence' for those statistical methods as well.

An independent samples t-test showed that the mean difference between the groups turned out to be significant at $t(27) = 2.4, p < 0.05$

Many students tend to skip the results sections of research reports, because they do not understand the meaning of sentences like the one exemplified above. The unfortunate consequence of this is that they will not be able to fully assess the merits and weaknesses of those studies. Moreover, they will not be able to report on their own studies in a conventional and appropriate way. This is not only unfortunate, it is unnecessary. Once the underlying principles are clear, understanding statistics is a piece of cake!

— Lowie and Seton (2013: 17)

In all cases, we will take particular care to explain the *reasons* why some particular measure works: many introductions to statistics, in our experience, fall into the trap of relying on technical aspects that are not introduced and which are consequently difficult to relate to what is being discussed—although, fortunately, there are now some good exceptions to this 'rule' which we will point to as further reading in the conclusion to the chapter below. We will also try to avoid this problem as far as possible as we proceed so that, subsequently, armed with these basics, you should also find it much easier to find out information yourself because most of the terms and their motivations will then be familiar.

The question of assessing whether your results say what you thought they did is addressed by several techniques and methods quite directly. Here we set out briefly how this is approached when an empirical study provides some kind of *quantifiable results*. Note that this does not mean that the methods we describe here are only relevant to so-called 'quantitative' studies (→ §5.1 [p. 140]); in many cases even qualitative studies can be usefully augmented by considering the proportions and frequencies with which their categories appear, often in combination with others. Such results can be anything that you can count, including differing numbers of categories that you may have found according to different groups or contexts, or some varying range of labelled phenomena in some body of data set, and so on—we will see examples below. For all such situations, it is beneficial to employ some basic statistical techniques specifically developed for empirical research. These techniques help show whether any trends or patterns found in your data are actually *reliably* indicative of what you thought (or not!).

A few of these basic techniques have very broad application and it is increasingly expected of reports on research results that any claims made are backed up with at least some statistical evaluation. As illustrations of what might occur, we can consider examining standard questions that have been taken up in multimodality research and recasting these as more empirical studies. In Kress and van Leeuwen's (2006 [1996]) widely used model of page composition, for example, it is claimed that information presented in the upper portion of a visual representation will depict some notion of an 'ideal', whereas information in the lower portion will depict something that is comparatively 'real'. Then, as they themselves ask: "how can we know that ... left and right, top and bottom, have the values we attribute to them, or, more fundamentally, have any value at all?" (Kress and van Leeuwen 1998: 218)

To move forward on concerns such as this, we need to explore them empirically as well, treating them as *hypotheses* to be investigated as argued in the previous chapter (→ §5.1 [p. 141]). One straightforward way of approaching this would be to collect a large range of image materials that appear representative (what this might mean we addressed in the preceding chapter also) and have a number of 'coders' independently make decisions concerning whether they think the top of the picture is more 'ideal' than the lower part or *vice versa*. We can then check using the statistical methods we present in this chapter whether those coders agree with one another to a 'significant' extent or not, thus moving beyond individual discussion of cases that may or may not be taken as supporting some claim.

By applying the methods set out here it becomes possible to use the term 'significant' in its *technical statistical sense*. This involves making sure that the results we present just could not, to any reasonable degree of likelihood, have occurred by chance. When we make this step, results can be presented in a much stronger fashion than had we just looked at a collection of examples in order to report whether they tend to agree or not with the original claim or that some of their 'averages' match. Academic journals and other avenues for disseminating research results increasingly expect that results be backed up in a stronger fashion and so it is beneficial to know something of these basics.

Obtaining some statistically derived figures is not sufficient in itself, of course. We need to work with the results that we obtain and also to consider carefully whether we are even asking the right questions under the right conditions. Any positive or negative statistical result is only as good as the use made of it and so must be seen as a *supporting* method for empirical inquiry; it is by no means an automatic way of guaranteeing good results, or even sensible results.

If we find substantial agreement between the coders in our example, then we might at least be satisfied that there appears to be some phenomenon that Kress and van Leeuwen's hypotheses are correlating with. If, on the contrary, we find that the coders cannot make this distinction consistently, then we must go back to see what might be happening in more detail. Perhaps the coders needed better explanations of what they should be looking at; perhaps the test materials were not so good after all; or perhaps Kress and van Leeuwen's hypothesis is not one that is usable for larger bodies of data. Until we find more refined questions and ways of testing them, we will not know. What we will have, however, is a progressively stronger understanding of just what people are doing, and can do, with this kind of material.

Let us now go through the most basic of the techniques that allow us to take these steps in a bit more detail.

6.1 Why statistics?—and how does it work?

The essential idea that makes the use of statistics appropriate and necessary is very simple, and that is the fact that we can rarely look at *all the cases* of any data or phenomena that we are interested in. We can only look at some selection. This rather obvious point has two rather important consequences:

- First of all, it raises the question of how we can know that the data we *did* look at tells us something more about how things work in general. If what we look at does not reflect, or is not representative, of the 'broader picture' then any conclusions we come to may be of much more limited validity than we were hoping.
- And second, whenever we do look at some selection rather than the whole, the precise results we get *will vary depending on our selection*. This point is critical because we need to know to what extent any differences we find are because of some real difference in the data we are exploring or is just due to this intrinsic, and unavoidable, 'sampling variation'.

Statistics provides a bank of methods for dealing with just these issues. And since issues of this kind are almost certain to occur *somewhere* in almost all studies that we undertake, we need to see that statistical methods are not something that we can ignore.

In short: we should never just 'hope'—even if implicitly and without conscious reflection—that the particular selection of data that we make for some analytic purposes will turn out to be in some way representative. We can instead set out tests and methods that can systematically increase our confidence that we are finding out real things about our data rather than perhaps random or irrelevant variation.

These two points—representativeness and intrinsic variation—can be used to explain and motivate most of the particular techniques that we will introduce here. They in fact share the same goal: they seek to measure to what extent any variation that we find in our analysed data is *significant* variation and not just the results of noise, that is, a consequence of the fact that analysed data will unavoidably vary by itself anyway. To achieve this, statistics relies on some very interesting properties exhibited by many natural phenomena, including the 'sampling' of some selected measures or values. Sampling will never lead to absolutely accurate results that are always the 'same' value: what will occur instead is that results will group themselves subject to regularities that can be used to characterise the situation that one has.

This all works by describing certain properties of any collection of results or data values that one has obtained. These properties relate to how frequently particular values occur in relation to others and can tell us a surprising amount. One basic result from statistics is that any values we 'measure' will tend to cluster around the true value and that this clustering itself has a particular 'shape'. Finding out about that shape is then the first step towards understanding what most statistical measures are doing and why they work.

Since we are interested in the shape of a range of values, it is useful to consider just how we can characterise such shapes succinctly. This is all to do with information and keeping hold of just enough to tell us something useful without requiring all the details. For example, if we find the average, or *mean*, of a set of numerical values, this gives us a shorthand way of describing the entire set. So if we take the five values {1, 2, 2, 5, 6} and work out their average (3.2), then we have achieved a succinct characterisation of the original set.

This characterisation of course loses a lot of information as well. The five values {-102, -8, 0, 25, 101} *also* have an average of 3.2! So we can see that describing sets of numbers just by their means still leaves a lot open. What is then most useful statistically for making the characterisation more informative while remaining succinct is *how much the values spread out* with respect to their average. If they spread out a lot, then many of the values will be very different to the average; if they do not spread out very much, then many will be close to the average. So we also need an account of their *dispersion*, or *variation*. When we know how much they spread out, then we have a much better approximation to the original set.

Working this out is straightforward. First, we might consider just adding up how much each individual value differs from the average. This does not quite work, however, because some differences might be greater than zero (if the value is larger than the mean), and some might be less than zero (if the value is smaller than the mean). In the present case, then, the first three values of our first set are smaller than the mean, while the last two are greater. Just adding these differences together would then lose

Table 6.1. Calculating the basic parameters describing variation for the set of values {1, 2, 2, 5, 6}, which has a mean of 3.2

Values	Differences from mean	Squares of differences from mean	
1	$1 - 3.2 = -2.2$	$(-2.2)^2$	$= -2.2 \times -2.2 = 4.84$
2	$2 - 3.2 = -1.2$	$(-1.2)^2$	$= -1.2 \times -1.2 = 1.44$
2	$2 - 3.2 = -1.2$	$(-1.2)^2$	$= -1.2 \times -1.2 = 1.44$
5	$5 - 3.2 = 1.8$	$(1.8)^2$	$= 1.8 \times 1.8 = 3.24$
6	$6 - 3.2 = 2.8$	$(2.8)^2$	$= 2.8 \times 2.8 = 7.84$

Sum of squares: 18.8   (SS_x)
Divided by the number of values: 3.76
Positive square root of result: $\sqrt{3.76} = 1.94$

information and, in the case of perfect symmetry, the differences would just cancel each other out—that is, if we have lots of negative values and lots of positive values, then, on the one hand, we have a lot of different values to the average but, on the other hand, this information is lost when we add the positive and negative numbers together.

We need therefore to measure the average *size* of the difference regardless of whether it is positive or negative; this is most commonly done by just taking the square of the differences instead, because this is always positive—e.g., $(-3)^2$ and 3^2 are both 9. So, first, we add the squares up. This is such a common operation in many statistics tests that it is given a shorthand label, SS_x, where the x picks out the individual values. Then we divide the sum by the number of values as before to get an average. Finally, we take the square root of this average to get back to the 'scale' of the original values. This is summarised for our first set in Table 6.1, giving an end result of 1.94. So together we know that a characterisation of our original set of numbers can be given by their mean (3.2) together with the 'average' of distances of those numbers from the mean, i.e., 1.94. When we do exactly the same calculation for our other set of five numbers above, the value we get is very different, i.e., 65.18. This average distance from the mean shows how the numbers of the second set are very widely dispersed indeed.

Sums of squares

Next, we have a bit of statistics magic. So far we just described the mean and 'spread' of a complete collection of numbers. Imagine now that we want to work out what the mean might be when we do *not* have all the numbers. More particularly, imagine that we have a large collection of *possible* data but practicalities stop us ever examining all of that data to work out what its mean might be: we could take as an illustrative example something that might occur in any statistics introduction, e.g., people's height in some country—we will move on to multimodal questions in the next section. Now we could notionally consider measuring the entire population's height, although this is hardly realistic. Alternatively, we can measure the height of a more manageable *sample* selected from the entire population.

Population vs. sample

A standard result in statistics is then that, when the sample is large enough, the mean of the sample will give us a way of deducing the mean of the population. Moreover, the *variation* that we find in the sample will have very particular and systematic properties. That variation can be calculated with the same scheme as above with one small difference: we divide not by the number of values in our set (which would be the size of the population in our height example), but by the number of values in the *sample* minus 1. The reason for this difference is subtle: since the overall properties of the variation in any sample are actually fixed—we will explain why this is the case in a moment—we do not need to look at all of the values. When we have looked at all but one result, we have enough information to work out the last one.

Degrees of freedom

It is exactly as if we take 10 numbers but already know that they add up to 25. After we have looked at 9 of those numbers, we already know what the next one must be—i.e., 25 minus the sum of those 9 numbers we looked at. The same holds for the calculation of the variation: we never need to consider the differences from the mean for all of the values in the sample, looking at one less than the number of values is sufficient. This is called the 'degree of freedom': in the present example, when we looked at all but one of the values, the last one is no longer 'free'. So if we call the size of the sample N, we have $N − 1$ degrees of freedom. The degree of freedom will occur in several of the statistical procedures described below. Sometimes we will need then to be slightly more general: if we have divided our samples into various groups in order to explore different aspects of what we are investigating, then the rule of reducing by one holds for *each* group. Thus, with 3 groups each consisting of 50 samples, the corresponding degree of freedom (or df) is $(50 − 1) × 3$, or 147. Alternatively we can just add up the number of samples and subtract the number of groups.

Standard deviation

This then gives the formula for working out what is called the *standard deviation* for any sample. If we carried out the calculation of the standard deviation for our 5 numbers above—now treated as a sample rather than the entire population—then the revised table is identical to that of Table 6.1 apart from the last two lines. Instead of 'Divided by the number of values', we have 'Divided by the number of values minus 1'; the result in this case would then be 4.7, and then the value of the square root naturally changes as well, to become 2.168. So if we got these numbers in our sample, we would expect the variation of values in the entire population to actually be a bit broader than what we had calculated before.

The reason for this is also straightforward: when we had the entire population, we already knew that we did not have anything higher than 6; but if this is a sample, and with our selection we already got a 5 and a 6, as well as a 1, which are all some way away from the mean, then it is likely that if we continue to sample we will get a few numbers that are even further away from the mean—therefore we can well expect that the standard deviation will be higher.

We have said at several points that when we select a sample, the properties of that sample are known to be quite systematic—that is we can expect that a particular shape of variation will occur and that this shape can be used to good effect for testing whether some results that we have obtained are in fact 'significant' or not. In the next section, we will set out just what this means and briefly sketch its consequences. These consequences will then let us set out the examples applied to multimodality research in the sections following.

6.2 What is normal?

Many statistical methods depend on the following basic phenomenon: if we select samples from some overall population, then those samples will tend (a) to cluster around the mean and (b) cluster with a particular shape. Let us look at this with the following example.

Imagine that we are performing some analysis of webpages and we are interested in the extent to which graphical or pictorial information is used. For the present we will not concern ourselves with making this question more interesting or multimodally discerning, we will simply count for some random selection of webpages how many times an image is used—this can even be automated to a certain extent just by looking at the internal representation of the webpage if that page is written in HTML, but we will ignore this here. All we want to do is investigate 'how many images are used?'. This is a construct that we need to *operationalise* to produce numbers that we can actually work with statistically. We will assume for current purposes that we can just take any webpage and get some number that says how many images there were on that page.

Now let us start collecting the data in our sample by picking webpages and seeing how many images there are. We will record the results in what is termed a *frequency graph*: this simply keeps a running total of how many times a particular number of images has been found so far. The graph usually has the numbers that we obtain running along the bottom, or the x-axis, of the graph, and the number of times we find that number shown on the vertical, or y-axis of the graph. Note that we could be doing this with *any* question that we have operationalised to produce data: lengths of words on the webpage, number of links, number of links with images, and so on. In addition, the numbers we report here and in other examples are *for illustration* and do not represent the results of actual studies!

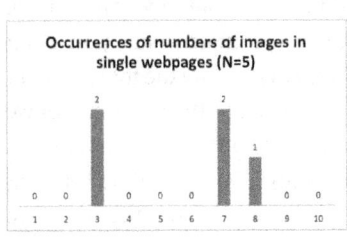

Fig. 6.1. Frequency graph for the first five webpages we look at

Our counting might proceed as follows. The first webpage has 7 images, the second has 3, the third has 8, the fourth has 7, and the fifth has 3 again. Our frequency graph is then mostly filled in with zeroes, but for 3 we have the entry 2 (i.e., we find two occurrences of a webpage with 3 pictures), for 7 we have again 2, and for 8 we have 1. This is actually just a simple task of piling up blocks: for each webpage with a particular number of images, we add a block to the respective column in the frequency graph. Our frequency graph so far then looks as shown in Figure 6.1.

What happens next lies at the heart of statistics and is an important result in its own right. Let us keep on counting webpages and look again after we have piled up the blocks for 1000 webpages. There is a very good chance that the result will look similar to the frequency graph shown on the left of Figure 6.2. This is exactly the same in construction as the previous graph, but just contains more data—our sample has been increased from 5 webpages to 1000 webpages.

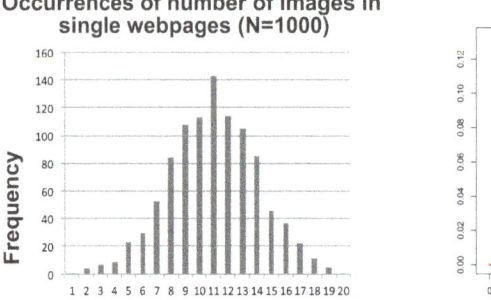 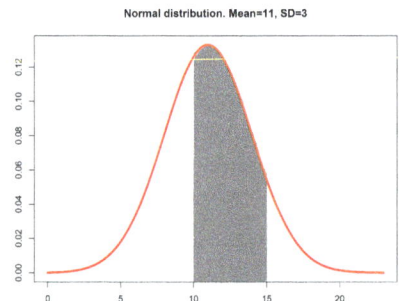

Fig. 6.2. Frequency graph for the first 1000 webpages we look at (left) and a stylised version of the 'shape' of the distribution shown in red with an example 'area' picked out (right)

This kind of shape of frequencies is called the *normal distribution* and is the usual way that values in a sample cluster around the mean. The graph on the right hand side of the figure shows the curve more clearly. The interpretation of the graph is that many values in the sample will be at or close to the mean, whereas rather few will be further away from the mean. Between these the frequencies of values will drop off in the manner suggested. The precise shape of this curve can vary—for example, it might be a bit more pointed or a bit more spread out—but we will ignore this for the present. Curves varying in this way can generally be 'transformed' back to a standardised version of the normal distribution which has several valuable properties.

The most important of these properties is that the curve allows *predictions* to be made about the likelihood of values that will be found in a sample. If we think back to the idea of piling up blocks for each webpage, then the likelihood of any particular value occurring is just the number of blocks in that column divided by the total number of blocks. For example, if we had 100 blocks and 50 of them are in the column that we are interested in, then the likelihood, or probability, of picking out one of those blocks (i.e., selecting a member of our sample with just that value) is 50 divided by 100, i.e., 0.5 or 50%. Half of the time we might expect to land on a block from that column. Considering the shape (and therefore the number of 'blocks') we can see in the normal distribution, we can infer that the likelihood of a value occurring a long way from the mean is much less than that of a value occurring close to the mean. There are far fewer blocks there. We can also work out the probability that some individual member of a sample will lie within some *range* of values—again simply by adding up all the blocks involved and dividing by the total number of blocks used.

Finally, this can be made far more precise as follows. If we keep the blocks all the same size, then we can see that counting blocks is exactly the same as (or, actually, is an approximation to) working out the *area under the curve* for some particular range of values. If we need to work out the probability that values between, for example, 10 and 15, will occur, all we need is the area under the curve shown in the right hand graph that lies on the 'slice' between those two values: in the figure this slice is picked out in grey. And, because the normal distribution is such a useful curve and has a precise mathematical definition, these areas have all long been calculated and are directly available in any computer software that lets you work out statistical measures.

As an indication of how we might immediately use this, consider now that we revise our look at webpages and classify them according to webpages before 2010 and webpages from 2010 onwards and we want to know if there has been an increase in the use of images or not. We do our sampling as before but now have two collections of numbers: we might find in one case that we have an average number of 6 pictures per webpage and in the other an average of 8 pictures per webpage. Are we then justified in saying that there has been an increase (or not)? Intuitively it should be clear that the answer here will depend again on the 'spread' of the data: if the data exhibits a large standard variation then it could well be that 6 and 8 are just not very different; conversely, if the data has very narrow variation, perhaps this is sufficient to say that there has been an increase.

We will show how this works in detail and how it can be calculated using appropriate statistical tests in the next section. For now, we can simply refer again to our normal distributions: if we work these out for our two collections of webpage data, then we not only know their respective means but also their variation. We can calculate (or, more likely, look up or have a computer program tell us) what the probability is of getting a 6 with respect to the data whose mean was 8 or the probability of getting an 8 with respect to the data whose mean was 6. If the probability is in any case quite high, then we cannot deduce that there has been any particular change. If the probability is very low, then this would suggest that something *has* changed.

This is all made possible by the fact that the normal distribution allows predictions of likelihood in the manner that we have suggested. If reporting the results of some study of this kind, one can (and, in fact, should) report not only that there was some difference but that that difference is statistically significant. Again, as we mentioned above, this means that the *likelihood of it occurring by chance* is sufficiently low that we can *rule out* chance as an explanation. Moreover, we can now understand 'is sufficiently low' in terms of just how far a value lies away from the centre of these distribution curves. The shape of the curves lets us calculate exactly how far some value needs to be away from the mean to be considered unlikely enough to meet rather stringent tests—as we shall now see.

6.3 Assessing differences

The topic that we ended the previous section on—ascertaining whether the difference between two samples of data selected according to some criteria is *statistically significant*—is one of the most common tasks that has to be done for many kinds of empirical analysis. In this section we give a detailed illustration of how exactly this works, introducing a few more of the basic techniques and their motivations along the way.

For this, let us go back to the mention we made at the beginning of the chapter about ways of testing Kress and van Leeuwen's proposals concerning the 'information value' of areas on a page or image. One scenario for this might be the following. First, we might ask whether it is possible to recognise the kinds of distinctions that Kress and van Leeuwen propose with any reliability in any case. We might test this by taking a collection of students or co-workers and training half of them in the application of Kress and van Leeuwen's notion of 'ideal'/'real'; the other half should have no specific training. We then prepare a test set consisting of 100 cases of pictorial or illustrated material and have these coded by some experts to collect the 'correct' analyses. We should be cautious here and throw out any examples that have multiple 'correct' answers or differing interpretations, only keeping those for which there is general acceptance of how they should be analysed. The resulting annotated collection we call our *gold standard* and we will use this to compare the performance of our trained and untrained groups.

There are various ways of measuring how some experimental participant fares when attempting to do an analysis of this kind—this is all part of the experimental design and producing an *operationalisation*. For current purposes it does not make much difference how we do this, so let us just assume that any participant has to decide whether a specific instance in question from the test set is vertically organised along the ideal/real dimension or not. The score for that participant is then the number of times they succeed in making the same selection as given in the gold standard for each instance in the test set. Again we need to emphasise that it is the general principles that we are concerned with here, so any other example that might be fitted into this scheme would do just as well.

Let us therefore take 20 untrained participants and 20 trained participants, have them classify the examples of the gold standard, and record their scores. We already have to be quite careful here in how we go about this testing. It is quite easy for analysts confronted with a range of data to 'get into a rut' and keep on producing similar analyses once they have decided on one way of doing things. We should counter this by some means—for example, by presenting the 100 examples in random order so that each participant sees the examples in a different sequence (as one example of how to do this, search online for 'Latin square' design). When randomised, we know that the order of the examples cannot then be playing a role in any results that we get.

As has probably become clear by now, we do not want to be doing many of the necessary operations on the data we collect by hand. It is therefore highly recommended that one practices and gains at least some basic competence in everyday computer programs that allow data of this kind to be easily manipulated. This begins with spreadsheet programs such as Microsoft Excel or Apache OpenOffice and goes on to include rather more powerful dedicated tools such as the open source and free program R, or standard statistical tools that need to be bought, such as SPSS or Mathematica. For the kinds of tests we are using here it does not make very much difference because all of these tools cover the basics. So we run the tests and then have two spreadsheets as a result, one for the trained and one for the untrained. Each spreadsheet includes the score for each participant.

As a matter of good practice, one should first look at the data for each set of participants to see if it looks 'normal'—one can do this by plotting a graph for the values as done in the previous section. As we explained above, the normal distribution commonly occurs when sampling data and so many straightforward statistical measures rely on this property. Cases can occur, however, where the curve looks very different and, if this happens, different statistical techniques are necessary. We will say more about this below in the last section of this chapter.

Fortunately, several of the tools for working out statistical tests automatically check that the data is more or less 'normal' before doing their main work in any case. This is valuable because if the data deviates for some reason then the test may not be valid. We can see this with a simple example: throwing dice. If we made a frequency distribution of what we get when throwing a normal six-sided die, and the die is completely fair, then we should *not* get a normal distribution since all outcomes are equally likely. There is no 'mean' around which the numbers will group themselves. More complex shapes can also commonly occur and these are all the subject of a variety of statistical tests. The basic *principles* do not change, but the precise calculation necessary does.

We remain with the simple case here. Assume we have the two data sets illustrated in Figure 6.3: this graph simply plots the scores for each participant in each group. The scores of the trained group are shown in small diamonds; the scores of the untrained group are shown as larger diamonds. Already from this data we might have some hope that the

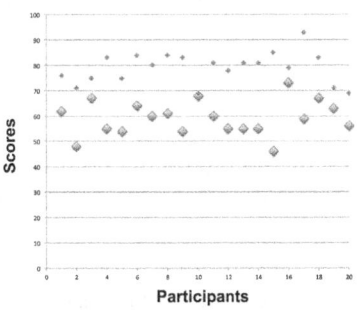

Fig. 6.3. Raw data from our two groups: 20 scores for 20 participants each. The small diamonds are the scores of the trained group; the large diamonds the scores of the untrained group.

groups are different as the occurrences of overlap between the two types of scores are rather few. But in general this may not at all be the case: the scores we obtain may overlap far more and it is then only with a statistical test that we can be sure about reporting results concerning possible differences.

This means that we want to answer the question as to whether the two groups performed differently to a statistically significant extent. The *descriptive statistics* of the two samples are given by their respective means and standard deviations. These can be worked out in the ways described above or directly using the functions that the respective tools that you are using offer. Let us do this for our two illustrative data sets: the untrained group has a mean of 59.1 and a standard deviation (SD) of 6.78; the trained group has a mean of 79.1 and standard deviation 6.13. This means that there is still quite a spread of results in each group and so we certainly need to check whether there is a difference between them. To do this, we construct what is called the 'null hypothesis': this is the minimal hypothesis that there is actually no effect, or no difference between the groups. We want to calculate the statistical likelihood of the null hypothesis being true. If it is sufficiently unlikely, then we can be reasonably sure that we have found a difference between the groups.

Null
hypothesis The null hypothesis is used because it is in principle impossible to *prove that a hypothesis is true*. Although one all too often reads statements in student papers such as 'these results proved the hypothesis' and similar, this is wrong. One can only obtain results that *support* a hypothesis (or not)—one can never demonstrate unequivocally on the basis of an empirical investigation that something is and must be true. The only things of this kind are theorems in logic, where the truth of a proposition may be reached by means of logical deduction. Empirical results and hypotheses are just not of that kind. So what we can do instead is to investigate whether we have found a situation that is so unlikely to have occurred by chance, given our empirical results, that rejecting the hypothesis that the situation holds leaves us on safe ground. In short, we work out the chance that we are rejecting the null hypothesis by mistake; if that chance is sufficiently low, then we are making progress.

The null hypothesis is written as H_0 and so in the present case we are going to check the hypothesis:

H_0: The two groups, trained and untrained, are both drawn from the *same* overall 'population', i.e., there is no real difference between the two and people do as well (or as badly) with training as without.

We want to be as sure as we can that we do not falsely throw this hypothesis out by claiming that there is a difference when actually there might not be.

The measurement of 'being sure' is then assessed in terms of the probability of incorrectly rejecting the null hypothesis. Typically adopted measures for this probability are 0.05 or the stronger 0.01. These figures give an indication of how likely something would be to occur by chance: clearly, if something is only likely to occur 5 times out of 100 (0.05) or even just once out of 100 (0.01), then it is extremely unlikely. So, if we find that the probability of falsely rejecting the null hypothesis is less than 0.05, we say that the results are significant at the 95% level; if the probability of falsely rejecting the null hypothesis is less than 0.01, then the result is significant at the 99% level. There are standardised ways of writing this: usually one uses the shorthand $p < 0.05$ or $p < 0.01$ respectively. Naturally, if we want to be even more cautious, then even lower probabilities can be selected as the hurdle that results will have to reach to be considered acceptable.

Now, we mentioned above that samples vary systematically, in fact they are normally distributed if the sample is large enough. And, if samples are normally distributed, then we can work out likelihoods of values occurring by looking at the area under the normal distribution curve as we explained. But what if the sample is *not* large enough? Actually the 20 participants making up each of our groups in our present example is not a very large number and we might well then have distributions that are not normal. In fact, if the sample is smaller than 'enough' then there may be more variation. Fortunately, it has been found in statistics that *even this variation is systematic*.

The frequency distribution in this case looks slightly different, however. When the sample size is small, the distribution of measurement 'errors' constituting variation is described more accurately by another distribution, called the *t-distribution*. This is actually a family of curves that gradually approximates the normal distribution as the sample size increases. For each sample size, therefore, we have a slightly different curve and we use this curve (or rather, the spreadsheet or statistical program that you select uses this curve) to work out the likelihood of something occurring just as we illustrated above with the normal distribution. In fact, the parameter we use for selecting the particular t-distribution curve to be applied is the 'degree of freedom' that we introduced above. This is the complete size of the samples minus the number of groups.

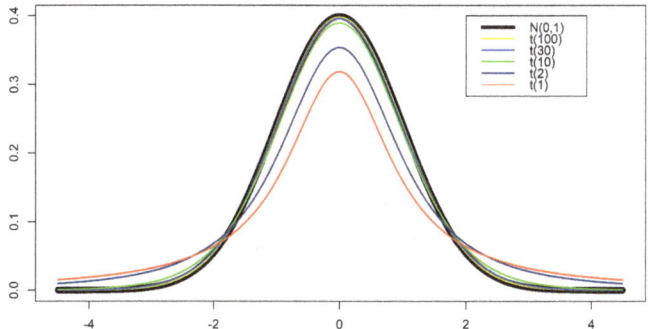

Fig. 6.4. t-distributions with varying degrees of freedom (in comparison with the normal distribution)

For the purposes of comparison, Figure 6.4 shows a 'standardised' normal distribution with mean of zero and standard deviation of 1—written as N(0,1)—and the family of t-distribution curves varying according to degrees of freedom—written as $t(x)$ where x is the degree of freedom. Remember that the area under the normal distribution gave us the probabilities for particular ranges of values occurring; the same is then true for the various t-curves, depending on the size of the sample.

So our final step here is to do the actual statistical test, which is called the *t-test* because it is using the t-distribution. To have this calculated we give our data for the two groups and specify how many degrees of freedom are involved. For our data, this is the total number of samples (40) minus the number of groups (as described above), giving us 38. We must also specify whether the test is what is called 'one-tailed' or 'two-tailed'—this refers to whether we look at both 'sides' of the curve around the mean or just one side. The reasons for this choice relate to what is possible in the data variation: sometimes variation is restricted (e.g., not being able to go below zero, as with weight or height) and so we only have to look at the area under half of the distribution curve. We will ignore this here and assume the generally safer 'two-tailed' condition: that is we look at the entire distribution curve and the probabilities it entails.

Independence Another assumption that we will make here is that the two sets of data that we are comparing are *independent* of each other. It should be intuitively clear that if the two sets of data are related somehow, then this might well throw the calculations off. So we need to rule this out to begin with, preferably in the design of the experiment. Sometimes we might have situations where we do *not* have independence—for example, we might ask the same subjects a question at two different times. In such cases, called a *within-subjects* or *repeated measures* design, it is likely that the answers given by any single subject would be related. This requires different calculations and has a different value for the degrees of freedom (e.g., one less than the number of *pairs* of linked results, i.e., 19 in our present example), but the overall logic remains. By linking data into pairs of values, we can then investigate to what extent the *difference* between each value in a pair shows the variation that the t-test picks out.

There are also statistical tests for exploring whether samples show dependencies, correlations or other relations which would enable us to check independence—for example, if the untrained participants in our present example were in fact cheating by looking at the trained participants' results! For the present case, however, we will continue to assume independence; this is called a *between-subjects design*. The data values collected are assumed not to have any dependence and so the degrees of freedom will be as indicated above. Calculating the t-test for the data 'by hand' is then broken into two tasks. The first step is calculating a 't-value' or 't-statistic'. The t-value is a score that reflects how far the means of the samples are from each other and, precisely analogously to the case with the normal distribution, the value obtained lets us work out in the second step what the probability would be of such a result occurring by chance. Here reference is made to the t-curve to allow calculation of the corresponding probability (just as with the normal distribution case in Figure 6.2). If we make our hurdle for statistical significance more strict, i.e., we want to be sure that the chances of the event occurring by chance are even lower, then the value we accept has to be even further away from the mean. As we saw above, values moving further from the mean are progressively less likely to occur, i.e., they have lower probability.

To find out what probability is given by each t-value, one would need to look at a table of pre-calculated results (like the old tables of logarithms that were used in schools before the advent of calculators) or, much more likely, use the calculations provided automatically by a statistics tool. Reports of results generally give both the t-value and the associated probability of falsely rejecting the hypothesis that the groups compared are drawn from the same population.

Calculating the t-value in full is straightforward when we already have the standard deviations (SD_1 and SD_2) and means ($\overline{X_1}$ and $\overline{X_2}$) of the two groups (of sizes N_1 and N_2 respectively) being compared, although the formula may look a little more daunting than it actually is. The t-value is the result of calculating:

$$\frac{\overline{X_1} - \overline{X_2}}{\sqrt{\frac{SD_1^2}{N_1} + \frac{SD_2^2}{N_2}}}$$

i.e., subtract the mean of one group from the mean of the other group and divide this by the square root of the sum of the squares of the respective standard deviations each divided by the respective sizes of the two groups. All this does is give a sense of how far the means of the groups differ from one another taking into consideration a kind of 'average' of the variation within each group.

For the example data used in this section, the result of calculation is -9.763, as you can readily check yourself given the standard deviations and means given above.

Nowadays, statistics tools most often combine all the above steps and so it becomes very easy to work out the statistical significance (or not) of our data. For the current example, and assuming a spreadsheet program such as Excel, this works as follows. First, put the data of our first group somewhere, say in the cells A1:A20 and the data of our second group somewhere else, say in the cells B1:B20. The probability of these two sets of data coming from 'the same underlying population', i.e., of not being different, is then given by the formula

```
=T.TEST(A1:A20,B1:B20,2,2)
```

where the final two '2's indicate that we have a two-tailed test and independent sets of data respectively. The equivalent formula for OpenOffice uses the name 'ttest' instead of 't.test'.

For our data here Excel and OpenOffice then give back the value 6.62×10^{-12}, which is rather small and much less than the 0.05 or 0.01 that we might normally set for rejecting the null hypothesis. In other words, we can be certain to a level of 0.000 (the number of zeros here indicate that we have achieved a level of significance so strong that it cannot be shown with three decimal places) that we are not incorrect in rejecting the null hypothesis. When you have a result from a statistical test, there are set phrases that you must use for reporting this in any article or report; these phrases tell the reader exactly what test was used on the data, the degrees of freedom, and what the result was—this is therefore very important for presenting results appropriately. In the present case, you would write that:

the mean difference between the groups was significant at $t(38) = -9.763, p < 0.000$

This shows that the training certainly has an impact on the recognition of the appropriate categories or classifications and so there is at least *something* that is being learnt. It does not tell us about why the trained group is doing better, but at least would suggest that there is some phenomena in the data (or in the participants) that is worth examining.

If our study had had different results, then the calculations would of course play out differently. Imagine that we had performed the experiment in the same way but our two groups had given us data with respective means of 62.7 and 65.9 and respective standard deviations of 13.4 and 12.9. In a case such as this, it is much more difficult just to look at the results and say that there is, or is not, a difference. Calculating the t-value using the formula above, however, gives us a value of -0.7561. The corresponding probability level returned by the t-test is 0.4542, which is well below (i.e., numerically greater than) the required level for statistical significance. This probability level would mean that if we were to reject the null hypothesis, i.e., we state that there *is* a difference between the samples, then we would quite likely be doing so wrongly.

6.4 Assessing similarities

A further simple calculation that can be done when we have a body of data is not how different they are but how *similar* they are—this is generally characterised in terms of *correlations*. If we have data that can be characterised in terms of two ranges of numerical values (we return to this issue of kinds of data below), then we can see how much the data exhibits a correlation between those two ranges. For example, if we have the number of images per webpage as above and the year in which each webpage was made available, then we can see what the degree of correlation is between these two ranges. If we expect that the number of images per webpages has increased with time, then we would see a *positive* correlation; if the number has decreased, then we would see a *negative* correlation.

Correlations   We can measure the *strength* of the correlation by plugging the individual values for the two ranges of numbers, call them x and y respectively, into a formula that is similar to the calculations for standard deviations given above. In the present case, though, we need a kind of 'combination' of the standard deviations and specific differences from the means in the two groups of data in order to see how the two ranges of values relate. We have seen the way of assessing *variation* in a body of data above:

essentially one looks at some 'sum of the squares' of deviations from the average, or mean, of the data examined. Here, since we are comparing two sets of data to see if their variations are similar to one another, we employ an equivalent calculation of *co-variation*, where each difference from the mean among the *x*'s (call this mean $\bar{x}$) is multiplied by each difference from the mean among the *y*'s (call this mean $\bar{y}$). In analogy to the shorthand used above (cf. Table 6.1), we can label this sum SC_{xy} to show that both sets of values are combined and that we are looking at co-variation rather than squares.

This gives us one of the most common correlation measures used, that of Pearson's Correlation Coefficient, *r*. This coefficient simply takes the combined sum of co-variation we have just described (SC_{xy}) and divides this by the square root of the sum of the squares of the *x*-values (SS_x: giving us a measure of the variation among the *x*'s) multiplied by that of the *y*-values (SS_y: giving us a measure of the variation among the *y*'s). The square root here is again to puts the individual 'divergences' from the means back in proportion to the overall variation found in the data. The coefficient then captures the ratio or proportion between the *co-variation observed in the data* and a combination of the *individual variations* found for the *x*'s and *y*'s. The formula for the entire calculation is therefore as follows, first on the left-hand side using our shorthand abbreviations and then filled out in full on the right:

$$\frac{SC_{xy}}{\sqrt{SS_x\ SS_y}} = \frac{\sum_i(x_i - \bar{x})(y_i - \bar{y})}{\sqrt{\sum_i(x_i - \bar{x})^2\ (y_i - \bar{y})^2}}$$

In such formulas, the sigmas, i.e., $\sum$, mean that we sum up all the values that we have for the *x* and *y* ranges. There has to be the same number of these since we are comparing each *x* with a corresponding *y* – in our example of images in websites, then, some particular *x* value might be the year and the number of images in a webpage might be the corresponding *y*.

The closer that the value of *r* comes to 1, either plus 1 or minus 1, then the stronger the correlation we have. A value of 0 means that there is no correlation. If the number is greater than zero, it means that the values of *x* and *y* both increase or decrease in sync with one another (i.e., a positive correlation); if the number is less than zero, then the values of *x* and *y* move in opposite directions—as *x* increases, *y* decreases, and *vice versa* (i.e., a negative correlation).

Whenever we are working with sampled data, it is also sensible to report whether the correlation is statistically significant by giving a *p*-value as done above. This is normally done by taking the *r* score from the formula, the number of *x*-*y* pairs that one has, and by selecting the significance level that one wants to investigate—typically 0.01 or 0.05 as explained above. There are several online tools that then produce a corresponding *p*-value, indicating whether your correlation was significant at the level of significance that you selected. What is going on behind the scenes here is similar to what we have described above for the t-test: to work out whether the correlation is significant, it is necessary to factor in the degrees of freedom—this is done by giving the number of pairs one has. The degree of freedom is then that number minus 2 (because we are looking at all the values as possible things that might vary and these are organised into two groups, the x's and the y's. But the online pages (search, for example, for 'calculating p value from Pearson r') will do all this for you.

As a concrete example of applying this technique in a multimodal context, we can consider a study of the changes that have occurred in the page composition of mainstream American superhero comics. We will hear more detail of approaches to comics and their layouts in our use cases below—for the purposes of the present chapter we are concerned just with the statistical technique employed. Pederson and Cohn (2016) examine 40 comicbooks with publication dates balanced across decades from 1940 to 2014, characterising the pages of those books according to several features describing their layout. They then worked out an average rate of occurrence of features per book by dividing the number of occurrences by the total number of 'panels'. This then gives a set of data where each book is classified for each feature according to that feature's occurrence and the year of publication—i.e., 40 pairs of x-y values for each feature. Pederson and Cohn then calculated correlations to see whether the features examined had decreased or increased over time. Results here could clearly be relevant for exploring possible changes in graphic styles and conventions over the years.

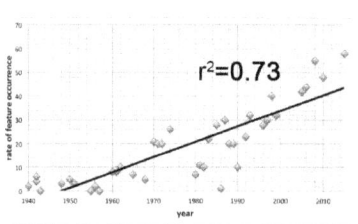

Fig. 6.5. Scatter plot of a comicbook page layout feature over time, with the calculated 'trend line' representing the correlation.

One usual way of showing data of this kind is as a 'scatter plot', which simply shows the x-y values on a graph. The correlation can then be indicated by a *regression line* running through the 'cloud' of data points that results. This is illustrated with constructed data in Figure 6.5 approximating the feature that Pederson and Cohn term 'whole row', i.e., the extent to which pages are made up of panels taking up the entire width of a page—for the actual graphs, readers are referred to the paper. Most spreadsheet programs will provide such graphs, the 'trend line', as well as the value of the correlation automatically.

Here, then, we can see that there is good evidence for a steady increase in the use of these wide panels, sometimes termed 'cinematic', over time—we might use such a result to support arguments that the visual style of American superhero comics has become more cinematographic. The graph also shows a further derived value that is often reported: simply the square of r, i.e., r^2, which is called the *coefficient of determination*. This gives a sense of the percentage of variation in the data that is explained by the correlation found. Clearly, the higher this value, the less unexplained, or 'unaccounted for', variation there is. In our example here, r^2 is 0.73, so we have 73% of the variation accounted for simply by taking the year of publication into consideration.

Results concerning correlations should then be reported with a phrase of the form:

the values were positively correlated according to Pearson's r, r(38)=0.858, p<0.001

The value of 38 given is the degrees of freedom as usual—so in this case, the total number of pairs (40) minus the number of 'groups', i.e., 2, because we have an *x* group and a *y* group. Therefore we can see here that there has been a statistically significant increase in the use of this particular page compositional feature over time.

Pearson's correlation coefficient assumes that the values being related are both distributed more or less normally; you may then need to throw out any obvious 'outliers'—i.e., values that are way off from the means of each of the ranges examined since these may well be errors or non-representative extreme cases. The coefficient also assumes that it can look for a 'linear' correlation: that is, both the *x*'s and the *y*'s increase and decrease in a constant proportion: this is the trendline shown in the last figure. If the relationship becomes more complex, then other, similarly more complex, measures of correlations would be necessary.

Once a correlation has been established between some set of data A and B, it may be useful to consider whether one *causes* the other—this would then place a direction-ality on the two groups, either A is the independent variable and B the dependent or *vice versa*. This question may be explored by experiments designed to gauge the ab-sence and presence of whatever variables it is that A and B correspond to. In all cases, however, one must always be on the lookout for further confounding variables that were not explicitly examined in the data but which may turn out to be far better can-didates for playing the role of causes. A classic example from the statistics literature is the strong positive correlation that exists between the number of fire-fighters called out to a fire and the amount of damage done: we should naturally be hesitant to posit a causal relationship between calling out some number of fire-fighters and the amount of damage that the fire will cause! There is a further variable that was not measured: the size of the fire itself.

We also see here that we can quickly run into ethical considerations. One way of ascertaining which of the two variables—the number of fire-fighters called out or the size of the fire—is the better predictor of damage would be to vary these independently of one another in experimental situations… Fortunately, another indicator of cause is temporal succession: causes come before their effects. This would hopefully be suffi-cient in the present case to tip the scales in a more sensible direction! It must always be emphasised, therefore, that just because two sets of values show a correlation, there is no sense in which we can then immediately talk of one 'causing' the other.

Finally, we have noted here an analogy between the kinds of calculations done for correlations and that for standard deviations above: in statistics, some kind of calcula-tion of how far a body of values differs from the mean for that data reoccurs again and again—it is the basic way of ascertaining the overall 'shape' or degree of 'variation' in some data. It will be useful in several of the tests given below and, particularly, for their formulas, to bear this in mind: whenever you see some multiplication of differ-ences between values and means, it shows that some indication of how a body of data is varying is being calculated. And this is usually just what we need to get at in order to see if that variation is different from chance to a statistically significant degree.

6.5 How much is enough?

When one has decided to take a more serious look at one's data and results so that it becomes relevant to consider various statistical measures, the first and most common question that arises is 'how much data do I need?' We raised this issue in our discussion of sampling in Section 5.2.1 of the previous chapter. We need to be able to estimate the sample sizes necessary to be confident that we can find significant results. This is an important question, of course; one does not want to work with unrealistically and, worse, unnecessarily large sample sizes, but on the other hand, one does not want to perform statistical tests on sample sizes that actually make those tests unworkable or unreliable.

The basic idea follows directly from the issues we have already discussed: we would like any data sample that we look at to be *representative* of the population, or entire body of potential data, that we are considering. If, however, that body of data contains a large amount of variation, then we are going to need a larger sample to make sure that we cover that variation. Conversely, when a body of data is relatively homogeneous showing little variation, our sample can be smaller and still remain representative. The problem, as usual, is that we do not have access to the full theoretical data set or population and must somehow estimate this to know how big our sample will have to be.

Fortunately, there are many established statistical results that can be employed for this purpose too. Here we will characterise very briefly the notion of statistical *power analysis*, which, quite simply, is a way of measuring how strong your conclusions can be given a particular size of sample. With this, you can work out how much data you will need to be able to make statements with particular selected degrees of statistical significance.

As a first step towards understanding this measure, let us step back and introduce two concepts that are often found in information retrieval contexts and so are useful in any case: *precision* and *recall*. If you are retrieving documents on the web using a search engine then 'precision' is a measure of how accurately the selection was made. Accuracy in this case is the number of retrieved documents that were actually relevant. Recall, on the other hand, is a measure of how many relevant documents were obtained at all when compared with how many *should* have been retrieved.

These measures are based on a combination of two notions: positive and negative cases, e.g., documents that were either relevant or not relevant, and true and false, i.e., documents that were classified correctly (as being either relevant or not relevant) and documents that were classified wrongly. Putting the two dimensions together gives four classes: true positives, i.e., the relevant documents actually selected, false positives, i.e., irrelevant documents that were selected, true negatives, i.e., irrelevant documents that were not selected, and false negatives, i.e., relevant documents that were not selected.

The formulae for calculating precision and recall, along with a further metric often used for the 'goodness' of some selection process, the F or F_1 value, which combines the two, are as follows (where tp is true positive, fp is false positive, tn is true negative and fn is false negative):

$$\text{Precision} = \frac{tp}{tp + fp} \qquad \text{Recall} = \frac{tp}{tp + fn} \qquad F_1 = 2 \times \frac{\text{precision} \times \text{recall}}{\text{precision} + \text{recall}}$$

This is directly applicable to our current task. If we consider the notion of the null hypothesis above, and remember that we wanted to make sure that we did not reject the null hypothesis wrongly, then such a situation corresponds to that of a false positive—i.e., we think we have a relevant situation but actually we do not. We can now turn this around and consider the other unwanted condition, that of false negatives. That is, we want to avoid the situation that we do not falsely *accept* the null hypothesis and so miss the opportunity of finding a significant result. These two type of errors are called α errors (or Type I) and β errors (Type II) respectively.

What is more, and this is the final step, we want to find out *what size of sample* would be sufficient to be confident that we can avoid the false negative and find some effect if it is actually there. We saw above that it is usual to select probabilities for Type I errors, i.e., of falsely rejecting a null hypothesis, of 5% or, if being very strict, 1%. There are similar standard values for the Type II errors: typically we want to be at least 80% sure that we do not wrongly accept the null hypothesis. This is the desired *power* of an experiment: we want to find a sample size that gives us this power, i.e., the likelihood of finding an effect if it is there.

Now: effects also vary. Some are stronger than others and this also has an influence. If the effect is strong, then we should already see it with a smaller sample; if it is weak, then we will need a larger sample. Lowie and Seton (2013: 49) suggest that in order to reach a power of 80% and with a confidence level (Type I) set at 5%, a strong effect would need a sample size of 28 in each group analysed, a medium effect would require a size of 85 in each group, and a weak effect would require a size of 783 in each group.

There are several ways of actually calculating effect sizes, falling into two main groups: measures based on correlations and measures based on differences of averages. Some calculations based on correlation make use of the t-value that we discussed above, and so if we have already done a t-test then it is straightforward to provide an effect size measurement as well. This in turn allows you to check whether the sample size used and the effect size are appropriate. If an effect size is small, and one used only a small sample size, then the results may well be less than reliable.

Power analysis

Sample sizes

Two correlation-based measurements commonly calculated from the t-statistic are r and η^2 (eta-squared). These can be calculated from the t-statistic as given above, df (the degree of freedom), or the size of two samples compared, N1 and N2, as follows:

$$r = \sqrt{\frac{t^2}{t^2 + df}} \qquad\qquad \eta^2 = \frac{t^2}{t^2 + (N1 + N2 - 2)}$$

If r is less than 0.3, then we have a small effect; if r is greater than 0.5, we have a large effect; for η^2 the corresponding values are 0.01 and 0.14 respectively (cf. Cohen 1988: 27–74).

Returning then to our example of the effect of training that we discussed at the end of Section 6.3 above, the t-test values we calculated make it straightforward to apply the formula for r. So, using the values for the statistically significant data in that example (t=-9.763, df=38), we now get an r value of 0.846. This tells us that the training effect was a very strong one, consequently only requiring a relatively small sample size to reach the desired power of 80%. This value is then added to the standard phrase used to report the results, i.e.,

the mean difference between the groups was significant at $t(38) = -9.763$, $p < 0.000$, $r = 0.846$

Another measure of effect commonly found in the literature is the difference-based measure Cohen's d; straightforward formulae can be found in statistics textbooks and online (cf. Woodrow 2014: 65–66). A slightly more technical overview containing practical examples, explanations and formulae for both kinds of effect measures is given by Lakens (2013).

A further reason for working out effect sizes is that it is not always beneficial to have ever larger samples. In fact, the larger a sample is, the more likely it is for even small effects to show up as statistically significant. Effect size gives a measure that is independent of sample size: i.e., if an effect is small, then it shows up as small even if the sample size is so large that the difference shows up as significant. It is, then, a good idea to take the effect size into consideration so as to balance any values of statistical significance obtained. A small effect remains a small effect, regardless.

6.6 Refinements and restrictions

In this section, we set out a handful of further comments and refinements that may need to be considered and introduce one of the statistical tests that is most frequently used in the kinds of studies relevant for multimodality: the chi-square test.

First, we need to be more explicit about certain assumptions that the tests described so far make; we always need to be careful in making sure that the data we collect meets such assumptions—if this is not the case then, even though we obtain numerical values when applying the tests, those values may not be meaningful. We mentioned above some of the assumptions that the t-test makes, for example: in particular, that the data sets that are being compared should be normally distributed. In fact, the t-test requires in addition that the data of the two groups compared should *have approximately the same degree of variation*. That is, if we compare groups where one group varies considerable and the other not, then the values given by the t-test are less informative and may even be misleading.

The huge diversity in statistical tests that one finds in the literature are largely the product of a considerable history of exploring how to deal with such situations—different statistical tests are able to deal with more or fewer of these kinds of changes in assumptions. Therefore, although in many studies the data *will* be normally distributed, it is still a good idea to check this first and to calculate the standard deviations to compare them. There are even special statistical tests for this: for example, the *f-test* and the *Levene* test. Such tests indicate the probability that the two sets of data are comparable in the appropriate way. Most statistical tools either run such tests when performing a t-test automatically, allowing you to check whether the results will be reliable, or offer a way of doing the tests by yourself.

Second, and very briefly, it may be the case that the data turn out not to be normally distributed after all. This is a more serious situation and generally means that the t-test is simply not appropriate. When this occurs, you can either consider changing the operationalisation adopted so that it is more likely to give results that *are* normally distributed, or adopt a different kind of statistical techniques which is less sensitive to this restriction. Tests of this kind include so-called *non-parametric* tests that do not rely on particular distributions in the data. These consequently contrast with 'parametric tests', such as the t-test, which work precisely by fixing certain parameters to give the 'shape' of some distribution as we have seen.

Most parametric tests have 'non-parametric' versions or equivalents that can be used without assuming particular distributions. Common statistical packages and computer programs generally provide both kinds of tests and so when one knows how to use one, it should be fairly straightforward to work out how to use the non-parametric versions too. A typical non-parametric test corresponding to the t-test, for example, is the 'Mann-Whitney U test'. The results of this test are used in just the same way as those of the t-test, i.e., by obtaining a probability that the null hypothesis holds. We will occasionally mention the names of non-parametric equivalents to the tests we describe as we proceed; full lists can readily be found online or in more complete introductions to statistics in any case. We will also see in a moment, however, one non-parametric test that is very commonly used: the chi-square test.

Third, the t-test can only be used to compare two groups with respect to one variable of interest—above we had the scores for our participants. If more groups are to be compared, or there are more variables of interest, then other statistical tests need to be employed. The most common tests here belong to the family known as *analyses of variance* or ANOVA, which we will return to below in Section 6.8. There we will see both how these tests work and how to read texts that report results using them; you

Parametric *vs.* non-parametric tests

will often encounter this in empirical studies and so it is useful to know what is being said.

Types of data

Fourth and finally, we have described how we can test whether there is a statistically significant difference between two sets of measures of a single variable. This works well when we have numerical values that we can add together and find means for. This kind of data is called *ratio* or *interval* data: this means that the values that are possible lie on a continuous scale. But what if we are concerned with data that can *not* be related to numerical values in this way? Then we have a different *type of data* (see: Krippendorff 2004: Chapter 8, 150–170). For example, another common type of data, often arising from questionnaires, is called *ordinal* data. This type occurs when one has a series of categories that are ordered but where the values may have different or varying 'breadths'—we can see this in cases where values are collected according to selection from scales of, typically, 5 or 7 categories, such as 'hate it', 'dislike it', 'don't care', 'like it' and 'love it'. Response sets of this kind are generally called Likert scales. Their categories are clearly ordered—that is, 'hate it' is worse than 'dislike it', which is worse than 'don't care', etc.—but it is *not* possible to say that the 'dislike' band is the same width as the 'like' or 'love' band. So we cannot perform the usual kind of arithmetic operations on the categories selected.

Deciding which type of data is being used is therefore very important because it helps determine which statistical tests can be used and which not. For example, data collected according to an ordinal scale is most likely not going to be normally distributed (because we do not have access to any actual numbers that might underlie the selection of categories) and so testing for statistical differences with a t-test would not be appropriate—this would be a situation where we would typically need to employ non-parametric tests.

Another very common type of data, particularly but not only for multimodal work, is called *nominal* or *categorical* data. In this case, we just have names (or categories) for the relevant values, which cannot be placed on some continuous scale and which cannot, even, be ordered with respect to one another. We already touched on cases which would require data of this kind above—for example: what if we want to compare not scores but values such as 'given' and 'new', 'male' and 'female', or 'picture', 'text' and 'diagram'? Here again we cannot use a test such as the t-test because, quite simply, we cannot work out what the 'average' of, for example, 'male' and 'female' would be. The question is nonsensical because we are again dealing with a different *type* of data.

Note that we cannot 'cheat' here in order to turn nominal data into interval data— that is, we cannot simply replace 'female' by 1 and 'male' by 2 and say that we can now use the tests we have seen so far. The reason should now be clear: it is because these numbers are not actual numbers, i.e., numbers which we can sensibly add and multiply; they are simply labels being used in place of the original categories. We can add the 1 and the 2 to get 3, but this is of course completely meaningless with respect to the categories we are considering; neither, in this usage, is 1 less than 2 or *vice versa*—the usual properties of numbers are no longer relevant and so we cannot use calculations that would rely on them.

Fortunately, there are a variety of statistical tests that can be applied to nominal data. One extremely common test for comparing groups on the basis of nominal data is *chi-square analysis* (or χ^2 if one likes symbols); most statistical tools provide this test as a matter of course and there are even several websites where one can simply enter data into tables and receive the corresponding statistic. There are therefore very few grounds for not performing a chi-square measure of significance if the data is appropriate. To conclude this section, therefore, we will briefly work through an example of its use selected from another potential multimodal empirical research question. This will also give us an example of a non-parametric test in action, where we will see that most of the principles that we have seen so far are equally relevant for understanding what is going on.

Let us take the following research scenario. We have noticed that a considerable range of emoji appear in various text messages produced by younger users of the medium, but it appears that there are differences. We want then to investigate whether there is a gender difference in specific emoji usage. For current purposes, let us take as categories the old favourite of traditional sociological studies: 'male' and 'female', probably answered by self-identification in a survey of some kind. We could naturally imagine a host of further categories that might be considered more ideologically sound for current research questions—but these would also all be nominal categories and so could be added to a study of the kind we describe here without problem; we will leave this extension as an exercise for the reader at this point however! Selection of categories of these kinds offers a typical case for chi-square as we are only investigating nominal categories: concretely 'male'/'female' is one dimension, while the distinct types of emoji form another dimension.

Chi-square then gives back a value that again helps us to decide the chance that we are mistakenly rejecting a null hypothesis; the null hypothesis, as before, will typically be that there is no difference between the groups we are examining. And, as with the t-test, there are two stages: first the calculation of the chi-statistic and then looking up where this falls in the corresponding frequency distribution. If we lie a long way from the mean, then the probability of the value occurring by chance is very small and we can be reasonably certain that we are not doing something wrong when we reject the null hypothesis and assert there is a statistically significant difference between the groups we are examining.

Since chi-square is working with nominal data, and we cannot add or multiply nominal data directly, the only thing that we can work with is the *frequencies* with which particular values occur—and this is essentially what chi-square is doing. The value indicates whether the respective frequencies of values we find in a study differ from each other significantly or not. The calculation of the chi-value is straightforward but somewhat tedious, and so again using Excel or another spreadsheet program, or pre-prepared tables on the web is recommended. Therefore, we will only sketch the

Chi-square analysis

calculation here as we proceed as it is always good to know a little of what is going on in the background.

We first need to decide how we are going to collect our data. There are various possibilities here and these need to be considered carefully as is always the case when designing a study. For example, we might just collect 100 text messages and in each one look to see how many cases of the individual emoji we are investigating occur. We then use the chi-square test to see if there is a difference in frequencies according to whether the sender of each respective text chooses to identify themself as male or female. This would be a classic case of needing to be wary of *confounding factors*: that is, some property of the data that gets in the way of investigating what we are really interested in. In the present case, for example, perhaps the male members of the sample happen to send texts that are 5 times longer than the female members of the sample: then they may have more emoji of various kinds but this is not necessarily related to gender, just to quantity. We always need to pay attention to such possibilities and come up with study designs that avoid them.

A good way to recognise such situations is to imagine more extreme variants of what might occur: imagine that the female texts were only 1 character long and the male texts 500 characters: clearly just counting these would result in a very unbalanced data set that may tell us something significant about the respective lengths of texts that are sent according to gender, but not much about emoji usage. Using what we have introduced above, we could now be far more sophisticated of course—for example by first demonstrating that there is no statistically significant difference in the lengths of the texts that we examine with respect to the gender of their senders—but, for present purposes, let us just take a much simpler solution and take not the texts as sent as units but the characters in those texts. We could then take one data set to be the first 2000 characters in texts sent by male senders, and the other data set to be the first 2000 characters sent by female senders. This would at least give us a fair chance of finding differences that are genuine differences in emoji usage.

ℹ️ The mentions we have made so far of manipulating bodies of texts to get them in more appropriate forms for doing our statistical tests—for example, picking the first 2000 characters of some collection of text messages—make it clear that you would be well served by having at least a basic competence in one of the computer languages available nowadays that support such operations with considerable ease. These computer languages range in difficulty but some are quite straightforward for these kinds of tasks and have a very low learning curve.

Python, for example, is freely available, runs on almost every type of computer and operating system (Windows, Mac, Linux, etc.) and would manage a conversion of the kind just described in a couple of lines of code. When considering empirical research, the investment of time in learning how to do basic tasks in Python repays itself extremely quickly. Imagine looking through 100 text messages to pick out the first 2000 characters: this is tedious and therefore error prone. Increase the size of the data set to 1000 text messages and 20000 characters and the task is considerably worse than tedious—such an increase makes no difference to the Python code that you would write, however. Doing the task manually might take a couple of weeks; doing it with Python, including writing the code, might take a half an hour or so—even less when you have become fluent at programming these kinds of tasks. Search for: "Python for absolute beginners".

This also turns out to be more than a simple quantitative difference that the task takes less time; having the computer code to do the task quickly also leads to *qualitative* improvements in your research. Imagine, for example, that you have just spent two tedious weeks preparing your data according to the assumption that you should count by characters in the text messages. Then you find out, for some reason, that it would actually have been better to count out words. The time and energy to go back and re-count all the data by hand might simply be lacking. If you had written Python code to do the selection, the change would be achieved in a matter of minutes and the data prepared anew at a click. This begins to make it possible to explore different ways of solving any task and of checking the results of those alternatives in a reasonable time. The chances that both better research and better research results can be produced are considerable.

So let us collect our data in this way—giving us 2000 characters from self-identified male senders and 2000 characters from self-identified female senders, and taking five popular emoji for present purposes more or less at random. To prepare the data for the chi-square test, we first construct a table that sets out the respective counts for the values we are interested in. In our case, the table will have two rows (for 'male'/'female') and 5 columns (one for each emoji we address). The choice of which dimensions to assign to columns and which to rows makes no difference for the final outcome.

Consider first the example data set, or 'contingency table', in Table 6.2. Here we might by inspection suspect that our data is not showing much difference between our selected dimensions of interest, our *independent variables* 'male' and 'female'. There is variation, as there always will be when we examine real situations of use, but it does not appear to be large. The chi-square test for this table in fact gives us a probability that we would be in error if we reject the null hypothesis that the two groups are not different of 0.494, which is of course way too high a probability to count for statistical significance. We could not, therefore, reject the null hypothesis for this data as doing so would be wrong *almost half of the time* (49%).

Table 6.2. Emoji usage: first data set

	🙂	😇	😍	🥴	😡
male	134	43	27	16	93
female	118	54	33	20	89

Table 6.3. Emoji usage: second data set

	🙂	😇	😍	🥴	😡
male	134	43	27	16	93
female	101	44	43	35	112

The value of the chi-square statistic—which serves the same role as the t-statistic or t-value we saw above—is 3.394 for this data set. This of course tells us little by itself: we need to relate the value to the corresponding distribution to see whether this indicates that something is likely or not. Broadly speaking, the higher the chi-square value, the lower the probability that is attached. Again, as with the t-test, this looking up of an area under a distribution curve is something that the standard statistics tools generally do for you, giving the associated probability directly.

Preparing data so that you can run a chi-square test and obtain the corresponding p-value either online from a dedicated website (there are several) or with a spreadsheet program is straightforward. Essentially all that you need to do is enter the values for the counts obtained – always the actual counts, never percentages or other indirect calculations – in a table of the appropriate size, click on 'calculate' or similar, and the result will be shown. If we do this with respect to the rather more varied data given in Table 6.3, for example, we obtain a rather different result to the previous case. The values in this table for the male usage set are the same as before, but the values for the female usage are different. Now, just by looking at these values it would by no means be clear whether this represents a *statistically significant* difference in gender patterns or not. So we would be well advised to carry out an appropriate test of significance, such as that given by the chi-square test.

Entering these values into an appropriate online form would then tell us that the chi-square statistic in this case is 16.414, firstly a much much higher value than we had with the first batch of data. The online form will then also usually inform us that this value gives us a probability of falsely rejecting the null hypothesis that male and female usage do not differ of 0.0025—which tells us that rejecting that hypothesis is certainly safe at both the 95% level (i.e., $p < 0.05$) and the 99% level ($p < 0.01$). All that remains to be done is then to report these results appropriately and, for this, we need to give one further piece of information. This is, similarly to the case with the t-test, again the degree of freedom involved.

This is necessary because, although we will not show it here, the chi-square distribution is actually a family of curves exactly analogous to the t-distributions we saw in Figure 6.4 above. The chi-square curves also vary according to the 'degree of freedom' and this is why we also need to report on this when presenting our final results. For the reasons motivated in general for degrees of freedom above, the degree of freedom value for chi-square tests is simply the result of multiplying the number of columns in our contingency table minus one by the number of rows in the contingency table minus one. For a 2 × 2 table, the degree of freedom would therefore simply be 1—i.e., (2−1)×(2−1); for our 2×5 table, we have 4 degrees of freedom—i.e., (2−1)×(5−1). Thus, following our now successful study, we would report the results using the following set phrase:

the groups were found to differ significantly at $\chi(4) = 16.414, p = 0.0025$

The 4 given after χ is the degrees of freedom value just as with the t-test above. Checking these figures is left as an exercise for the reader!

With fewer conditions, and consequently smaller degrees of freedom, the corresponding chi-square distributions change their shape quite considerably, and so there are a number of further related tests to balance this: particularly *Fisher's exact test* and the *Yates correction*; the calculations are broadly similar. If expected frequencies appear below 10, then it is advisable to use Yates's correction—some tools might perform this automatically if the values warrant it. A further alternative for low values (but not only low values) is the *log-likelihood* (LL) ratio; online calculators and tools usually give back a LL value (analogous to the chi-statistic) and the corresponding probability.

Calculating the chi-square statistic by hand (or, usually, using a spreadsheet program) is also possible but, although straightforward, is rather tedious and so we will not illustrate this in detail here. Essentially one forms several tables within the spreadsheet program, beginning with the actual data and then several derived tables of values. These follow from the underlying motivation for the chi-square test quite directly. In particular, we (or the program) needs to compare the variation that actually occurred in the data with what would be expected 'by chance'. To do this, it relies upon comparing the data that actually occurred with the values that might be expected if the data had been fairly, i.e., proportionally, distributed among the categories. Sometimes we might already know the expected values on other grounds: for example, if we throw dice, then we would expect values to turn up more or less equally. Usually, however, we do not know the expected values and so we work this out on the basis of the proportions actually found in the data.

For example, if we find some number of emojis of a particular kind altogether and twice as many females as males participating, then we might expect that there should be twice as many cases of that emoji being selected by females as by males. So, for each gender and for each emoji, we would divide the total number of the gender considered by the number of males and females together (to obtain the proportionality) and multiply this by the total number of the emoji. Performing this calculation repeatedly for all the cells in our original contingency table gives then a table of 'expected values'. A spreadsheet program such as Excel or OpenOffice can then perform the comparison of the occurring values with the expected values directly. Thus, if our data is collected into one portion of our spreadsheet (say B2:F3 for our 2×5 emoji data) and we have worked out all the expected frequencies in another portion of our spreadsheet of the same size (say B12:F13), we can use the spreadsheet formula:

```
=CHITEST(B2:F3,B12:F13)
```

This returns as a result the final probability associated with the chi-square value. More sophisticated tools will go further and allow you simply to provide raw data, such as questionnaire or survey answers, and save you the trouble of working out the immediate expected values—again, it is well worthwhile looking at such tools as they can save you a considerable amount of work.

When performing the chi-square test, regardless of which method of calculation is used, it is also appropriate to calculate effect sizes as we introduced in the previous section. However, because the chi-square test is concerned with the frequency of nominal data and not with numerical data, we cannot use methods based on means and their differences. What we can do instead is measure strengths of associations, for example by using Cramer's V. This is calculated in a very similar manner to the measures we gave above for the t-test; i.e.:

$$\text{Cramer's V} = \sqrt{\frac{\chi^2}{n(r-1)}}$$

where n is the sample size (i.e., the total number of observed occurrences) and r is either the number of rows or the number of columns, whichever is the smaller. The interpretation of the values is also similar: less than 0.1 is little or no effect, more than 0.3 is a medium effect, and more than 0.5 is a strong effect. For our emoji data, Cramer's V is 0.1591, i.e., quite a weak effect. An alternative measure of effect size for nominal data that you will occasionally see is the *odds ratio* (OR), generally only calculated for 2 × 2 tables.

Finally, and just as was the case with the t-test, things get more complicated if we need to compare more groups or variables; ANOVA was mentioned above as a possibility for interval data, but this does not apply to nominal data. A more sophisticated collection of techniques applicable to nominal data is then offered by various *regression* methods, which look for relationships between bodies of data (cf. Woodrow 2014: 85–95). Jaeger (2008) gives more, if somewhat technical, details.

6.7 Inter-coder consistency and reliability

In the previous chapter we noted at several points the value of annotating or enriching data with categories or labels that help one carry out a research task more effectively (→ §5.2.3 [p. 146]). The idea is that one codes the data so that it is easier in subsequent stages of analysis to find regularly reoccurring patterns, which can then be investigated as meaning-bearing in their own right. This is particularly important for all corpus-based approaches to multimodality research. The typical use case is that one has decided on some body of data to be studied and will then subject all the individual materials to the type of analysis selected. Bell (2001) sets out as an example of such a study an imagined analysis of the front pages of women's magazines coded according to the multimodal expression of social distance, 'modality' (i.e., truth-claims); we will address some similar examples below.

A question that then arises is how can one have confidence that the codings or annotations of the data on which one builds the analysis are correct or, better, 'reliable'? Various answers can be pursued. One form of answer simply relies on the analysis being done 'well'—i.e., if one is sufficiently well trained in a scheme of analysis then one should be able to apply it in a consistent fashion. This method is often selected when the analysis in any case includes more interpretative effort on the part of the analyst, taking into consideration aspects of context or other features that might call for one coding rather than another.

Although it is certainly to be hoped that analyses performed in this way are, indeed, 'reliable', measures have been sought for evaluating annotations or codings of data in a more transparent fashion. For many tasks of annotation or coding, there may be a considerable body of hand-coded data and it is by no means clear that this will always be sufficient for building solid research. There are also many studies where the coding of data according to some criteria may be being done by specifically trained 'coders' who do not necessarily have all the theoretical knowledge required to fully engage with difficult cases. Particularly in these kinds of studies the question of the accuracy or reliability of the data annotations or codings given to the data reasserts itself quite strongly. Such situations demand that some evaluation of the coding quality be undertaken and academic journals and other scientific publication forms increasingly expect that articles report on just how reliability was assessed.

In the fields of content analysis and linguistic analysis, schemes have been developed for calculating what is termed *inter-rater reliability* or *inter-coder consistency*. These presuppose that one can have a body of data coded or annotated with respect to a given set of categories by more than one coder or annotator independently of each other. The measures then allow you to calculate to what extent these independently produced codings agree more than might be expected by chance. It is very important that the codings *are* produced independently. Reliability of coding cannot be measured unless this condition is fulfilled. This does not mean that coders should not discuss their disagreements and consider how best to analyse some data, but this phase of work should not be mixed with a reliability study. Care also would need to be taken in such discussions that the results achieved are genuinely driven by engagements with the data rather than well-known processes of social interaction and dominance.

When a set of codings has been produced independently and those codings agree significantly more than by chance, then one has good grounds for stating that the coding has been performed consistently or reliably. Moreover, different degrees of agreement can be very valuable in indicating places where there are problems of coding, where perhaps more instruction or training needs to be given, or where even the theoretical constructs might need revisiting to see if they are really applicable in the manner or to the extent originally thought.

As the mention here of 'significantly greater than by chance' indicates, we are dealing here with some further applications of statistics. The motivations for the actual measures available are therefore to be found in general statistical principles such as those we introduced in the previous sections. We do not need to consult these principles every time we calculate consistency measures, however—standardised formulae already exist which can be taken more or less 'off the shelf'. This section will describe how to calculate one of the more common of such measurements, explaining a little of the background as well.

Coding independence

> *Reproducibility* is the degree to which a process can be replicated by different analysts working under varying conditions, at different locations, or using different but functionally equivalent measuring instruments. Demonstrating reproducibility requires reliability data that are obtained under *test-test* conditions; for example, two or more individuals, working independent of each other, apply the same recording instructions to the same units of analysis.
>
> *— Krippendorff (2004: 215; original emphasis)*

Before beginning, it has to be emphasised that demonstrating reliability in terms of agreement or consistency can never be seen as a 'shortcut' to the truth: all that is being shown is that the analyses performed can be performed again, by different people, obtaining the same analytic characterisation within certain limits. As with all statistical tests, if one has not selected a sensible research question or sensible data to collect, no manner of statistical testing will improve on the situation! The converse does provide useful information, however: if an analysis cannot be shown to be reliable, then the status of that analysis must be given particular scrutiny and justification in order to ascertain just why that is the case and to limit that unreliability as far as possible.

Nevertheless, there will always be a tension between making results as reliable as possible and making results as interesting as possible. As Klaus Krippendorff, a prominent researcher in the area of content analysis who has contributed significantly to the discussion and measurements of reliability, describes it:

"*In the pursuit of high reliability, validity tends to get lost.* This statement describes the analyst's common dilemma of having to choose between interesting but non-reproducible interpretations that intelligent readers of texts may offer each other in conversations and oversimplified or superficial but reliable text analyses generated through the use of computers or carefully instructed human coders." (Krippendorff 2004: 213; original emphasis)

We will always need to be searching for appropriate compromises in this regard when pursuing empirical multimodal research.

Calculating agreement or consistency can be done in several ways and there is no single method which clearly outperforms all the others. Each of the standard metrics has its own quirks and may respond to particularities of the data in different ways. This is the reason why there are several measures at all: each tries to iron out some of the deficits of the others. When reporting consistency or agreement measures, it is now therefore common to give a collection of scores, each calculated with a different method—if one manages to obtain high scores on all the methods, then one is probably on safer ground! Several statistical tools, including some online webpages, consequently provide a collection of scores for different methods all at once as a matter of course.

The first point to understand when considering measures of consistency is that what would probably be thought of as the most straightforward way of showing 'agreement'—i.e., just seeing what proportion of the codings produced by some coders are the same as each other—is actually insufficient. This is usually expressed in terms of *percentage agreement* and is calculated in the obvious way: take, for example, two sets of codings and measure what percentage are equal. The reason why this is insufficient can be seen from our discussions earlier in this chapter: any sample of data is going to show variation and so we need to know not that two or more sets of codings agree, but that they agree *more than they would by chance*.

If different codings are the same simply because of random choices, this is not an indication that the coding was performed reliably. If there are few categories to be chosen, the chance that the same choice will be made is in any case higher; or there may be an implicit assumption that categories are all equally likely to occur, which can also be wrong. Our description of the role of variation in measuring data in previous sections tells us that there will always be 'error' in the technical sense. This will then effect not only codings that are considered 'correct' but *also* those that are 'wrong'. Measuring the percentage of correct codings does not get at this basic property of data and so may produce quite inaccurate results. Measures that take such properties of the data into consideration are therefore necessary.

Note that this leads to a few seemingly paradoxical properties of reliability measures. First and foremost, it is possible to have what looks like very good agreement but very low reliability scores. Imagine, for example, a set of data where two coders code 8 units of analysis and agree on 5 out of 8 choices. This looks good but until we know more about how the choices distribute, the specific value may not be indicative of reliable choices after all. It is useful to remember here that 0% agreement is about as unlikely to occur as 100% agreement and so we need to find ways of factoring agreement according to likelihood of chance outcomes as well.

This is usually done using calculations that provide indices of reliability. Although we only introduce one relatively simple method here, working through the examples shown should make it easier to follow up on more complex variants by yourself if necessary. Reliability metrics are often designated by Greek letters, qualified by their developers. The reliability metric we illustrate here is consequently called Scott's π— this is the index used in Bell's discussions of reliability mentioned above (Bell 2001). Scott's π can be applied whenever we have two independent coders who analyse data by assigning data to designated categories—this is therefore a test for *nominal* data as we introduced the term above.

Scott's π

Many of the reliability indices have similar formulae because they are trying to get at similar properties of the data. There is going to be a measure of what actually occurred and a measure of what was likely to happen by chance. The index then calculates some ratio of these figures to give the final metric. The formula for Scott's π is consequently as follows:

$$\pi = \frac{Pr(a) - Pr(e)}{1 - Pr(e)}$$

where $Pr(a)$ is the agreement that actually occurred and $Pr(e)$ is the agreement that would be expected 'by chance'. Whereas working out the actual agreement is straight-forward, it might sound curious that we also need to know what would be expected 'by chance'—how could this be estimated? Fortunately, it is not at all complicated and is very similar to the calculations that support the chi-square test. It is, by and large, simply a matter of seeing how the scores *would be distributed proportionately* on the basis of the totals that we actually find.

We will work through a simple example to show this in action. First, we consider how many categories we are coding for: this gives us an indication of how distributed our results could in principle be. Second, we look at the *total number of instances* of our data that were placed in each of these categories. This gives an indication of how both coders behaved and shows the proportion of the data that was placed in each category. This is what would happen, therefore, if the data were distributed 'proportionately' without regard for possible differences. The required value of $Pr(e)$, the expected agreement by chance, is then simply the sum of the squares of these proportions.

For some actual figures to calculate, let us imagine that we have done a study where we had two coders independently classify 20 examples according to a simple coding scheme made of three categories: 'Setting', 'Means' and 'Accompaniment'—these are grammatical categories that have also been extended for describing visuals by Kress and van Leeuwen (2006: 74) and we might want to see if some data is being reliably annotated with the categories for further investigations. Normally we would have given the coders some training, presented them with the definitions and some examples, discussed any problems that they may have had, and then let them loose on the data to be coded. We could, of course, take any other categories that might be of interest and many more instances. As long as the categories are *disjoint*, i.e., are not subject to overlap by having single examples classified simultaneously according to more than one category, then we can use the reliability metrics discussed here.

Table 6.4. Data formed by two independent coders working on 20 examples

Mary	Fred
Setting	Setting
Setting	Accompaniment
Means	Means
Accompaniment	Accompaniment
Setting	Setting
Means	Accompaniment
Means	Means
Means	Accompaniment
Accompaniment	Accompaniment
Accompaniment	Accompaniment
Means	Means
Setting	Setting
Setting	Means
Means	Accompaniment
Setting	Accompaniment
Means	Means
Setting	Setting
Accompaniment	Accompaniment
Accompaniment	Accompaniment
Setting	Setting

After the coders—let us call them Mary and Fred for concreteness—have done their coding, we end up with two tables, one from each coder, each showing the categories they selected for each example. We will assume here that they always make a choice so that there are no 'holes' or 'gaps' in the data—one can always introduce an 'other' category to enforce this; one should also make sure, however, that coders do not feel obliged to use it too often! If the quantity of data being ascribed to 'other' goes above

10%, then the category scheme probably needs reworking. We can then combine Mary and Fred's tables into a single table consisting of rows for each example classified (20 in our present example) and two columns, one for each coder. Each cell in the table then contains the category that the respective coder selected for the respective example. This might then look like the table shown in Table 6.4. Krippendorff (2004: 227) calls this a 'reliability data matrix' and it already contains all the information we need for calculating reliability metrics.

First, we calculate the actual agreement from the table. This is simply the percentage formed by dividing the number of times the coders agree on a classification (which happens 14 times in this table) by the number of data examples categorised (i.e., 20). The agreement, $Pr(a)$ in the formula above, is then 0.7 (i.e., 14/20). Next, to work out the expected agreement, we divide the total number of times that a category was used by the number of times a classification was made at all, which is 40 because each coder coded all 20 examples, square all the values obtained and then add them up.

If we do these sums, then we find that we have 13 occurrences of 'Setting' in the table, 12 occurrences of 'Means', and 15 occurrences of 'Accompaniment'. Dividing each of these by 40 gives the proportions 0.325 (13/40), 0.3 (12/40) and 0.375 (15/40) respectively. We then work out the squares of these proportions and add them together to give us the value of $Pr(e)$. This weighs in at 0.33625 for the present case. We now have all the numbers we need—i.e., the two values for $Pr(a)$ and $Pr(e)$—to plug into the formula for Scott's π as follows:

$$\pi = \frac{Pr(a) - Pr(e)}{1 - Pr(e)} = \frac{0.7 - 0.33625}{1 - 0.33625} = 0.548$$

This value is actually rather low; we would not normally want to accept reliability scores below 0.7. So, even though a *percentage agreement* rate of 70% may sound quite reasonable, correcting for chance with Scott's π shows that this is not a convincing result after all. The intercoder consistency needs to be improved.

We can also illustrate the importance of correcting for chance by modifying our data somewhat. Imagine that Fred was having a bad day and actually just chose 'Setting' *all the time*. This can easily happen to a less extreme degree in real coding situations—some categories may, for example, not have been well understood and then a coder may systematically avoid using them. In this case with Fred selecting 'Setting' 20 times, note that we would still have a 40% (0.4) agreement rate!—this is simply because Mary happened to pick 'Setting' a few times as well. When we do the calculation of Scott's π using the revised data, however, we get a very different result. The proportions for the three categories are now 0.7, 0.175 and 0.125, which when squared and summed give 0.5362. Plugging these into the formula we have:

$$\pi = \frac{0.4 - 0.5362}{1 - 0.5362} = -0.294$$

Now, reliability metrics are usually designed to yield a value between 0 (meaning no reliability) and 1 (meaning very high reliability or perfect agreement). So when we get a negative value we know that something is very wrong! This can occur only when the sample is too small or when there is a *systematic* skewing of the selections, i.e., a repeatedly occurring mistake or an avoidance of some options in favour of others by individual coders, so that reliability is not even possible. This is the case here, again showing that calculating a suitable reliability metric is very important for a variety of reasons. The values that one obtains can function as useful diagnostics concerning problems with the codings that we have received, which may in turn indicate problems in the coding *schemes* that have been used.

As noted above, there are several measures of reliability that are commonly used. In linguistics, Cohen's kappa has been popular for some time although, in qualitative studies more generally, Krippendorff's α has been argued to apply to a broader range of kinds of data—including cases where there are gaps in the responses given, more than two coders as well as smaller amounts of data—and so might be preferable (cf. Krippendorff 2004: 222). The values returned by the metrics are, however, often broadly similar. For our two illustrative cases above, for example, the values for Krippendorff's α are 0.559 and -0.2615 respectively, quite similar, therefore, to Scott's π. Moreover, also as noted above, several online services and computational tools produce measures according to all of these metrics as a matter of course. Consequently, as long as one aims for values in the region of 0.8 and above, the reliability of the codings produced can safely be assumed.

6.8 What affects what? Looking for dependencies

We have seen so far how to find out whether there are statistically significant differences between collections of data, whether there are correlations between collections, and how we can calculate levels of agreement between coders annotating data. When conducting more detailed studies, however, perhaps the most common goal is to find out not only whether there are differences but also to look for the possible reasons any differences found—that is, we often want to know what factors might be responsible for the observations we obtain. This leads us to the extremely useful collection of statistical methods that we mentioned at a couple of points in passing above: analysis of variance, or ANOVA.

ANOVA is appropriate whenever some measure is produced in an experiment or study (i.e., interval data), there are one or more qualitative variables that classify the data into groups, and we have *three or more* groups of interest; we will see examples below. The calculations involved in performing ANOVAs are essentially similar to those that we have seen above, particularly for the t-test, but there are many more of them because *comparisons* have to be made across any data sets we are considering to see just where effects are coming from. One can get an intuitive sense of this by thinking of one of the most basic uses of ANOVA, that is: looking for differences when we have *more than two groups of data* to compare. We explained above how the t-test provides a straightforward test for interval data with two groups but that it did not, for example, operate if we had three or more groups. We might then seek to work around this difficulty by running several *separate* t-tests, one for each pair of groups. This is generally not a good idea, however, because each individual test brings along its own chance of *mistakenly* rejecting the null hypothesis: i.e., assuming there is a difference when there is not. And, each time we repeat the test on the same body of data by picking out a new pair, we also 'add together' (not literally, but approximately) the chances of making mistakes. What would then be a reasonable level of significance for a single t-test then quickly becomes unacceptably low due to this snowball effect.

ANOVA is the technique of choice for circumventing this problem but, as a consequence, it is rather more careful (and intensive) in its calculations. This means that we will not go through how to do these calculations by hand—an appropriate computer program should be used instead. The main issues to be dealt with when performing ANOVAs are consequently: first, to organise your data in a format that lets it be 'consumed' easily by such a computer program and, second, to read (and understand) the results obtained. This latter skill is a useful one even if you do not intend to run ANOVA tests yourself because it is common (and increasingly so) to encounter papers and reports of research results where analysis of variance results are given. As with all statistical tests, there are set phrases and methods for phrasing the results of ANOVA tests and so, once you know what these are saying, it is much more straightforward to follow articles where empirical results of this kind are being built upon.

To know why certain values are being given by a software package whenever you run an ANOVA test, it is useful to have a basic idea of just what the ANOVA test is doing. We will describe this as simply as we can first. Some of the concepts are a little subtle, but they all build nonetheless on what we have introduced in this chapter so far. We will refer back to concepts we have used before as necessary. We then proceed to the practical use of ANOVA tests with a multimodal example.

ANOVA's basic approach is to compare how much variation there is in several groups of data, both *internally* within each group and *across* the groups considered as a whole, regardless of how many groups there are. Any differences that occur are considered significant whenever the variation found (and we know that there will always *be* variation in collected data) in the groups is (a) more than that which would be expected by chance *and* (b) sufficiently different across groups. This is all captured in what is called the *F-ratio*, a value which is calculated by bringing together all the variations worked out for the groups involved and their combinations in a single measure.

We explained the idea of variation and how it can be calculated when we introduced standard deviation above ($\rightarrow$ §6.1 [p. 174]). Essentially, this was a matter of looking at how far each data value differed from the mean, or average, of the data values, squaring the result, adding them together (giving the sum of squares: SS_x) and dividing by the sample size minus one (the degree of freedom). The F-ratio does a similar job, building on the 'aggregate' degree to which the means of the considered groups differ and placing this in relation to what might be expected by chance. An 'aggregate' is required at this point because we cannot just work out a difference in means as would be the case for two groups (and the t-test); this would lose just the information that we are trying to measure. For example, while the difference between 10 and 5 is straightforward, the 'difference' between 10 and 5 and 2 is less clear; if we simply subtract 5 and 2 from 10, we are left with 3, which of course tells us nothing very useful about the *individual* differences between 10 and 5, between 5 and 2, and between 10 and 2.

To avoid this, the aggregate employed in the ANOVA test instead forms the squared differences of the individual group means to the total mean, multiplies these by the sizes of the groups involved to keep those differences 'in scale', and then adds up the results. Working out this score by hand would involve a lot of repeated calculations of finding out means, doing subtractions, squaring and dividing and so on, multiplied by the pairwise combinations at issue. While this is not particularly complex (once one knows what to calculate), the benefit of using a computer program to do the work should be obvious. This is just the kind of task that computers were (originally) designed for: many repetitions of more or less the same sequences of operations.

Once the final F-ratio has been reached, it is used just like the t-test and chi-square in order to see how unlikely that value would be to occur by chance. This works in the usual way, by placing the particular value calculated against a corresponding distribution (in this case, logically, the 'F-distribution'). This delivers a final *p*-value that is reported for indicating statistical significance as usual. When an F-ratio shows a statistically significant result, this indicates that we may reject the null hypothesis that there is no statistical difference between the means of the compared groups (regardless of how many there were). It does *not* tell us, however, between which particular groups the difference is significant and between which not. This is done by invoking other tests that focus in on pairs to search out the sources of the significant different. Since these tests are generally done after the main ANOVA calculation and are required for finding specific sources of effects, they are called *post hoc* tests. The most commonly used post hoc test is the Tukey HSD (honest significant difference) and, fortunately, most software packages that perform ANOVAs for you will also commonly tag on post hoc tests as well. Thus what would formerly have been a considerable effort is now generally rendered run of the mill.

Another similarity with the situation with t-tests and chi-square is that the actual shape of the F-distribution that has to be used to gauge significance depends on the degrees of freedom involved—and so this has to be given (or calculated) as well. However, since we are comparing *two* kinds of information—the variation within the groups and the overall variation—there are now *two degrees of freedom* in play. For the number of groups compared, we calculate the 'between-groups degree of freedom' by taking the number of groups and subtracting one; for the 'within-groups degrees of freedom', we take the total number of samples and subtract the number of groups—

i.e., if there are 4 groups, each with 5 values, then the corresponding degree of freedom is $(5 - 1) \times 4$, i.e., 16. These are both, therefore, variations on the degree of freedom calculations that we introduced above.

Finally, and also as explained above for the t-test, it is important to decide in the experimental design whether the measures gathered are independent or not—for example, if completely different groups are being compared, then they are independent, but if you are looking to see whether some single group's behaviour shows differences with respect to some varying condition that you have subjected them to, then one must use a 'repeated measures' (or 'within-subjects') ANOVA instead. This is one of the options that statistics programs will offer and so is not further problematic. The repeated measure variant can in fact be extremely useful for factoring out individual variation— if you measure some value for each participant both before and after some intervention, for example, then one can be more confident that any difference found is due to the intervention and not to pre-existing individual differences. This could have been applied to our t-test example above as well, testing not two groups of trained and untrained analysts, but one group before and after training. This would then 'factor out' the possibility that some coders might just have been intrinsically better (or worse) than others.

When we then report any results, or read any results, we use or see phrases such as the following:

> a one-way independent samples ANOVA showed a statistically significant difference in the group means, F (3,16) = 6.42, p = 0.005

'One-way' means that one is looking at one main variable, whose values are divided across three or more groups. 'Independent samples' means that the samples were independent, in contrast to a 'repeated measures' ANOVA. And the two numbers after the F are the two degrees of freedom as just described. The F-value calculated is 6.42 and, referring this back to the F-distribution that is appropriate for this particular *pair* of degrees of freedom yields a p-value of 0.005. This background information should help understand why particular values are produced and reported by the computer when we actually perform an ANOVA on some data.

To show this, consider a similar set-up to the example that we used above for the t-test, where there was one dependent variable (the score that participants obtained) and two groups: trained and untrained. Lets change the situation so that we now have three groups determined not by training, but by a different variable with three possible values, or 'levels', one for place of publication: newspaper, website, graphic novel. We can then see if there is a difference in performance according to whether the material analysed is drawn from one of these sources rather than the other. We keep the other details of the experiment the same, with the exception that we now only have one body of annotators, which we will presume to be comparable. This is then suitable for a one-way (we have one independent variable: place of publication), independent measures ANOVA.

The first step is to set out the data in a form that it can be read by a software package. This is generally straightforward: most packages that perform one-way AN-OVAs simply accept a spreadsheet with the values that were measured organised in columns, one column for each of the three values of the independent variable. Spreadsheets for this can be found by searching online. Performing the ANOVA test then requires that the columns be identified and then the program can run.

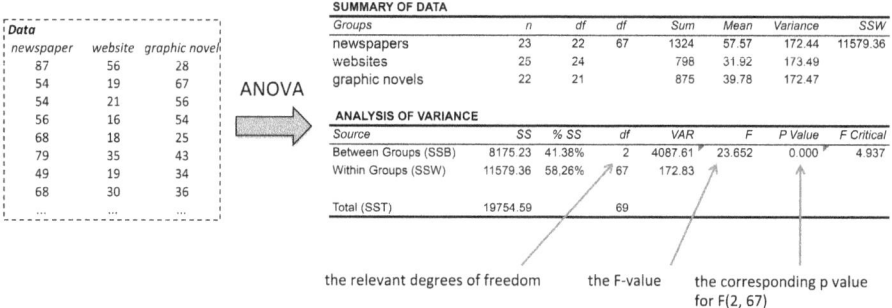

Fig. 6.6. Illustrative output of a one-way independent samples ANOVA performed with respect to three groups

Illustrative output is shown in Figure 6.6. An extract from the table containing this (made up) data is shown on the left of the figure. The tables on the right were produced automatically and contain many of the values we discussed above (sums of squares, etc.). The top table labelled 'summary' gives the descriptive statistics of the data, including the size of each group (n) and the calculated degrees of freedom for each group (one less than the group size); we also find here sums, means and variance, which is useful for checking that your data is not strangely behaved—e.g., exhibiting extreme variances. One of the conditions for applying ANOVA is that the variances across the groups are approximately equal, which is the case here and so we can have some faith in the result reported. The most important values are then the ones picked out in the figure by arrows, which show us that the difference in means across the three groups is, indeed, statistically significant (because the F-value is much greater than the F 'critical value', giving a very low value for p, the probability of mistakenly rejecting the null hypothesis). This result would then support an argument that the kind of pages analysed has a significant effect on the accuracy of any classifications produced.

Performing ANOVAs opens the door to a host of powerful empirical methods. We can consider, for example, more complex scenarios involving 'two-way' ANOVA where there is more than one independent variable in play at once. For example, we could then *combine* the original trained/untrained variable with the source of publication variable to explore both at once. Or we might examine whether the duration of shots in films varies systematically according to genre and 'shot scale' such as close-up, long shot, etc. Here the shot duration is taken as the dependent variable, and the genre categories and shot scales as two independent variables. Then, when reading the results of such a two-way ANOVA, we would generally first be told whether there has been a *main effect*: this reports whether (or not) a statistically significant difference between the means of the identified groups under consideration has been found.

Post hoc tests would subsequently characterise just which values of which variables have effects on the measured dependent variable. Such tests can also reveal *interaction* effects between the variables. Again, when such results are reported, the kind of post hoc test used is given together with any levels of statistical significance reached by the *combinations* of independent variable values. For example, we might be told that a particular genre category (e.g., 'comedy') and a particular shot scale (e.g., 'close-up') show an interaction (i.e., exhibits differences with the other values of the variables that are statistically significant) whereas another (e.g., 'thriller' and 'long shot') does not.

These are, again, standard methods offered by (now slightly more sophisticated) statistical software packages. Space precludes us going into more detail at this juncture, but much of the background you would need to take the next steps yourself are now behind us. The fact that such tests will run through all combinations of your selected variables to search for significant interactions should suggest just how useful and powerful these techniques can be. Several introductory statistics textbooks, such as Lowie and Seton (2013: especially Chapter 6 and the practical units) and Woodrow (2014: 73–84), provide further details and there are several online tutorials.

6.9 Summary: many types of tests and possibilities

We have now discussed some of the reasons why performing statistical tests is a Good Idea when carrying out empirical multimodal research and shown how some of the most frequent tests that are required in such research are calculated. The essential notions underlying the operation of these tests has also been sketched and, as should now be clear, there is generally little difficulty involved. The most time-intensive component of the work is making sure that the experimental designs tested are adequate and actually allow access to the phenomena under study. This is something that should be done in any case, however, and is not an overhead of employing statistics. It is just that the statistics makes it more visible when perhaps less than appropriate analytic decisions have been taken.

The actual calculation of the various measures is generally straightforward, but should not be done by hand. Gaining the necessary familiarity with some spreadsheet programs or basic statistical tools is more than recommended. In addition, any time spent gaining basic competence in a more straightforward computer programming language for manipulating data and placing it in the form required to support statistical testing will repay itself many times over.

In the space available here we have, of course, only been able to scratch the surface of what can be done using statistical methods. For those who would like to get a more thorough understanding of the basics set out in a manner that requires almost no mathematical background, we can recommend Field (2016) for both its excellent intelligibility and its novelty of form—an introduction to a wealth of basic statistic concepts and methods is embedded within a graphic novel. Also very approachable, with many detailed examples showing how to calculate each of the tests and methods introduced, is Richard Lowry's online introductory textbook (Lowry 1998-2015). Somewhat more technically, Woodrow (2014) provides a no-nonsense description of how to calculate and report on a broad range of tests relevant for empirical research. Lowie and Seton (2013) also set out a good overview of statistical methods and, in particular, the reasons for using some rather than others. Finally, Krippendorff (2004) offers a detailed introduction to many aspects important for empirical research in general, while Gwet (2014) works through methods for inter-rater reliability studies in particular.

Even without exploring further, however, the information given in this chapter should already permit the most commonly occurring empirical research questions in multimodality to be addressed and potential problems identified. Nevertheless, it will always be beneficial to discuss your experimental design with someone who has a background in running experiments and carrying out statistical evaluations. Maintaining good lines of communication with disciplines where this is the usual methodology employed is thus quite valuable and should always be considered when putting together a research team. Such team members not only bring with them expertise in calculating statistics but also, and just as importantly, experience in designing experiments in ways that are more likely to lead you to interesting results. Coming up with the appropriate designs is by no means straightforward and needs to be thought through very carefully in advance. Here, as in many areas, experience plays an important role and can save significant time and effort.

7 Multimodal navigator: how to plan your multimodal research

Orientation

In this chapter we change gears to practical analysis. We have now seen an introduction to general areas of interest for multimodality in Chapter 1, overviews of approaches and methods in Chapters 2, 5 and 6 and basic theoretical foundations in Chapters 3 and 4. Now it is time to see how to apply theory and method in the service of analysing diverse examples of multimodality in action. The specific focus of this chapter is to bridge between the theoretical categories introduced above and particular analytic decisions of just how to approach and then deal with any particular multimodal phenomenon of interest. We do this by setting out a 'navigator' that directs attention to just those aspects of a multimodal artefact or performance that will aid in its analysis. The chapters following then present use cases and example analyses that our navigation guide points to. It might then be useful to spring backwards and forwards between the example use cases and the account of diverse media in Chapter 3 as necessary. The navigator in this chapter will make it clear just how these are related so that you can gain a hands-on feel for just what 'work' the theoretical categories can do for you.

7.1 Starting analysis

When you are first confronted with the data that you want to analyse, it can be a daunting experience. It is common for complexities that you did not consider beforehand to suddenly arise and, worse, to keep arising as you work your way into the kinds of analytic categories and methods that you think will be necessary. Much of this difficulty can be coped with by systematically decomposing the particular 'places' where multimodal meaning making is going on 'within' the overall situation or entity that you are looking at. There are various ways to do this: some are helpful, others may raise more problems than they solve—so it is to be recommended very strongly that this is approached with a clear methodology that can guide you to focusing on just those facets that are the targets of analysis.

We have set out some aspects of a theoretical framework that can help make sense of what is going on in previous chapters, and particularly in Chapters 3 and 4. In this chapter, in contrast, we approach the task from the other end—not from general theory but from concrete situations, artefacts or performances that you may wish to address. This will, in the last resort, always draw on the theoretical foundations, but we recast the task here from the 'use' perspective so that it can be applied as needed for concrete cases. And, in the immortal words of Douglas Adams' *Hitchhiker's Guide to the Galaxy*, the most important thing of all is: "Don't panic!".

To show how to go about performing an analysis, or rather to get into the position where analyses can be done, we will start right away with a relatively complex and realistic example. This is useful because too simple examples will leave you floundering when confronted with the 'real thing'. The method we describe is completely gen-

eral and having the confidence to apply it to situations no matter how complex is perhaps one of the most important skills to be learnt and practised when becoming competent in multimodal analysis.

Example:
classroom
situation

Here, then, is the first example—some very different examples from different areas
will follow below and in considerably more detail in the use cases of Part III of the
book. Imagine you are wanting to analyse just how a teacher in front of a class uses
a blackboard and miscellaneous supporting material (such as maps) to explain some
geopolitical state of affairs. The precise details of the content will not be our focus
here; we are concerned solely with how to go about organising the 'data' that such a
situation might give rise to. Once the organisation of the data is clear, then it is considerably easier to go about performing the particular analyses that will be required.

As described in Chapter 5, for any analysis it is necessary to get hold of the 'data'
that is going to serve as the basis for analysis. The form of data adopted will, of course,
have considerable consequences for what can be analysed. Unless what is being analysed exists in a completely digital form already (e.g., an electronic newspaper, a social network message, or a 3D computer simulation in a computer game), the object
of analysis will need to be 'recorded' or 'captured'. The more that the recording medium differs in its properties from the properties of the situation being recorded, then
naturally the greater the degree of 'loss'. There will be some things that will just not
be available from the data. This is always a problem—but this is also unavoidable:
so the decisions about data need to be made in a way that is aware of potential loss.
Clearly, if you want to study intonation patterns and only have a written transcript of
the words that were said in a situation, then there are going to be problems. We will
see a considerable variety of ways of dealing with this situation in our use cases below; the questions of 'transcription' that were discussed in Chapter 5 are all equally
relevant here too.

For our current example, then, lets assume that we have a video taken from the
back of the class showing the pupils, the teacher and the blackboard being used. A still
from this video might look like the example shown in Figure 7.1. This gives us access to
the teacher's talk, the teacher's gestures and actions when writing on the blackboard,
possible interactions with the children, as well as possible interactions with other elements in the scene. This is already a lot of material and so in any real study it would
be beneficial to have decided in advance just what aspects of the situation are going
to be the main focus of the study. Clearly, if the focus is on how the teacher interacts
with the pupils, then this might require a slightly different emphasis than a question
concerning how the teacher uses the blackboard—although these may well then come
to overlap. Having a good idea (or as good as is possible beforehand) of what is to be
examined is also useful to avoid making mistakes, or less fortunate decisions, concerning the data collection itself—such as, for example, putting the video camera in
the wrong place so that important interactions or components are not captured. This
can itself become complex in its own right if participants or the location are moving!
In the most general case this comes to overlap with 'blocking' out positions and cam-

eras as commonly undertaken when producing films; this degree of sophistication will rarely be required for basic data recording however.

Fig. 7.1. A classroom situation

For the classroom situation as recorded, and indeed for any situation, there are two components that we can usefully distinguish prior to further analysis. The first relates back to the question of **genre** (→ §4.3.1 [p. 129]), i.e., which activities are being performed and which social actors those activities define (cf. van Leeuwen 2005*a*, 2008); the second relates back to the question of the **medium** that is being used for those activities. As we set out in detail in Chapter 3, 'medium' is itself a complex affair and considering it carefully will save us a lot of work (and potential confusion) later in the process. The medium being employed in a communicative situation provides the (possibly nested or even overlapping) *virtual canvases* (→ §3.2.5 [p. 87]) that the activities being addressed are being performed 'on'. Sorting out the media and its canvases is the first key to being able to bring organisation even to very complex situations and so this is the place to start.

Genre and medium

7.1.1 Media and their canvases

The approach to media and canvases to be followed when preparing for analysis is to go from the most inclusive to the most specific. It is necessary to take the most inclusive because this establishes the most 'powerful' or expressive virtual canvas that will be available. All further canvases used within the overarching situation will be limited by their embedding within the most general canvas available. We can see this most straightforwardly by, for example, turning the lights off in the classroom when it is dark outside—since the canvas then no longer supports simple visual access, all canvases requiring visual access that do not bring their own light sources with them can no longer be carried. So this is the sense of 'most inclusive' intended: any embedded canvas will be dependent on the most inclusive canvas provided.

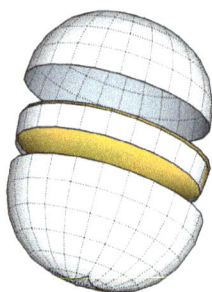

Fig. 7.2. A 'perceptual slice' through a more inclusive canvas

Canvases,
sub-
canvases
and slices   A useful way of thinking of this is to see the most inclusive canvas as a 'space' of possibilities for perception and action. Any canvases employed within that situation are then 'slices' through the overall space. This is suggested graphically in Figure 7.2. Here the overall sphere or ball represents the complete space of possibilities that some canvas provides. Then, any subcanvas used within that situation may select some restricted range of possibilities within that. As many slices (along any angle and of any 'thickness') though the overall space may be taken as there are relevant subcanvases being employed in the situation. This may also 'recurse' in that some particular subcanvas (such as a flat 2D form that is visually accessible and writeable) may itself support a range of further forms within it. Pulling apart these configurations of canvases may take some practice, but will always lead to a better understanding of just what communicative possibilities might be relevant.

Within our classroom situation, therefore, we begin with addressing the situation as a whole: this gives us the most inclusive space of perceptual and action-related possibilities corresponding to the ball in Figure 7.2. This space is three-dimensional and unfolds in real time. It affords free movement and placement of objects (including semiotic objects) and allows free line of sight and of hearing within it. At the most general level, agents within this space must create their own communicative contributions in order to bring that situation into being. The medium is then, in the terms introduced in Chapter 3, mutable ergodic, dynamic and involves participants *performing* communicative activities of various kinds. In principle, any communicative activities that fit within the canvas may be carried out.

However, as with all such multiple agent situations, there may be (and in this case are) constraints of a social nature on who in the situation gets to perform which roles. The classroom situation is typically asymmetric, with the teacher (even when facilitating) guiding and driving the communicative situation forward. Moreover, situations also never exist in a vacuum and so further constraints will apply from the relative cultural and social context in which the classroom situation is itself embedded: particular room layouts, particular architectural details, particular relative positions of the teacher and the pupils and so on may then work within such configurations and be given by those configurations (Laihonen and Szabó 2017). We discussed this in Chapter 3 using Kress and van Leeuwen's (2001) example of the child's bedroom (→ §3.2.7 [p. 97]); in the present case, one might see the entire classroom as expressing certain pedagogical positions—this then may motivate (to a greater or lesser degree) the bundle of properties making up the canvas.

The first step when considering multimodal research of any kind is one of fixing the analytic focus. This picks out just what is going to be examined and places it against particular *contexts* that were involved in the object of study's creation. There is no one fixed 'correct' answer to this decision: it has to be made with respect to the particular research questions and interests being pursued. Picking a focus can also be usefully related to what in several other accounts has been addressed in terms of the *higher-level actions* that are being performed (e.g., Norris 2011; Bucher 2011) as well as to potential *tiers of materiality* (Pirini 2016). Here we bind all of these aspects together in terms of the canvases within which actions are unfolding. This orients us quite explicitly to the affordances available for meaning-making and also opens up the framework to apply across the board to all multimodal 'situations', not just those involving interaction.

For the present, we will focus within the classroom and consider how this needs to be decomposed further. And, at this point, it is useful to start bringing considerations of genre to bear. Genre is one way of characterising the various kinds of communicative activities that might be performed in some situation. Genres relate 'upward' to social context—for example, a range of 'teaching' activities—and also 'downward' to the various communicative forms by means of which those activities get to be performed. Here we might profitably draw upon existing work on classroom discourse: how do teachers structure their lessons, what kinds of strategies are drawn upon and so on. Note, of course, that 'communicating some content' is just one of a broad array of tasks that might occur. Others range from making sure that inattentive pupils are either attentive or, at least, not too actively disruptive, on the one hand, through to explaining particular materials that may be made relevant for the class, such as something that has been written on the blackboard, on the other.

Each of the activities undertaken may draw on its own slice through the perceptual possibilities provided by the canvas. Directing the attention of pupils may draw on spoken language (including pauses, intonation, loudness, etc.), gesture, body position and orientation, gaze and facial expression. Explaining something on the blackboard will similarly draw on all of these plus a rather specific range of gestures typically employed for indicating particular places and areas on the blackboard, produced at the same time as 'deictic' language expressions ('here', 'in this part', 'there', etc.) that synchronise information producing verbally and information available visually.

In this latter case, the perceptual slice employed brings some more interesting properties—in particular, the visual representations employed are not only mutable ergodic, since the teacher will generally write them during the class, but also 2D and *non-transient* (→ §3.3.1 [p. 104]), since they remain on the blackboard until erased. These two distinct slices through the overall space are depicted graphically in Figure 7.3; the left-hand picture highlights the interaction with pupils, the right-hand picture highlights the use of the blackboard. These are, of course, only suggestive—the main point is the different activities and combinations of media characteristics that the two slices (and their carrying canvases) create.

Fig. 7.3. Two illustrative canvas slices though the classroom situation: on the left, we see the face-to-face interaction between the teacher and some pupils; on the right, we see the interaction of the teacher with the blackboard

We could readily imagine several further such canvases, each of which would then support its own area of multimodal investigation. For example, the teacher might interact with the map on the right-hand side of the classroom: this is similar to the blackboard situation but also different in that the map is a 2D, static, observer-based, micro-ergodic canvas of its own with its own conventions of representation and depiction. Understanding this canvas requires experience of the medium and the purposes for which it is used and would support a range of studies on its own—extensions and related artefacts will be seen in our use cases. Important for the present example is the way in which this medium is embedded within the classroom activities: access to the medium will again, typically, be managed by the spoken language and gestures of the teacher, which thus act as a kind of 'bridge' into this particular embedded subcanvas. This particular communicative task would then have its own combination of linguistic and gestural indicators directing attention to the map.

Alternatively, as a further example, if the pupils are also using textbooks, an entire constellation of possible slices is created. First, there is the interaction of each pupil with the pages of the textbook they are reading; this interaction will again be 2D, static, observer-based, micro-ergodic (like the map) for single pages and potentially 2D, static, observer-based, immutable ergodic if they have to search and navigate in the book themselves: here a range of literature on reading comprehension and the value of particular textbook designs would be relevant. Ideally the composition of the pages of the book would support the learner's re-creation of composition for their own purposes. Second, the teacher may include reference to the textbook as a further resource for sharing and making meanings: in this case, the interaction is similar to that found within the canvas including the map but without (usually) supporting gestural attention guidance—here (unless the teacher goes physically to the pupil and book in question) verbal cues will be required. Third, there is the internal design of the book itself: this subcanvas relies exclusively on the possibilities of page composition, and so focuses at a still finer granularity at the semiotic artefacts in question.

Each subcanvas slice will invite the application of knowledge that we bring to the situation from other contexts and studies, including knowledge of how semiotic modes may operate in each specific case, as well as setting up tasks for analysis that we need to perform to push that knowledge further.

i

Of course in today's classrooms the traditional textbook may well be pushed out of the attention of pupils by their smartphones, which opens up a further range of embedded canvases. To the extent that these activities—although subcanvases of the entire situation—are not embedded within the communicative activities under study, they might reasonably be ruled out of consideration. Whenever these are incorporated into the intended teaching situation—for example, by running searches for material relevant to the class—then they might well need to be included as well.

Figure 7.4 sets out in graphical form more of the embedded canvases that might be considered relevant for the classroom teaching situation. The diagram is to be read as if we could 'lift' each embedded canvas out of its embedding within the 'base canvas' that supports the activity as a whole. Suggestive labels are offered for each of the slices illustrated; we will build on these further below. A range of further possible slices are not included here on the assumption that it is the teaching situation that is in focus. As is the case with the smartphone, however, there is nothing to stop such slices being recruited for the teaching activity as well, and then they would indeed become relevant.

Once we have decomposed a situation of interest along the lines of embedded canvases as illustrated for the teaching situation here, then we move on to pursuing analyses. The reason for having the decomposition of canvases is that we can make more explicit just what kinds of analysis are going to be relevant. In general, this will lead us on directly to the use cases of the following chapters. In the next section, we sketch how this works, again focusing on the classroom situation.

7.1.2 From canvases to analyses

We have suggested that clarifying the media involved in a communicative situation can function as a solid methodological step for disentangling the often very complex range of communicative behaviours that might be present. This is to be seen as a launching point for subsequent analysis where the specific properties of the media identified can be drawn upon and, indeed, refined further by virtue of the analysis

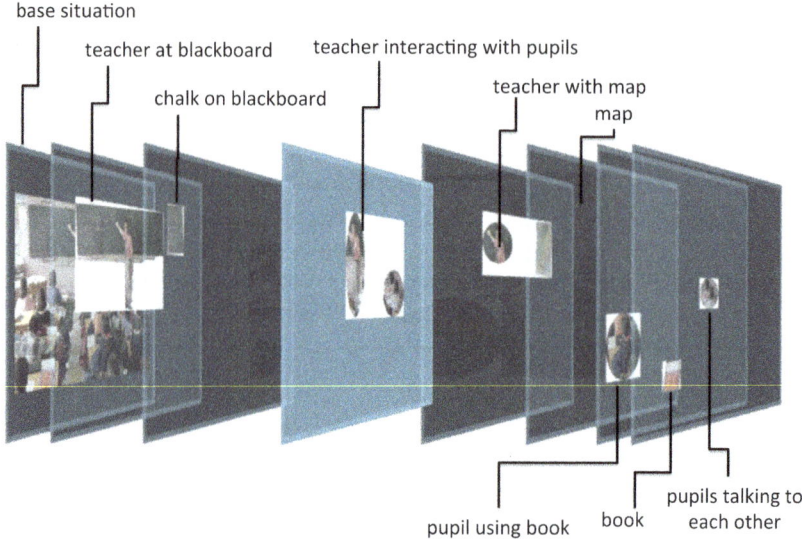

Fig. 7.4. A selection of potentially relevant canvases taken as slices with respect to the base classroom situation

performed. Several forms of expression already have very well developed schemes of description—such as, for example, the close analysis of verbal forms offered by linguistics or the classification and separation of styles of painting in art history; many other forms of expression find themselves in less well developed positions, and improving on this situation is one of the motivations for carrying out multimodal analyses at all. Remaining open to the kinds of resources that are employed for making meanings within a situation is one important further reason that we take the trouble to carefully separate out the various canvases and media involved.

The next step in our methodological process is to further refine the slices and their subcanvases so as to be more clear about just what varieties of meaning-making are in play. This then leads us towards the kinds of detailed multimodal analyses that are necessary to investigate just how communication is working in those situations. We suggest this next step graphically in Figure 7.5, where we begin to use more actively the categories and terminologies that we introduced in Chapter 3. Many of the media we see employed in this situation are mutable—i.e., they are created by the interactants themselves rather than being external to the situation—and transient—i.e., once they have occurred they are no longer accessible.

As we suggested in the introductory chapters and will see in more detail with respect to particular case studies in the analyses to follow, canvases of this kind have particular properties that leave traces on any forms of semiotic resources that are employed using them. Being clear about these properties of the media therefore puts us in a position to expect certain kinds of semiotic resources rather than others. Not all

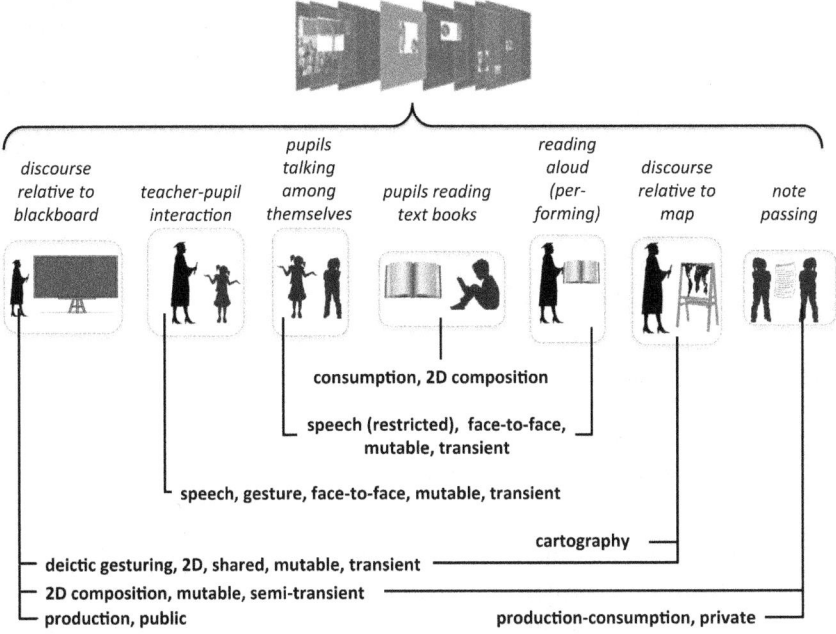

Fig. 7.5. Characterising subcanvases in terms of the media that they support

of the canvases involved here are mutable, however. Both the textbooks that may be read by pupils and the map that is used for discussion and background are immutable and rely on 2D composition in order to be able to make their own contributions to the meanings being created in the situation as a whole. Moreover, some of the canvases employed are 'public', in that the entire group can access them (at least for reading)— as is the case with what is written on the blackboard or reproduced in the map—others are 'private': as in, for example, the case of any notes passed surreptitiously among the pupils (assuming that this situation has ruled out the use of texting for the time being!).

Finally, it needs to be understood that each of these subslices can themselves usually be further segmented, each such segmentation revealing the canvases that contribute to their immediate configurations. It is at this stage that we usually begin to make contact with work undertaken in other disciplines concerning the organisations and meanings carried by specific types of semiotic artefacts. Again, we will see this carried out in detail in the use cases that follow—but for current purposes we illustrate this with respect to just one of the slices shown in Figure 7.5: that of the teaching using a displayed map as part of the teaching situation. This is already potentially complex, as we can see illustrated graphically in Figure 7.6.

Although complex, it is also at this level of detail that we begin to uncover strategies and semiotic resources that can be re-used across many different situ-

Strategies and semiotic resources

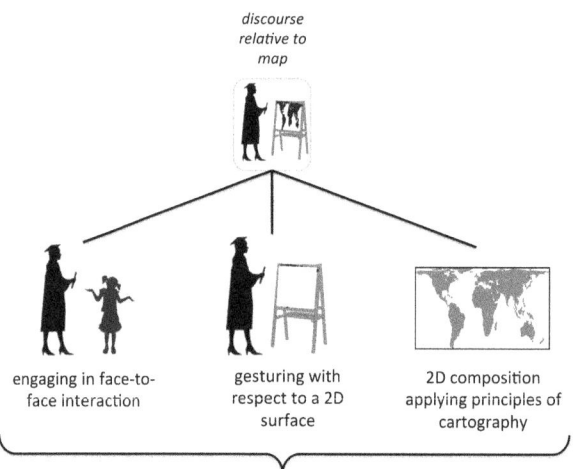

ations. For example, as shown in the figure, we can characterise teaching with the map in terms of the kind of language that is used for describing and directing mutual attention to some publicly displayed 2D artefact, in terms of the kinds of gestures that may be used to synchronise verbally constructed meanings with the spatial information present in the 2D artefact, and the conventions of the 2D composition of a map itself. Each of these will involve communicative strategies that will largely re-occur whenever we have the task of direction attention in this fashion. There will, therefore, begin to be existing work that one can beneficially draw on; moreover, the results that one achieves concerning how this particular communicative situation operates begin themselves to be relevant for others. In this way, the study of multimodality moves beyond the description of single cases and can target more general and more widely applicable results.

With the kind of characterisation of canvases and the properties of the media in which they are employed that has now been set out, we can launch into the particular methods and descriptive apparatus appropriate. This is then the 'navigation' function that we referred to above. The specific media uncovered support their own selection of semiotic resources, and those themselves involve their own conventions and styles of presentation. The individual sections listed under the use cases in subsequent chapters are indexed in precisely this fashion. For example, taking up again what may be written on the blackboard in our running example, only certain kinds of semiotic artefacts can be produced in this medium—they cannot show films, for example (although a film might be shown *on* a blackboard, or better, a whiteboard, but this is a different medium: we return to this in Section 7.1.4 below). In our descriptions to follow we will highlight particular forms of expression and make explicit in what media they occur.

We will also make explicit just what particular media *add* or *subtract* to other media that they support. For example, anything that can be drawn or written can in principle also be produced on a blackboard, although practical constraints might limit this. If, however, there is the additional constraint that the writing and drawing be done in 'real-time', e.g., performed accompanying the spoken language of the teacher as a form of illustration, then this naturally restricts what might be presented (due to time constraints), while also *extending* what can be drawn because of the synchronised verbal accompaniment. Thus, while a 'stand-alone' drawing of a house might need to be sufficiently accurate as to function iconically by resemblance (→ §2.5.2 [p. 59]), a simple circle or other squiggle may serve when accompanied by a verbally given label: 'here's the house'. There are certain communicative situations that rely extensively on this joint co-construction of meaning with online drawing and verbal accompaniment, such as, for example, architectural planning meetings where even non-physical, abstract information may readily be drawn—e.g., 'here's where the flow of people through the entrance hall will go' accompanying a roughly drawn path across a building plan.

A further possibility offered by the blackboard is that it readily supports *changes* in what has been written or drawn: this is different to, for example, a printed book, a map, or a PowerPoint presentation. The teacher might then use this affordance of the medium to further effect, and not just for correcting mistakes. For this reason, we referred to this medium above as both mutable and nontransient—that is, things are created and stay there until they are changed.

In Section 7.2, we set this out at the next level of detail, showing how the kinds of media distinctions introduced can be used as part of a triangulation process leading to the selection of appropriate analytic methods and previous work. The use cases that follow in subsequent chapters then lead on directly from the navigational directions given here.

7.1.3 Selecting a suitable analytic focus

Before proceeding, however, let us briefly emphasise again that not all analyses demand exhaustive slicing of their communicative situations in order to get going. This was described above in terms of selecting an analytic focus. The multimodal analyst needs to be able to follow how meaning-making is operating wherever it happens— and for this the quite detailed slicing we have illustrated so far is a necessary step—but the analyst also needs to be able to make sensible decisions about where to stop and where to begin: and this depends on the research questions being pursued.

For example, let us consider the at first glance simple case of analysing an illus- Context trated book. This will emphasise further the point we made above that the same steps in method must always be applied, even if we are looking at static artefacts such as books or other documents, images and so on. While it might seem obvious that we look at the book itself and presumably what is presented on its pages, this is still a decision. Any semiotic artefact, the book included, is the consequence of a web of enveloping contexts, any of which may have played significant roles in the production and appearance of the final 'object of study'. We might then, in the terms introduced above, have a more 'basic' situation of producing printed works and break this complex canvas of activities down progressively, only reaching 'the book' within a host of other embedded and overlapping canvases. Some approaches, such as ethnography (→ §5.2.2 [p. 144]), place particular emphasis on 'contextual embedding' of this kind—here we see this as an inherent component of method and employ the same methodological steps at all levels.

Discourse interpretation

Moreover, regardless of whether an artefact is static, or a performance is live and unscripted, or a participant is immersed within a designed environment, there is always the constellation of an interpreter who is interacting with these situations and engaging in abductive discourse hypotheses concerning their 'meaning' or signification (→ §4.1.1 [p. 116]). We will see this in practice in each of the use case scenarios we work through in subsequent chapters.

This kind of approach also opens up the field of investigation to encourage general questions of media history and structures of production. Here, for the book example, one might decide to fix analytic attention instead on the interaction between illustrator and text author (if there had been one), or on how the publisher's production department had imposed particular limitations on the size and quality of images and how these then played a role in the selection, or in the production, of the illustrating material used. Any of these interactions may have occurred in real-time, in meetings, in discussions over the phone; or may have been 'legislated' in advance, in company policies (which would themselves have emerged over time out of other meetings, discussions, written reports and so on). Any of the widening circles around the production of any specific artefact or performance may have left traces in the particular, concrete artefact or performance being analysed—and so, consequently, may also need to be allowed to play a role in the explanations given for appearance and design.

Resemiotisation or remediation

In some cases the effects of such contextualisations might be thought to be more imminent and consequential for individual analysis than others—but the flexibility to *decide* what the analytic focus is should always be remembered, at least in principle. Different methods and approaches may be more useful for different selections because they may, in turn, play out with different media. One common term for this is **resemiotisation** (cf., e.g., Iedema 2001, 2003). The process is typical of almost all complex semiotic productions, from buildings to operas, from museum installations to films. Concentrating more on the media involved, Bolter and Grusin (2000) term this *remediation*, although their net is cast somewhat less widely than ours is here.

Nevertheless, it is always still possible to focus attention on the particular type of artefact or performance that raised our interest. Returning, for example, to a focus just on our illustrated book, the medium of the printed book only supports particular kinds of possible expressive resources, and so we could sensibly restrict attention to just these when carrying out an investigation. This might not allow us to explain all the choices made when putting together such an artefact, even if we may not be able to offer detailed characterisations of just why something finds its way into the book. We will be in a better position, however, to see how any particular ensemble of distinct modes is contributing to a whole.

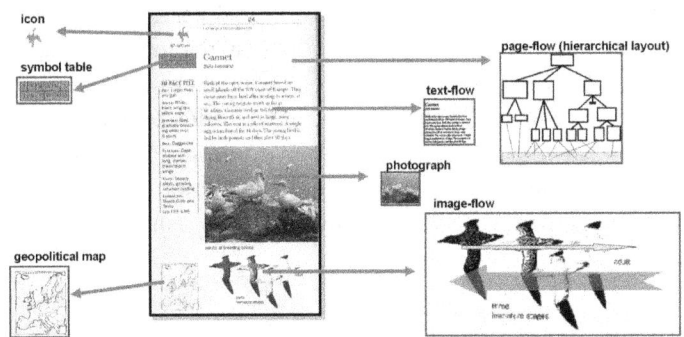

Fig. 7.7. Some sub-canvases claimed by the contributing semiotic modes in a printed page

Consider as an example the page shown in Figure 7.7, describing a page from a bird field guide as discussed at considerable length in Bateman (2008). Here we can see that it is possible to pull out particular 'subcanvases' that are used by rather different modes—flowing text, page layout, diagrams of various kinds, photographs and so on. These may all co-exist because of the depictive medium of the printed page. But, and this is crucial, they can also be separated and considered in the light of their own sociohistorical development. The meanings they are called upon to make and just how they are made are each a matter of practical, empirical and often historical investigation. Being able to separate them, just as we saw with the diverse activities contributing to the classroom situation described above, is of enormous importance for simplifying and structuring the activity of multimodal analysis; several more examples of this kind of analysis are given in our use cases.

We can also, however, open up the focus to take in design decisions made in the planning stages of the book, then the methods and approaches relevant there are going to be those concerned with interaction, possibly also in relation to discussions of how the book should look. Pulling apart the individual contributions in a multimodal analysis may then be complex, which is another reason that these decisions should be made carefully: bad analytic decisions at this point can leave a hotchpotch of phenomena that do not naturally fall together under *any* scheme of analysis.

More specifically, then, each selection of analytic focus brings to light particular 'canvases' within which, or on which, the semiotic artefact or performance plays out. Identifying those canvases provides crucial input for the next step of considering particular analytic schemes for the media products produced with them, as the next section will illustrate. The analyses of individual stages leading to an artefact (production) or away from that artefact once created (distribution, consumption) are then generally concerned with different kinds of multimodal ensembles: there is no straightforward relationship holding all of these activities together.

This is one reason why we do not suggest following Kress and van Leeuwen's (2001: 20–21) inclusion of 'design' simply as a stratal component of the analysis of any specific artefact or performance. The performance of a piece of music and the score of that music are not related in a relation of signified to signifier; neither does a building 'signify' its architectural blueprint. Semiotic artefacts that play the role of 'designs' or 'blueprints' are best seen as objects of analysis in their own right, employing their own media and conventions and embedded in their own circles of development. Their status as 'designs' lies in their particular purpose—i.e., to constrain and correlate other behaviours in particular respects in order that further objects or performances be created. We will say more about this particular kind of configuration of semiotic activities in subsequent chapters.

7.1.4 Media convergence revisited

The decomposition of media and canvases that we have seen is necessary simply because it is natural for communicative situations to be made up of 'smaller', or 'component', communicative situations that draw on only aspects of the full possibilities available. These form 'slices' through more complex spaces as suggested in Figure 7.2. There is, however, another way for canvases to be contained within other, more inclusive, communicative situations. This is commonly discussed nowadays in terms of **media convergence**—i.e., the ability of certain media to carry or show semiotic artefacts and performances that come from *other* media (→ §1.1 [p. 14]). Media convergence is generally attributed to the 'digital' turn in media of all kinds and to the fact that screens of various devices can now 'show' both traditional print media forms, such as newspapers, and more time-based offerings, such as films, videos and so on.

As a consequence, some of the artefacts that you will want to analyse may have this rather different type of canvas complexity. It is just as important to pull these media contributions apart appropriately as was suggested above for the 'naturally' complex communicative situations. Failure to make the necessary differences can leave both the process of analysis and any results that might be obtained confused. Actually, the idea of media 'showing' or 'suggesting' other media is not at all new, although it has received a new emphasis or centrality due to the advent of digital media. But, as Newall (2003) observes:

> "Pictures regularly depict other pictures. Paintings or drawings of galleries, studios, and other interiors, for instance, often depict pictures hanging on walls or propped on easels." (Newall 2003: 381)

Whether we are dealing with an iPad showing a newspaper or a painting showing another painting, therefore, we are in the same broad area and need to consider carefully the implications for the practice of analysis.

As set out in Chapter 4, we will refer to media that are being employed in such situations as being **depictive**. Much is currently made of such complex situations and it is indeed important to be aware that each of these 're-mediations' may introduce (or subtract) possibilities that have consequences for their multimodal analysis. In general, however, the principles at work overlap with those we have already seen. In particular, when moving across media, a depicted medium will only be able to make use of the canvas possibilities supported by its depicting medium. So, again, in our method of analysis we need to see such situations as moving from more inclusive communicative situations to less inclusive situations, picking apart the canvases involved at each stage, but this time being aware that *further, more restricted slices* may be being enforced by the depicting medium employed.

The important boundary function of depiction which we have described in Chapter 4 in fact applies across the board, even when depictions begin to construct correspondences between more distantly related communicative forms. For example, the kind of visual styles found in comics and many animations 'depict' the semiotic modes bundled together around facial expression and gesture; here it is even more evident that that depiction need never be exact or 'photographic'—all that is required is that sufficient distinctions are being drawn in the available material to allow application of the other semiotic modes being evoked as interpretative schemes. These are *not* seen, however, as cases of codes becoming free of their materials as suggested in some approaches to semiotic modes, but rather relies again upon structured mappings across semiotic modalities, although here more in the 'lower' semiotic strata near to form and material ($\rightarrow$ §4.1 [p. 112]).

This also re-emphasises that depiction should not be seen as involving 'resemblance' in a shallow sense because we are typically concerned with *structural* correspondences, i.e., drawing again on Peirce, iconic relations are not just images (cf. Peirce 1931-1958: §§2.276–2.277). It can, however, probably be assumed that the more 'abstract' the correspondence, the more effort a community of users will need to invest in the depiction for it to take root. Just which semiotic modes are to be considered as evoked may then also be strongly conventionalised.

Other cases of medium depiction are useful to discuss. One relatively simple class includes *notations*. For example, the use of braille as a printed form of representation for written language involves a medium that is different to that of regular print. However, the distinctions that are drawn in the material of that medium are sufficient to cover the distinctions drawn in at least the written alphabetic form of verbal language and so allow a straightforward transference to another material carrier with usefully different affordances. Whether or not any further semiotic modes have grown with respect to this medium would require empirical investigation—involving established practice and communities of users. Certain correlates, for example, of typographic

Depictive media

Notations

layout could be expected to serve a function, just as positions and divisions operate within the 'page' space visually, similar segmentations can function by means of tactile perception. There is in this case, then, no reason not to consider application of many of the semiotic modes related to the use of printed language to the medium of braille publications.

Other examples of notation would be the use of a light source for Morse code or the representation of music that occurs in sheet music—again, the question of whether additional semiotic modes specific to these medial forms have emerged is always an empirical question. For Morse code, this appears unlikely—for sheet music, almost definitely. We will return to some more interesting cases of 'notations' in our use cases below.

7.2 Undertaking multimodal investigations of phenomena

In this section, we build an explicit bridge between the media of any investigation and our dimensions for organising the use cases. The idea here is that any communicative situation to be investigated needs to have its canvas categorised according to the steps given in the decision procedure that we set out below. The divisions there then link into our use case areas so that analysis can be compared and contrasted with the analyses of similar kinds of communicative situations. We also then give more of an indication of approaches that have attempted to provide analyses of similar communicative situations either from within multimodality research directly or from other research traditions.

The decisions that we run through should be relatively simple to make for any communicative situation under consideration. The classification is not at this stage intended to be exhaustive and there are several other combinations that we will not explicitly differentiate at this point. Some of these will be discussed when we move to the use cases in detail, however. Moreover, the decisions need to be made for each communicative situation that is taken to be at work: that is, if there are embedded subcanvases, then these need to be characterised in exactly the same way as all other canvases. The results of these characterisations then point to how these subcanvases may be approached. Similarly, if there are depicted media, then *their* canvases and subcanvases must similarly be classified. By these means we build up not only a highly differentiated view of the situations that we have to analyse, but also a set of successive bridges into other analyses elsewhere that may have already concerned themselves with such canvases.

The first decision to make is then whether the canvas of the communicative situation at issue intrinsically includes spatial dimensions that can be used for signifying meaning differences. That is to say, is spatial distance, spatial shape, spatial connectedness and so on available for manipulation? This can also be in any number of spatial dimensions, although we are typically only concerned with two- and three-dimensional canvases. One-dimensional canvases are possible but certainly limited. More than three-dimensional canvases may also be possible, but would be difficult for us to deal with. If no spatial extent is available, then we know that the canvas has to be organised intrinsically in time, since otherwise there would be nothing to make meaning from—no 'manipulable' material would exist. Media that only make use of variation in time would include, for example, those supporting music and (monophonic) radio programs.

It is beneficial to distinguish between 'time-based' canvases and the question of dimensionality as it is generally raised with respect to space. For example, music, as we set out in Chapter 2, is a time-based mode of expression, but it makes little sense to talk of this as 'one-dimensional'. What is more the case is that time is *constitutive* for all of the variation that we find being exercised in music. Changes in pitches within harmonic spaces, multiple instruments playing simultaneously, the experiential responses to different kinds of material sounds (as in distinct forms of percussion) and so on—all of these unfold in time without collapsing to a single 'dimension' of variation or possibilities. Without time, and temporal extent, however, there would be no unfolding of this kind. And when a canvas supports time-based variation, then different kinds of perceptual responses become possible.

Alternatively, if there is a possibility of manipulating spatial extent, then the second decision to make is whether intrinsic temporal variations may be employed or not. If temporal variations are available for signifying meanings, then we have dynamic spatial situations; if there is no temporal variation available, then we have static spatial situations. Dynamic spatial situations would include film, dance, spoken face-to-face conversation and many more; static spatial situations might include traditional printed page-based documents (2D) or buildings (3D).

As the classifications suggested so far indicate, there are still substantial variations within the three broader categories discovered: temporal, static spatial and dynamic spatial (or spatiotemporal). We can provide a further breakdown of these situations in a way that will be beneficial when we set out the contrasting use cases as follows. First, we can consider whether the medium is being used *freely* or according to a previous *design*. This will be useful for distinguishing, for example, spontaneous telephone conversations (without visuals) and performances of previously composed music. More subtly, it can also then allow us to distinguish between spontaneously produced dialogue and a scripted performance of a conversation, as in theatre. We consequently label this distinction as that between *scripted* and *unscripted* canvases. And second, we can consider whether the canvas supported interaction or not, where interaction is seen as the ability to influence how the situation unfolds in the sense of mutability introduced in Chapter 3 above. The distinction drawn here is then between *interactive* and *non-interactive* situations.

The interplay between these categories can also be interesting. For example, we might group together listening to music and listening to a radio play. Both are immutable, or non-interactive, since the listener cannot influence how the communicative events unfold, and both are scripted. In contrast, playing an active role in the radio play as a performer invokes a canvas that is scripted but mutable—that is, the individual contributions act as if they influence the unfolding event, even though those changes have been planned out, or designed, in advance. In contrast again, taking part in a live radio discussion would be unscripted and mutable, whereas *listening* to that radio discussion is immutable, because the listener cannot normally influence the course of the discussion (unless it is a phone-in, which changes the canvas again), and unscripted. Each of the situations may bring influences to bear on how communicative resources and signifying practices evolve and play out in specific cases (cf. Martin 1992: 508–546). Figure 7.8 summarises once more the decision procedure and shows the decisions we have just introduced and the various paths through the outcomes of those decisions to arrive at the five use area cases we have just listed.

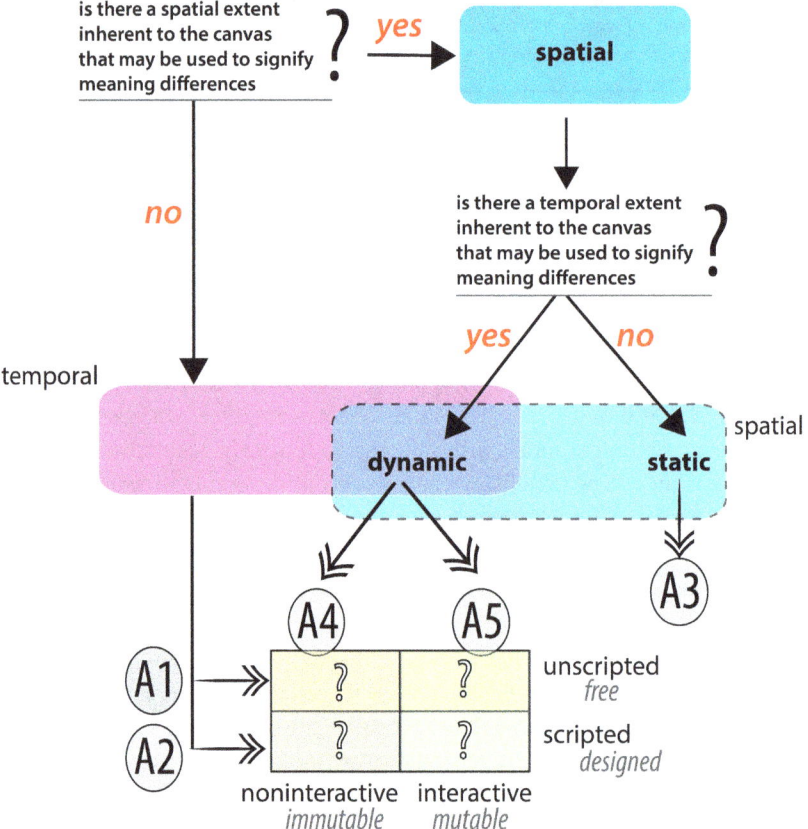

Fig. 7.8. Our proposed decision procedure for working from the canvas of any medium to the particular use case areas that we discuss and illustrate in the final part of the book

We can then group the use cases that we illustrate in Part III of the book to cover the most significant of these various possibilities. The decision procedure we have now sketched can then be used to help find those cases that are most appropriate for any particular investigation or study that we wish to perform. Specifically, at a first level of detail, Part III addresses the following five areas:

- A1: temporal unscripted
- A2: temporal scripted
- A3: spatial static
- A4: spatial dynamic
- A5: spatiotemporal interactive

In each area we will see that there is a rich variety of possibilities for communicative situations that bring along a host of multimodal signifying practices—these practices we then cover by setting out the semiotic modes that are employed with respect to the identified canvases or subcanvases.

7.3 The basic steps in multimodal analysis reviewed

Performing a piece of multimodal analysis can be beneficially divided, as with many such kinds of studies, into the eight steps we summarise below in the info box. We organise these steps in three main analytical phases that we now describe closely to allow them to be followed for any piece of analytic work.

The **first phase** addresses the overall design of the study and considers which materials are going to be relevant and how those materials are going to be acquired. These decisions are all subordinate to a particular set of research goals or questions. Naturally, one can also proceed more exploratively by examining a selected area of multimodal communication that is interesting in its own right—for example, for its complexity or because it has not been sufficiently studied previously. In this case, the forms of analysis to be performed may well emerge from looking at the data. Here the steps of the second phase may turn out to offer intriguing leads to be followed up and turned into research questions.

The **second phase** proper then collects a body of data conforming to the research questions or exploratory interest and defines (or selects) the analytic frameworks that are going to be applied to that data. We suggested above that the best way of gaining control of complex data is by determining the range of media that are employed and considering the communicative activities that are played out using those media in terms of the genres involved. The media identified then provide pointers to potentially applicable analytic frameworks that can tease apart the ways in which meanings are being made and communicated as required for achievement of the genres involved. We provide many examples of just what those pointers lead to in the use cases that follow in the chapters below. In each case we will begin by identifying the media and genre that may be involved and link these directly to frameworks that can be used for analysis.

The **third and last phase** of investigation then examines the results of analysis, drawing out patterns and interconnections that were not evident beforehand. There might also be experimental work where the analyses lead to predictions that can then be tested and evaluated as explained in Chapter 5 above. Alternatively (or in addition), there may be corpus-based work, where sufficient data has been analysed to support the application of more statistical approaches to finding patterns. For all of these, however, the basic steps of having a well analysed body of data, applying the analytic frameworks relevant for media, genres and research questions, are essential preconditions. The third phase then ends with writing up the results obtained in a fashion suitable for your intended audience—be that a term paper for your instructors or a submission to be subjected to international peer review for a journal. There are many articles offering advice on how best to set out the results of a research project: these are all equally relevant and necessary when presenting multimodal results and so should be followed as well.

---

i The steps identified for a complete piece of effective multimodal research can be summarised as follows:

1. Select a class of communicative situations to be studied and the particular focus that is to be adopted within this.
2. Decompose the media of the communicative situation to derive a hierarchically organised (although sometimes overlapping) range of canvases and subcanvases.
3. Map out the multimodal genre space in order to position your target(s) of analysis as activities being performed 'in', 'on' or 'with' the canvases in play.
4. Select from the data the activities that are being performed and identify, or hypothesise, the semiotic modes that may be using the canvases for those activities.
5. Triangulate your research problem with respect to other work on those modes, genres and situations.
6. Perform the actual analysis using the analytic frameworks relevant for the media and genres (our use cases below give illustrations of this).
7. Search for patterns and explanations for patterns in the data analysed.
8. Write up the results.

7.4 Conclusions and lessons for effective multimodal research

In this chapter we have considered some of the fundamentals of getting your multimodal research going. This can happen at several levels and so we distinguish between several complementary aspects. First, if you are undertaking multimodal research for some particular research question that you have already identified as relevant and interesting, then the main questions are those of methods and approach. We deal with this first. Second, if you are still looking for research questions and need to produce an example of multimodal research, such as might be the case if you are selecting to do a Bachelor's or Master's thesis or project, then most of the usual considerations when setting out on this task apply and you can find several suggestions and guidelines in other publications or online. We will address here only those particular aspects that arise as challenges of multimodality—issues such as finding a research question, selecting data, writing up results etc. are not that different from what is done with research questions in other areas that are not the sole preserve of multimodality.

The form of your description of your project will also vary according to whether you are primarily performing a piece of empirical research or a more theoretical piece, or something in between with theoretical discussions drawing on examples—again, these considerations are not particular to multimodality and so you should familiarise yourself with general advice in this regard. Perhaps one of the greatest challenges in both performing and reporting multimodal analysis turns out to be a surprising one: long descriptions of what one is analysing are *not* results!

The analyst has to avoid **relapsing into running commentary of what is going on or what is shown.** Much previous work falls foul of this tendency. The challenge of analysis is not to describe what one is analysing, but to *analyse* what one is analysing!

In fact, any descriptions being offered should always take a backseat and be employed sparingly, either to set the scene for the analysis that is offered, allowing the reader to understand just what was being analysed, or to provide illustrative material for the analysis that is given. Analysis then has to go further: its purpose is to increase our knowledge about how some multimodal communication is 'working', i.e., how it is doing what it does—just which aspects of this process of sense making are involved will depend on the research questions raised, but this must be more than just saying what is 'there' in the material placed under the microscope.

Avoiding 'running commentary' can be difficult when faced with the complexities of some multimodal artefact or behaviour. One always needs to be aware and attentive, looking for signs that an analysis is falling into this pattern. The following three 'traps' can be used as diagnostics: whenever you feel that your work is exhibiting any of the properties described, it is time to take a long look at the analysis and its methods in order to see where improvements can be made:

- **the description trap:** focus falls more on 'describing' the objects of analysis than 'analysing' the objects of analysis,
- **the pseudotechnicality trap:** the technical terms of the description can be largely removed without changing the results overmuch,
- **the circularity (or 20/20 hindsight) trap:** the technical description relies on the proposed results of analysis in order to be applied.

The circularity trap is particularly pernicious and is one of the main reasons why the kinds of more empirical methodologies that we saw set out in Chapter 5 are so important. Here one has to think very honestly about the analytic categories that have been applied and how they have been matched to the data. Would another analyst produce the same analyses? Would the categories apply the same way to other data?

! Changing the definition or application of categories to fit your data (and, worse, to produce the results you want) is like trying to measure the length of something with a rule made out of an elastic band: if you want to measure it longer, then you can stretch the elastic—this gives you the result you want, but tells you nothing of how long the thing you are measuring actually is!

Changing the definitions of categories so that they fit your data is always to be avoided because any analysis that is then produced using those categories is, unfortunately, often worthless. Beneficial results arise only when there is a productive traction between well defined categories and the particular necessities and uses found in instances of their use. It must be possible for the data *not* to fit the categories! Without traction, results are unlikely.

Finally, there is one last situation to be aware of, one which we would not describe as a 'trap' but where we would nevertheless advise caution—we call this one the **re-description syndrome**. This relates to the description trap and arises when one considers a translation of some results into a specific frame of description as a result in its own right. Re-description alone is not usually a result, however: one needs instead to show that one has taken understanding of some artefact or performance further, regardless of the framing of that understanding. Often this can best be shown by translating any description back into the terms and constructs used in other disciplines or approaches—which will also increase considerably the chances that results will be noticed and engaged with.

Triangulation   For this reason we would actually argue against one of the most commonly adopted 'methodological' stances taken towards interdisciplinary work: the step of 'achieving a common vocabulary' or getting to 'speak the same language'. This is often doomed to failure, particularly when pursued reductively. We consider being able to *speak each other's language(s)* infinitely more important. This has repercussions at all levels of multimodal work. Consider the following position suggested by Kress and van Leeuwen: "In our view the integration of different semiotic modes is the work of an overarching code whose rules and meanings provide the multimodal text with the logic of its integration" (Kress and van Leeuwen 2006 [1996]: 177). This is another manifestation of trying to 'speak the same language'. We suggest here that other models of inter-relationship may take us further, particularly the step of *triangulation* listed above in our points for effective multimodal research. One needs to check (i.e., triangulate) that one's re-description is not simply saying something that has already been said, and perhaps said better, in another form of analysis or community concerned with the kind of artefact or performance at issue.

Part III: **Use cases**

In this part of the book we present a diverse but organised range of areas where multimodality is present, showing how these have been addressed in the past and how the foundation introduced up to this point in the book can now move them forward in several ways.

Organisation of this part of the book

The chapters in this part of the book present particular kinds of artefacts and performances raising challenges for multimodal analyses. They are also among the kinds of artefacts and performances for which we are beginning to see a growing number of analyses. In general, we organise these so as to work gradually from less to more complex media or canvases, although occasionally it will be advantageous to adopt a more general perspective and focus in on variations within that.

The overarching structure of our presentations will follow the dimensions for characterising canvases that we motivated in Chapter 3. First, the canvases will be decomposed into particular areas of attention as explained in our Navigator chapter (Chapter 7). We then use the affordances of the identified canvases to structure ensuing analyses, showing how the kinds of semiotic modes that may be identified rely on both the canvases available and the communicative purposes being pursued. We will also give substantial references to other work that may have addressed the artefacts or performances at issue so that comparisons and other sources of input can be considered.

When carrying out your own research, it should be possible to locate your object(s) of analysis with respect to the dimensions of classification so that you can jump to the respective use case areas below. The use case areas will then show examples of multimodal approaches that have addressed similar concerns as well as bringing out how the the distinctions we have introduced in earlier chapters can guide research further.

We group the individual use cases that we discuss into 'use case areas' according to the dimensions of classification and decision tree that we ended Chapter 7 with. All the use cases presented in this part of the book can therefore either be read alone or combined together according to the requirements of a particular more complex communicative situation. We will see many examples where various canvases can be combined profitably to build up more complex descriptions.

Use case areas

The groupings are nevertheless important because a reoccurring theme throughout all of our discussions will how to avoid compartmentalisation. It will often be the case that the treatment of one area of multimodal meaning-making can beneficially draw on related or neighbouring areas. This is another reason why we have proposed the foundations set out in previous parts of the book: only when we have a sufficiently broad and robust foundation for multimodality can we make such transfers across media and modes and at the same time avoid the constant danger of making one medium look too much like another. Descriptions must always be able to bring out what is specific to any particular multimodal communicative situation while still drawing on generalisations when possible.

Space restrictions prevent us presenting all of the use cases in the detail we would like and so some should be seen more as 'use case sketches'. Nevertheless, we will

Use case sketches

also seek at least to identify the properties of the canvas(es) involved, mention previous work, and pick out aspects that are particularly challenging or illustrative of the integrative account that we are proposing.

Our grouping of use cases is then as follows:

- The first use case area groups together canvases that unfold in time and which are generally 'unscripted', as in face-to-face interaction. We take this quite complex canvas first because it is in many respect the primordial mode of meaning-making within human social groups. Facets of its organisation return in almost all other forms and so it is always beneficial to have a broad understanding of how forms of multimodal communication operate.
- The second use case area turns to scripted, or semi-scripted performances, such as music, theatre or dance. This moves beyond the previous group by adding planning and design as an affordance of interaction with the canvas.
- The third use case area turns to static canvases that employ space and spatial extent, thereby adding 'layout' to the affordances of the canvas. Typically this includes 'externalised' media such as print publications, webpages, schoolbooks and other 2D canvases as in comics or graphic novels, for example.
- The fourth use case area adds temporal aspects and movement to the affordance of a 2D spatial canvas, taking in a range of media, audiovisual presentations and so on. Such media are generally 'page' or 'screen' based and involve time but are at most micro-ergodic.
- The fifth and final use case area adds actions and reactions to the affordances of the canvas, ranging over webpages, digital texts in 2D as in social media and virtual or augmented reality in 3D and games. These are therefore intrinsically ergodic.

The specific use cases themselves are presented as individual chapters for ease of reference, although there may well be 'cross-over' areas, or transitions, where properties are shared across areas. We draw attention to these situations of cross-over as we discuss them. In general these will further clarify just how to deal with complex media and the emerging combinations that make up much of our current media landscape.

Use case area 1: temporal, unscripted

8 Gesture and face-to-face interaction

We noted in Chapter 2 that work on spoken language has always been pushed in the direction of multimodality (cf. Pike 1967; Birdwhistell 1970). It is for this reason that the area is the most common direction explicitly addressed in terms of 'multimodality' at all. In fact, when engaging with the work of several of the communities working on natural, contextualised occurrences of spoken language, one would gain the impression that this topic and 'multimodality' are synonyms: 'multimodal communication' in this context *means* the study of face-to-face interaction situation! As we can see from this book, however, this is far from actually being the case: multimodality is a far broader topic. Nevertheless, it is still fitting that we begin our use case discussions and examples with the face-to-face situation—not only is it in many respects really a 'primordial' situation for multimodality, but also and as noted earlier, many important facets of a broader theory and practice of multimodality research can be drawn from work and methods developed with respect to spoken face-to-face interaction.

Despite the inherent multimodality of spoken language, much earlier work on face-to-face interaction still took 'language' as its point of departure, grouping all other components together as the 'non-verbal'. This was (and is) a misleading characterisation because it lends itself far too readily to seeing language as where the 'real' communication occurs and the rest, the non-verbal, as additional inflections or colourings, flavourings, etc. of what was (actually) said. Although there are occasions when this may make sense—i.e., accompanying activities are in some strong sense supportive and subordinated to the ongoing language events—it is not an adequate description of face-to-face interaction *in general*. We could equally well imagine a situation where, for some reason, interactants have to be quiet and so, for a while, the interaction might be carried forward using quite different expressive resources, with language taking a backseat. A multimodal account of interaction needs to address these and any other situations that occur as well.

Researchers in face-to-face interaction now mostly accept that many expressive resources apart from those narrowly considered as 'linguistic' need to be seen as contributing to meaning-making not parasitically, but as *intendedly carrying communication forward in their own right*. Burgoon et al. (2011), for example, offer eight coding systems, suggesting that all receive their own descriptions during analysis: kinesics, vocalics (paraverbal, prosody), physical appearance, proxemics, haptics, chronemics (use of time: waiting time, promptness, 'polychronics'—i.e., doing several things at once), environment and artefacts (arrangement of environments, design or objects, selection of objects) as well as olfactics. All must be considered "essential ingredients in the interpersonal communication mix" (Burgoon et al. 2011: 270). The appropriate take-up and production of such messages plays an important role in identity perception, relationship management, persuasion, sincerity signalling *and* often constitute basic components of the semantics being communicated as well.

Any characterisation of the 'carriers' of such phenomena as simply 'non-verbal' is then not particularly useful—the term in the end only expresses what we already know, i.e., that other expressive resources are being used than 'verbal language' considered narrowly as streams of sounds being emitted by speakers. In contrast, the sheer complexity and internal richness of the canvas involved in face-to-face interaction is impossible to avoid and quickly demands our focused attention. This is very much more than the 'sound signal', taking in expressive resources working with body codes, contact codes and spatiotemporal codes, all manipulating their materials along many simultaneous dimensions. Understanding how these resources combine and interact in the service of of organising interaction and exchanging meanings is the principal multimodal challenge.

There is a huge range of literature in this area, with diverse approaches, research assumptions and methods. For our present goals, we will focus primarily on the expansions that have been, and are being, made to address language use as involving the entire body and, particularly, the role of *gesture* within this. We will also, however, place this far more strongly within our overall model by bringing out the 'canvas' (→ §3.2.5 [p. 87]) within which face-to-face interaction plays out. As we explained in detail in Chapter 7, we can use this to build connections across a far broader range of communicative actions.

8.1 Previous studies

As mentioned above, a very substantial body of work exists on face-to-face interaction from a very broad range of orientations and disciplines. Several approaches begin from the *situated social interaction* itself, i.e., the situation where the interaction is actually happening in all its complexity and contextual-embeddedness. Most of these can be characterised as 'interactionist' in some sense, which means that interaction is seen as a primary site of making meaning. We saw some of the broader motivations for this in Chapter 2, when we discussed the relation between multimodality and society and media (→ §2.6 [p. 64]).

CA   The longest established approach of this kind is *Conversation Analysis* (CA)—we offered an example of multimodal transcription in the style of CA in Chapter 5 above as well. Conversation analysis began its life addressing the fundamental question of how social order emerges from the moment-by-moment acts of participants in everyday social interactions (cf. Garfinkel 1972 [1967]). 'Meaning' was identified consequently as the moment-by-moment *achievement* of a sense of successful interaction that can be directly observed by analysts and participants alike. Thus, when interactants respond smoothly and unproblematically to some utterance, this is seen as *constructing* that utterance as having been understood. When there are problems, then these are also generally manifest in the observable behaviour of the participants. Relying on the *directly observable behaviour of those involved* is adopted as a strategy for understanding how social order is constructed in action *without* assuming that that order is already in place. The final motivation in CA for accepting any particular aspect of spoken interaction as relevant for their description is therefore that the participants themselves show an 'orientation' to that aspect in their behaviour.

One particular example of this kind of process is the phenomenon of *turn-taking*, for which CA research offers a detailed account (Sacks et al. 1974). People in interactions were observed to try and 'take the floor' in a conversation only at particular points that could be described in terms of the grammatical and intonational phrasing of the turn being produced by the current speaker. Violations of this tendency would be marked explicitly with other observable linguistic signals as well—such as hesitations, apologies, repeated words and so on. This observable behaviour is taken in CA as demonstrating that the *speakers themselves* are orienting to the fact that they are attempting to 'break into' the ongoing flow of discourse. In other local contexts, speakers might designate other speakers as being required to take the floor by asking a question or raising his or her hand to signal that a change of turn is being aimed at.

Turn-taking

Turn-taking takes on an even broader utility when properly combined with the varying affordances of canvases. Consider, for example, air-traffic control discourse, where an extremely restricted canvas must be used, or computer-mediated discourse, such as chats (→ §16 [p. 355]), or computer games (→ §17 [p. 366]), where turn-taking must be managed in other ways. The canvas may also need 'artificial' or designed restriction—for example, in airplane *flight deck interaction*, where spoken interaction may be regulated for safety reasons. All of these applications of turn-taking can be beneficially tracked as media *depictions* (→ §4.2 [p. 126]) of the spoken language situation.

!

CA-inflected work is now commonly extended to include a broad range of multimodal phenomena. As we saw in Chapter 1, it is rare in more real-life scenarios to have only language to consider—many situations, ranging from leisure time to highly specialised professional activities, directly and irrecoverably involve visual (and other) forms of communication and these can always both directly impact on any language that is occurring and act instead of such language. CA work demonstrates how adequate descriptions of communicative behaviour will often not be possible without giving due attention to these broader contexts (cf. Goodwin 2000; Poncin and Rieser 2006; Mondada 2007; Deppermann 2013). Multimodal extensions of CA are consequently growing in breadth and have also been described in several introductions to multimodality (cf. Jewitt et al. 2016); another detailed overview of the field and many references can be found in Stivers and Sidnell (2005) or Machin and Mayr (2012).

When one is considering analysis that attempts to address more of the full breadth of the canvas within which communication is unfolding, the challenge of 'acquiring' the necessary data returns with considerable force. A recording of the spoken language with notes concerning hesitations or other properties is not going to be enough and many approaches, including CA, have turned to working with extensive video recordings as their prime method for fixing interactions for analysis. Mondada (2007, 2016), for example, suggests the use of the video camera as an ideal tool for making a range of embodied interactive practices of this kind accessible to research—a move now being followed in several disciplines, including visual anthropology, visual sociology, education and more. Many of the premises here are similar to those we have been emphasising throughout this book, treating embodied action as the central starting point for discussing and analysing varied communicative situations *without* assuming that verbal language automatically plays the most central role.

Mediated discourse analysis A further area of research focusing on interaction is that promoted for many years by Ron and Suzie Scollon (Scollon 2001; Scollon and Scollon 2003), which developed through several broadly 'ethnographic' variants with names such as 'nexus analysis', 'geosemiotics' and *mediated discourse analysis*. Common to all of these is the commitment to see interaction as happening in, and involving, their material contexts and environment, taking into consideration, for example, the situatedness of signs (or other semiotic resources) in the world. Language is here again seen as just one resource within a material world in which gestures and body movements play an equally important role (cf., e.g., de Saint-Georges 2004).

Norris This general approach is taken the furthest towards explicit accounts of multimodality in the work of Sigrid Norris (Norris 2004*a*). Norris offers a detailed methodological approach to analysing natural interaction in context, allowing the attention and actions of the interactants to determine just which aspects of a situation are to be drawn into the analysis. As she writes:

> "One challenge for the analysis of multimodal interaction is that the different communicative modes of language, gesture, gaze, and material objects are structured in significantly different ways. While spoken language is sequentially structured, gesture is globally synthetically structured, which means that we cannot simply add one gesture to another gesture to make a more complex one. [...] Gaze, however, may be sequentially structured, and during conversation it often is. But, during other interactions, gaze can be quite random. [...] There are other communicative modes, which are structured even more differently. [...] [F]urniture is a mode, and when thinking about it, we find a functional structure. Chairs are usually located around a table, or a reading lamp is located next to an easy chair. Thus different modes of communication are structured in very different ways." (Norris 2004*a*: 2–3)

This position is in line with several proposals made in the literature that even material objects may need to be considered during some particular analysis (e.g., Kress and van Leeuwen 2001: 31–32).

Nevertheless, to counter the danger that an analysis might become too broad and lack direction, Norris (2004*a*) relies as well on the essential conversation analysis tenet that it is what *participants in an interaction themselves* show attention for that counts. This remains the ultimate motivation for including some facet of a situation in an analysis and others not. In addition, Norris refines this further by taking the 'mediated actions' that interactants themselves see themselves as performing as the primary analytic unit. This allows a responsiveness to what it is that the interactants themselves construct themselves as doing without needing to place language at the centre of attention. Norris sees this as an essential strategy for working with communicative modes exhibiting very different kinds of organisation within a single framework (cf. Norris 2016*a*,*b*).

A further broad direction of (originally) spoken language research can be characterised more under the rubric of *Discourse Analysis*. This has quite distinct origins to those of CA and is generally far more accepting both of abstract models of the kind widespread in linguistics and of accounts that focus on individual, often cognitive capabilities of language users. Considerable work in this area has addressed the specifics of *gesture* as one of the most developed 'additional' expressive resources accompanying spoken language in all natural interactive situations (Müller 1998; Duncan et al. 2007). Adam Kendon, one of the central protagonists in gesture studies characterises gesture in the following terms, very much in the spirit of Pike (1967):

Discourse analysis

> "[…] this bodily activity is so intimately connected with the activity of speaking that we cannot say that one is dependent upon the other. Speech and movement appear together, as manifestations of the same process of utterance." (Kendon 1980: 208)

Kendon's position is an extremely multimodal view. Face-to-face communication is not seen as primarily linguistic at all but rather as the mobilisation of a rich array of synchronised and integrated resources. The challenge from a multimodal perspective is to characterise the communicatively effective integration of this diversity that regularly occurs.

8.2 Describing gesture and its functions

Several approaches attempt to characterise gestures with the same kind of analytic attention traditionally allocated to language. This involves finding the basic units at work, their distributions and possibilities for co-occurrence, their meanings (or contributions to meanings), and so on. Constructing individual classifications of gestures has proved to be a complex endeavour, since the meanings they provide most commonly *work together* with meanings being made with other expressive forms—this has been, therefore, another one of those areas where compartmentalisation can hinder progress.

Kendon (1972) proposes that gestures exist along a continuum in which the presence of speech is reduced when language-like properties of the gestures increase. Spontaneous gesticulations which are often accompanied and co-expressed with language, but which do not share linguistic properties, are therefore at one end of the continuum, whereas full sign language where specific manual signs have specific meanings and are combined according to rules—which then have linguistic properties of their own—are at the other end of the continuum. 'Language-like' gestures form a further category in this continuum closer to spontaneous gesticulations and can generally be described as iconic or metaphoric. Their meaning is mostly revealed, or refined, in combination with speech.

McNeill (1992, 2006) then picks up these descriptions and offers a classification scheme oriented more towards the semiotic categories of Peirce (→ §2.5.2 [p. 56]). His characterisation includes gesture types identified as *iconic, metaphoric, deictic* and *beat. Iconic* gestures "present images of concrete entities and/or actions", while *metaphoric* gestures "picture abstract concepts" (McNeill 2006: 59). *Deictic* gestures are 'index' fingers or other body parts which entail "locating entities or actions in space vis-à-vis a reference point"; *beats* are "flicks of the hand(s) up and down or back and forth" (McNeill 2006: 59). McNeill emphasises that these types need to be seen as 'dimensions' of variation because they often combine in one and the same gesture: "We cannot put them into a hierarchy without saying which categories are dominant, and in general this is impossible" (McNeill 2006: 59).

Work on the metaphorical use of gestures and their function for the multimodal construction of extended discourse has recently come particularly into the spotlight. Figure 8.1 shows an extract from a dialogue analysed in Müller (2008: 100) and Müller and Cienki (2009) where this phenomenon is very evident; we will also consider this here from the perspective provided in this book. In the example a young woman describes her teenage love relationship with her boyfriend as "clingy'. The transcription used for this example is simplified to indicate intonational and other speech-related properties with 'iconic' punctuation—such as slashes and backslashes for intonational contours, etc. (→ §5.2.3 [p. 146]). Quite sophisticated annotations schemes are now available for gesture (Bressem 2013; Bressem et al. 2013). The present example starts with a description of the gestures (G1 and G2), which continue while the person is speaking; the speech is then separated into the German original text and an English translation.

The analysis that Müller, Cienki and colleagues offers shows that gesture can not only accompany speech, but also *pre-figure* (i.e., occur before) and *reiterate* extended metaphors that add considerably to the meanings being expressed. In order to illustrate the communicative situation, Müller (2008: 100) uses a line drawing showing two women talking to each other on a sofa. We can therefore assume that this is a situation with a dynamic representation of speech and gesture (and probably other resources which are not transcribed), which—as we set out in Chapter 3—is three dimensional and in which the sign consumer is an observer of, and listener to, the gestures and their accompanying speech. It is therefore a mutable ergodic canvas in that both the speaker as well as the viewer/hearer are engaged in constructing the emerging discourse within the situation (→ §3.3.1 [p. 106]).

Following the approach provided in our navigator chapter, we take the transcription as representing our analytical canvas under consideration, which is already a precise and detailed sub-canvas of the much larger canvas of the entire environment

Gestures as metaphors

G1: open palms touching each other … repeatedly

1. [also da hab ich schon gemerkt naja\ das is: ganz
    well there I did already realise well\   this is really

     G2: hands clap repeatedly

2. schön (-) (mh) (-) klebrig\ oder heftig]
    pretty (-) (mh) (-) clingy\ or heavy]

           PUOH   point upwards

3. un[dadurch dass er so depressiv war] denke ich letztendlich /]
    an because he was so depressive   I think finally

Fig. 8.1. Extract of interactional transcription from Müller (2008: 100). Spoken language is the main numbered line in each segment; transcription below the line gives an English gloss and transcription above characterises the simultaneous gestures performed by the speaker.

in which the example took place. The transcription is then helpful in order to narrow down the focus on the semiotic modes operating in this specific case, looking at the descriptions of gestures and their manners of combination with speech. An analysis can then naturally draw on the characterisations available in the literature already mentioned above.

These apply fairly straightforwardly. The first gestures, G1 and G2, present a performance of the action of 'clinging together' and are therefore iconic gestures. Müller and Cienki explain further that "the speaker's flat hands repeatedly touch each other, moving apart and then back to 'sticking' together" (2009: 309) . Interestingly, the speaker only uses the term "clingy" at the end of the second gesture when the hands have already been clapped repeatedly. Gesture G1 is thus, according to Kendon's distinction, a language-like gesture that is accompanied by language which, however, expresses a different meaning. Gesture G2, in contrast, directly co-expresses what is also said on the verbal level as "pretty clingy": the clapping hands echo visually the meaning of clingy. It is thus an iconic gesture, which, again following McNeill, also has a metaphoric function in that it describes the *relationship* as being clingy.

We can characterise this in a similar manner to Müller and Cienki, focusing on the *discourse* construction that combines the expressive forms in play: the speaker is concerned with describing a relationship and can do this in several ways simultaneously, each making use of specific affordances of the expressions deployed. Following Kendon (1980, 2004), Müller and Cienki develop the premise that "gesture and speech are visible and audible actions that form one single utterance" (Müller and Cienki 2009: 298–299). With regard to the first gesture, they argue that "the gesture enacts […] the verbal metaphoric expression, indicating that metaphoricity of this expression was activated or in the foreground of the speaker's attention" (Müller and Cienki 2009: 308). The movement of the hands repeatedly touching each other thus not only underlines or supports what may be expressed verbally as being 'clingy' but also, by virtue of the fact that it is uttered *before* the verbal use of the metaphor, demonstrates that this metaphor is conceptually 'active' in the formation of the discourse prior to its use in selected expressive forms. The overall combination then constructs a multimodal metaphor in which both modalities share common source and target domains. The domain from which the metaphorical expression is drawn is that of physical substances being sticky; the target is that of interpersonal relationships. As Müller and Cienki (2009: 309) underline, the gestures here then are not only individual occurrences or spontaneous gesticulations, but have the function to structure an entire segment of the discourse.

The affordances of gestures can be used in other ways as well. Consider, for instance, the other two gestures in the example. The first, PUOH—'Palm Up Open Hand'—is drawn from a class of contrasting gestures that has now received substantial study in the literature Kendon (2004); Müller (2004). PUOH contrasts with several other similar but differently used gestures, such as PLOH: 'Palm Lateral Open Hand'. The second gesture, pointing upwards, is similarly identifiable as coming from a broader set of related, but quite distinct, possibilities. Several such 'gesture families' have now been explored (cf. Fricke et al. 2014). The gestures here accompany the verbal speech but do not appear directly related. We see from this that they may be depicting actions by resemblance and so operate iconically, but they are not metaphoric with respect to the content of the speech. Nevertheless, they still function as discourse-structuring elements, drawing attention to something else. Müller characterises this kind of gesture as 'performative', "that is, gestures, that perform a communicative activity [presenting] a discursive object on the open hand and focusing attention by pointing" (Müller 2008: 101). We might consider this further in terms of functional metafunctions (→ §2.4.5 [p. 49]) since the current gestures appear to carrying both textual (i.e., discourse organisational) and interpersonal (evaluative) meanings (cf. Martinec 2004).

The work of Kendon, McNeill, Cienki and Müller represents one slice through the broader landscape of approaches currently targeting the systematic categorisation of gestures. Considerable effort, especially from a corpus-analytical perspective, now aims to generalise the meanings of gestures further. We have seen in our example analysis that it is important to look at the combination of speech and gesture together as expressive resources in order to say more about how they perform a variety of discourse-relevant tasks, ranging from anchoring the behaviour into the current situation to constructing metaphors that may structure extensive stretches of interaction.

We do not yet know, for example, at just which 'levels', or locations, in the linguistic system such synchronisation takes place. There appear to be some kinds of language-gesture use which favour a very tight intertwining of the physical manifestations of language and those of gesture—so much so that the syntax and semantics of utterances may need to cover both forms of expression—and there are other usages which appear looser.

Fricke

Examples exhibiting a tight degree of structural integration across the phonetic and gestural material being manipulated in the interaction are easy to find. Consider, for instance, the following two examples discussed by Ellen Fricke translated from the original German of her corpus and modified here to pick out the gesture for ease of discussion (Fricke 2013: 748):

(i) "I want to buy [G↺] such a table" (+ *iconic gesture of a circle*)

(ii) "I want to buy [G↗] such a table" (+ *deictic gesture to a circular table*)

The linguistic element 'such a' acts as a synchronisation marker across the distinct material forms: it is present in the syntactic structure of the corresponding grammatical phrases, but also 'involves' a gestural realisation accompanying the phonetic form. Gestures of this kind are contributing to the semantic information communicated and so cannot be characterised as parasitic on information already given linguistically (cf., e.g., Kendon 1980; McNeill 2000; Poggi 2007; Müller and Cienki 2009). Several researchers are now consequently attempting to expand the notion of 'grammar', 'grammatical constructions' and the like to include gestural elements to cover such phenomena (cf., e.g., Steen and Turner 2013; Mittelberg and Waugh 2014; Bergs and Zima 2017). Fricke herself gives a particularly far-reaching proposal in this regard when she argues for a 'multimodal' view even *within* syntax (Fricke 2013). She describes the integration of gestures in the syntactic structure of utterances as either occupying a syntactic gap, overlapping chronologically with the verbal utterance, or by instantiating specific utterances.

Broad *vs.* narrow multimodality

The extent to which this could offer a sufficient foundation for the diversity of multimodality we are concerned with in this book is, however, unclear. Fricke also discusses the notion of multimodality more generally from this perspective. In a series of publications (cf. Fricke 2007, 2012, 2013), she proposes that it is useful to distinguish between a 'broader' sense of multimodality and a 'narrower' sense. The broader sense is when different communicative forms appear alongside one another, as loosely coordinated activities or performances—which Fricke describes as *code manifestation* or 'multimediality'. The connection between 'text' and 'image', which we return to in subsequent use cases, is suggested to be an organisation of this kind. In contrast, the narrower sense is when one individual semiotic 'code'—for example, a specific human language such as English, German, Finnish, etc.—*itself* employs expressive resources that draw on different sensory channels. Both possibilities are covered within our general model of semiotic modes presented in Chapter 4 above.

Gesture allows a clear illustration of this. If gesture and spoken language were simply 'accompanying' one another as independent, albeit correlated, activities, this would be an instance of multimodality in the broader sense. If, however, gesture and spoken language are so strongly coordinated that the description of specific linguistic organisations or constructions are themselves dependent on both phonetic and gestural material, then this would be a genuine case of *code integration*, i.e. the close intertwining of two or more resources, and so constitutes multimodality in the narrow sense. Many approaches have in fact already trodden this path for spoken language with respect to the phonetic material introduced by syntax and the lexicon on the one hand, and intonational contours on the other. Grammatical descriptions have consequently been proposed with both expressive forms fully integrated—gesture may then need to be seen in this way as well.

8.3 Conclusions

Gestures are evidently far more than redundant signals that can be ignored or considered subservient to more explicit meanings made in language and there remain many open questions involving issues of multimodality. The most general form of these questions is, as we have suggested, how the diverse sources of communicative information in face-to-face interaction are to be integrated within larger models of communication and language.

As suggested throughout this book, therefore, it is clear that there is a need for considerably more *empirical* work here to unravel the descriptions and theories necessary. Systematic accounts that incorporate gesture (and other body movements) established on the basis of large quantities of natural data must receive a high priority. Current empirical work of this kind includes both fine-grained descriptions and corpus-oriented investigations of synchronised gesture and language for human-computer interaction (→ §2.4.2 [p. 41]), as well as a rapidly growing body of work employing computational tools of the kind discussed in Chapter 5 aiming at the automatic detection and description of gestures, facial expressions, posture and more (cf., e.g., Kipp et al. 2007; Allwood, Kopp, Grammer, Ahlsén, Oberzaucher and Koppensteiner 2007; Kopp and Wachsmuth 2009; Sowa and Wachsmuth 2009; Escalera et al. 2016).

It seems unlikely, however, that the inherent richness of spoken language and face-to-face interaction data will be fully understood without employing more finely articulated positions on multimodality. Extending theoretical and descriptive interest to deal with this richness does not long allow the researcher to remain within any 'narrow' linguistic boundaries. And it is for this reason that we have emphasised in this chapter the challenges raised by *combinations* not only of gestures and other bodily movements with verbal and other resources, but also of the embedding of interaction in broader, also multimodal, communicative contexts. This emphasises the basic assumption, fundamental in the context of multimodality, that language is no longer the central semiotic resource in all types of communication (cf. Kress and van Leeuwen 2001).

Use case area 2: temporal, scripted

9 Performances and the performing arts

As a major category for the description and exploration of a wide range of social and cultural activities, the term *performance* has been used in diverse disciplines, including psychology, sociology, philosophy, linguistics, literary theory, history, anthropology and many others—competing with the heterogeneous field of *performance studies* with its own on-going analyses of a multiplicity of situations all seen as performative. Schechner (1988), for example, sees actions, behaviours and artistic practices as the main objects of interest in performance studies, including all kinds of intended and often, but not always, scripted works deploying resources such as language, image, gesture, sound, music, etc.

These resources are not only used in environments for the fine arts, as in theatre or opera, but also in more or less spontaneous situations in the street or semi-controlled situations in classroom or consulting situations—all of these constitute moments of live performance. More inclusive approaches today also consider recorded performances of the fine arts, such as filmed rehearsals and opera or ballet films. Similarly, the internet with its multiple opportunities to create quasi-live or as-if performances offers considerable material for discussion (cf. Schechner 2013). Any basic understanding of performance in its very broad sense thus includes:

> "a 'broad spectrum' or 'continuum' of human actions ranging from ritual, play, sports, popular entertainments, the performing arts (theatre, dance, music), and everyday life performances to the enactment of social, professional, gender, race, and class roles, and on to healing (from shamanism to surgery), the media, and the internet." (Schechner 2013: 2–3)

This definition reflects a radical reconsideration of the underlying frameworks of analysis within performance studies that has taken place over the last three decades. We also find here approaches paying particular regard to semiotic perspectives—these will be of considerable help below as we build bridges to multimodality more broadly. Gay McAuley, for example, acknowledges the diverse activities subsumed under the label of performance and then goes further, arguing for the presence of an artist, performer or social actor as well as a public or spectatorship as the lowest common denominators defining a performative situation (cf. McAuley 2007, 2010). A characterisation of this kind closely echoes, therefore, the essentials of communicative situations as we set them out in Chapter 3 above. Performance is seen as establishing a semiotic space which brings about both an act of performing as well as acts of viewing and understanding. This paves the way for approaches to performance that ask for their meaning-making potential on several canvases and with regard to the specific affordances involved just as set out in our navigator chapter above (Chapter 7).

This chapter engages with situations of this kind, considering performances as multimodal interactions exhibiting productive interplay of verbal and musical discourse together with facial expressions, gestures, body postures, head movements, spatial positions and other resources.

9.1 Performance and scripted behaviour

At its most general, the notion of performance is already well anchored in the context of multimodality studies. Ron Scollon and Suzie Scollon (→ §8.1 [p. 242]), for example, argue that while earlier approaches to the study of performances that are more language-based have been interested primarily in active real-time dialogic performance by humans, multimodal studies must be concerned in addition with "the design of objects, the built environment, works of art and graphics, film, video, or interactive media productions" (Scollon and Scollon 2003: 171).

Fine arts   This perspective casts its net far wider than performance in the sense of the performing arts, which is generally taken to subsume traditional genres of fine art which are performed in front of an audience, including theatre, dance, opera, music and so on. One could then include here characterisations of teacher-led classroom interactions, for example, as a case of a semi-scripted performance in which the social interaction about topics and texts similarly involves a range of semiotic resources in a planned and constructed configuration (see Chapter 7).

Healthcare communication   Indeed, the door is opened to many further semi- or fully scripted performances, such as controlled performances in healthcare communication in the operating theatre where, as Bezemer et al. (2012, 2014) describe, specific functions of holding the scalpel may also serve communicative purposes. In Bezemer et al.'s case, the research focuses mostly on questions of how performances in such specific environments can be generalised according to the social roles involved, for example in surgery. Mondada (2007) and Heath et al. (2000) address similar contexts from a conversation analytic perspective, also showing particular regard to the interrelationships between interaction patterns and gestures and other body movements. These kinds of performance may be characterised in terms of more or less formalised restrictions on the free mutability of the canvas involved (→ §3.3.1 [p. 106])

For the purposes of the present chapter, however, we will consider a different kind of 'restriction' and the challenges it raises. The concern here is with the space opened up between fully mutable canvases, where the contributions of interactants are driven by their immediate 'communicative' concerns, and more finely structured canvases which support 'free' contributions only along some specified dimensions while keeping others fixed. Thus, performing a song, for example, might fix the song text and melody, while still allowing considerable variation in voice and tonal qualities, movements and postures, gaze and so on. It is the configuration as a whole that then constitutes the 'performance'. This brings about an important difference to the kinds of performances we have described so far in this book. As Schnechner emphasises: "[p]erforming arts frame and mark their presentations, underlining the fact that artistic behaviour is 'not for the first time' but enacted by trained persons who take time to prepare and rehearse" (Schechner 2013: 52). The potential tension between fixity and freedom is a locus of considerable 'intended' creativity and it is this that marks out most particularly the difference between performance in general and the types of performance we address here.

9.2 Previous studies

The performing arts subsume several traditional genres of fine art which are performed in front of an audience, including theatre, dance, opera, music, the circus and so on. Many traditional studies of these arts, particularly within the context of performance studies, have followed the model of the disciplines characterised in Chapter 2 above; that is, researchers have given primacy to one resource or expressive form over others—music in opera, for example, or verbal language in theatre plays. Moreover, it is also commonly the case that arts such as theatre or opera are each studied within their own, more specific disciplines, such as theatre studies or theatrology (as a particular area of performance studies) and opera studies (as an area of musicological research), bringing together both the practical production perspective as well as a reception-oriented or text-based perspective, although with less explicit focus on the actual processes of meaning-making. The performing arts in these contexts are approached more from the angle of cultural and historical criticism, or with a specific focus on genre, historical periods, playwrights, composers and so on.

However, when the quantity and diversity of resources involved in opera, theatre, ballet, circus, or mime do receive attention, then there is also a tradition of work that is strongly oriented towards semiotics (cf., e.g., Helbo 1987; Fischer-Lichte 1992; de Marinis 1993; de Toro 1995; Kornhaber 2015). Among these, the position set out in André Helbo's *Theory of Performing Arts* from 1987 is in many respects well aligned with the direction we are pursuing in this book. Helbo situates the role of semiotics for the performing arts "not only at the levels of the expressive phenomenon, of the message, and of the relation text/representation, but especially at that of the analysis of codes and of the matrix of codes" (Helbo 1987: 35). By including a particular view of *semiosis* and the complexity of the semiotic objects in question, he paves the way for a comprehensive analysis of all elements involved that is similar in many respects to the account of multimodal analyses given in our navigator chapter (Chapter 7).

Helbo summarises further three points of interest which have influenced the development of a semiotics of the performing arts over time: the place of externality and the question of the materiality and mediality of the body, which are the basis for any theory of embodiment in the performing arts; the issue of the researcher and his/her position towards the performing arts, which activates a reception-based perspective, supported by other disciplines such as psychology, sociology, pragmatics, etc.; and the 'scientificity' of the methodology, which questions the universalistic claim of

Semiotic approaches

traditional semiotics and calls for a re-orientation of performance studies towards a broader perspective "from the lived experience of meaning to its conceptualisation" (Helbo 2016: 349).

Following Sindoni et al. (2016), therefore, we will see the multimodal analysis of performances as an investigation of:

> "how different social actors (auteur, artist, producer, performer, etc.) produce artwork (object, practice, event, display, etc.) and grapple with tradition (e.g. continuity and stability of codes vs. innovation and experimentation), also in relationship with other social actors, be they spectators or collaborators (e.g. audience, listeners, and co-artists) and/or providing other forms of professional or financial participation (e.g. producers, agents, directors, event organisers, etc.)." (Sindoni et al. 2016: 333)

Experiential semiotics

Helbo himself takes these questions as the starting point for the development of an *experiential semiotics* that puts the focus on the *experience* of the spectator watching a performance and of the creator doing and showing the performance. This brings with it research questions that are similarly important for a multimodal approach to the performing arts: "How can one record a set of complex and protean practices (performing arts: theatre, dance, opera, circus, street performances, concerts, etc.)? What paradigms and mechanisms are used? Does the spectacle exist as objects?" (Helbo 2016: 349). These questions resonate with several of the methodological concerns we set out for multimodality in Chapter 5 and are essential for a multimodal analysis of the semiotic resources that different genres in the performing arts produce—especially with regard to the status of their materiality, mediality and, in particular, textuality.

The performing arts are then explored in these terms as dynamic processes of multimodal semiosis, taking into consideration both production-oriented as well as reception-oriented ways of thinking, focusing throughout on the meaning-making patterns of texts and performances. Recently developed frameworks within the context of multimodal performance studies (cf., e.g., Fernandes 2016; Sindoni et al. 2016, 2017) also follow these aims in order to tackle the challenges of unpacking the various semiotic resources involved as well as their patterns and regularities. In the following, we illustrate how a systematic multimodal analysis can be pursued with regard to the canvases and subcanvases involved in examples of the performing arts, drawing particular attention to their respective affordances.

9.3 Example analysis: theatre and its canvases

Theatre studies

In order to narrow our focus on the performing arts to more concrete examples, we take a classical theatre play as a case study, bringing in comparative perspectives on other performing arts as necessary. Theatre and performance studies in fact often go hand in hand in study programmes and in the designation of university departments; Schechner (2013: 6–11) offers a comprehensive list for these programmes in the USA, the UK and beyond. The long-standing traditions of theatre research are consequently combined with broader approaches to all sorts of performances. In general, a shift is observable "from oral interpretation and theatre to performance studies" (Schechner 2013: 21), which has opened up the field of interest to engage with several further disciplines. Performance studies today is therefore a highly interactive and interdisciplinary field. This extension also offers further illustration of the phenomenon of 'discipline-broadening' that we discussed at some length in Chapter 2.

This broadening notwithstanding, theatre still constitutes a dominant art form discussed and analysed in performance studies and so we will begin with this here too. For a multimodal analysis of theatre that takes into consideration the communicative situation and the media involved, it is important to note that, in contrast to many traditional analytical approaches, the source text of a theatre play cannot be the medium in question. Even though David Birch, for example, states that the actual performance of the play is the "stage-enacted" equivalent "with all of its written meanings kept intact" (Birch 1991: 190), it is exactly the process of enacting and performing with all possible deviations from, and extensions of, the written script and the ways of representing and realising that script that is important for the construction of meaning *qua performance*. As Tan et al. (2017) highlight, "stage performances have a greater range of semiotic resources available, arising from the contribution and interaction of linguistic and non-linguistic modes", which all come into play in the description.

The communicative situation is thus not one of reading the script or the different roles in the play; it is rather one in which several artists or performers act on a stage (or a similar, limited and framed place) and present actions and events, in theatre often creating a particular fictional world, in front of a live audience. This audience, i.e. the spectators or recipients, watch and interpret the actions performed on stage as a community of consumers. As an example, consider the image in Figure 9.1, showing the stage and auditorium of the Globe Theatre in London, where both actors (in the image centre) as well as spectators are shown in their current communicative situation.

Since a performance is always a live event, the signs and actions performed are always made once and do not last—the canvas is *transient* with respect to the ongoing action, although that action may also play out against a (locally) more stable 'backdrop' constituting the setting. This is a clear challenge for every analysis, since the genre of theatre is in general "eminently ephemeral [...], different at each performance, and construed differently by different audiences" (Sindoni et al. 2017). For related performing arts such as opera, ballet or music performed in concerts, the communicative situation is similar in that spectators listen to or watch the respective show in a specific building, such as a concert hall or opera house, or in front of an open-air stage which localises and restricts the semiotic space of the performance within a certain context. The performers, i.e. artists, singers, players, musicians (also orchestral musicians in the orchestra pit) have a limited space to perform their actions following the given and previously rehearsed script.

Fig. 9.1. A photograph showing stage and auditorium of the Globe Theatre in London, UK. ©Tracy Ducasse, Flickr Creative Commons Licence, https://www.flickr.com/photos/teagrrl/17574687/

Mise-en-scène is therefore a central feature of any performing art, since it includes the specific stage design with the composition of the set, the various props, costumes and further objects and details used on stage or the lighting used in the theatre hall. The general canvas for the theatre situation as a whole includes all these details. It is thus a 3D environment in which the actions and performances unfold in time and with semi-free, scripted movements by the actors and props. Not all movements and actions actually happening in the play might be fully pre-scripted, however, since actors are often free to move around in a limited space or change the respective actions individually, as long as they follow the 'general intention' of the piece.

The situation of a theatre play is therefore in many respects similar to the teaching situation we described in Chapter 7, with the important distinction that a theatre play is typically fully scripted and based on a written screenplay or dramaturgical descriptions. The actors and performers follow specific instructions, which are often known in their written form. The planning and design of this script in comparison to the actual performance are important aspects of the analysis and in particular the former must be seen as a specific property of multimodal theatrical performances. Also similar to the teaching situation is the fact that the actors within the theatre space must create their respective communicative contributions to bring the play into being. The medium of the theatre play is therefore also mutable ergodic and dynamic and involves actors performing communicative activities that all serve to reconstruct the pre-scripted and now newly evolving play.

Other participants in the communicative situation, the spectators or recipients, are—similarly to the pupils in the classroom—relatively inactive in that they (usually) do not perform specific activities that influence the communicative situation of the play. In contrast to situations in the classroom, however, where pupils might react to

Mise-en-scène

the teacher and enter into discussion, the spectators in a theatre are most likely addressed by the performers only in a 'one-way', highly restricted (and often ritualised) form. Performers may, for example, engage in eye contact with members of the audience and an audience member may, by means of facial expressions, etc., offer some response. The primary role taken up by the audience in their interaction with the play in general and the different activities and performances in particular remains, however, understanding the performed events and the relationships between the represented characters as constitutive for the story or plot of the drama.

Note that it is possible, and generally desirable, to describe each of these activities—both production and reception-oriented—within their own slices provided by the broader canvas, and with regard to the various semiotic resources involved in the meaning-making processes. We can also 'open up' the discussion to consider larger-scale encompassing canvases as described in Chapter 7. For example, even though the script itself in its written form is not part of the immediate canvas of performance, its realisation in monologue or dialogue by the actors plays an important role for the performance of the play. In fact, much information about the characters, their names and relationships as well as their emotions are conveyed by the spoken language and resources such as intonation or loudness, for example, since no other verbal information is generally available within the performance itself. In addition, theatres may offer brochures or booklets accompanying any respective play and giving a summary of the story as well as information about actors, the director or other details.

While often relevant for interpretation, from a multimodal perspective such information clearly belongs to distinct communicative situations. For the performance itself, specific gestures, body movement and positions as well as gaze and facial expressions remain the most important mediators of meaning. Similarly, the interaction of actors with other actors or objects on stage represent slices of the canvas which bring about different activities. Any of these slices can then be approached with a more detailed analysis and on the basis of a more specific research question.

In the case of theatre, for example, we might address questions of genre or the creation of tension and suspense, as analysed in Tan et al.'s (2017) treatment of two different performances of the Gothic horror play *The Woman in Black* (Hill 1983; Mallatratt 1989). Tan et al. approach the specific genre of this play and demonstrate that its characteristics of tension and suspense are created by the combination of several semiotic resources, classified as speech/narration, mise-en-scène and soundtrack. From these resources, the specific dialogue and voice-over narration by the narrator of the play create a dialogic space, whereas lighting and the placement of characters on stage construct a prominence and foregrounding of the protagonist or antagonist. Different sound types such as the human sound of whispering or the object sound of creaking doors interplay with these resources and together realise certain genre conventions that can be described as typical for the Gothic horror genre (Tan et al. 2017). The authors examine key scenes of the two plays and show how different performances use different combinations of choices from the semiotic systems available to produce dif-

Suspense in theatre

ferent meanings. As they write: "The story is the same, but the spine-chilling experiences that are thus constructed for audiences in each performance are unique" (Tan et al. 2017: 35).

Theatre vs. opera

Further examinations of plays from different genres or cultures, such as the Chinese sung theatre or the Polish Laboratory Theatre in the 1960s mentioned by Schechner (2013: 183, 226) as examples, might reveal further genre patterns—both for contemporary theatre as well as in historical overview. Relatedly, comparisons to other performing arts such as opera are possible, and become increasingly revealing the more a rich ground for comparison can be drawn on. Analogically to Tan et al. (2017), for example, Rossi and Sindoni (2017) describe and analyse the sociosemiotic systems of verbal language, music and mise-en-scène as providing resources such as libretto, score or staging to produce meaning in opera. In a comparative slice of both the canvases of theatre and opera, it then becomes interesting to ask how these semiotic systems and their respective resources differ from each other and which medium-specific details are available for each canvas. Moreover, modern theatre plays often also incorporate music or video installations interacting with the actions and activities on stage, adding further potential slices for analysis which, in turn, demand more inclusive and stronger accounts of multimodal meaning-making.

9.4 Example analysis: Berlin Philharmonic concerts 'live'

Live vs. recorded performances

We emphasised above that any original performance is by definition 'live'. Nevertheless, it is still quite usual to analyse *recorded* versions of theatre plays, operas or music concerts, as these are widely available on YouTube or exclusively produced as theatre films. This is an understandable response to the difficult task of analysing live performances in their non-reproducible and fleeting character—with a recording, one has at least secured access to one specific performance of a play or other performance in its respective situation of production. A further blurring of boundaries then occurs with the advent of live transmissions or online streaming of operas and concerts to movie-theatres to make the performances available to a broader audience. The Metropolitan Opera and the Berlin Philharmonic, for example, are pioneers in providing access to video-streamed concerts in high definition and excellent sound quality.

Characterising this kind of access in the terms we have developed here will further illustrate the utility of cleanly separating out the canvases and resources at play during any analysis. We will see that the video itself, as well as its possible embedding in websites or other applications, necessarily involves several aspects of media convergence (→ §1.1 [p. 14]) that also need to be taken into consideration in analysis. In particular, for any video showing a performance, we need to ask about the specific *filmic techniques* used for the recording as well as the process of editing in post-production. It is relevant whether only one single camera, for example, was simply recording the full stage and everything that happened on it from the back or some other privileged position, or whether the director of the recording used several cameras and their different positions to record the play from several perspectives and to edit, in post-production, a variety of shots showing the same event. Here, the question of remediatisation plays an important role in that films "change the semiotic system of a live performance" (Sindoni and Rossi 2016: 399). As McAuley emphasises:

> "[a] video recording of a theatrical [or music] performance is necessarily already an interpretation of that performance: it involves choices of what to record, what position to record from, what point of view (in both senses of that term) to adopt, and the video recording results in the creation of a new artifact." (McAuley 2007: 187)

Let us consider a concrete example to make this clear. The Berlin Philharmonic and their digital concert hall[1] provide free access to several concerts, including one from September 2014 performing Robert Schumann's Symphony No. 1 in B flat major, *Spring*, under the baton of Sir Simon Rattle as the first concert in a Brahms/Schumann cycle. Considerable further detail is offered on the dedicated website[2] (itself inviting further multimodal, multi-canvas analysis), including information about the recording and production process. Seven cameras were installed above the orchestra's stage to film the performance from different perspectives. These cameras were controlled by computer to avoid any performance disturbance by camera crews or assistants. *(Digital concerts)*

The video begins with a long shot of the whole concert hall showing both the auditorium as well as the orchestra tuning up. As soon as the conductor enters the stage, the video cuts to a medium long shot focusing on the musicians and the conductor on his platform taking a bow and starting the concert. Since the first movement of the symphony starts with a solo of the trumpets, the video then cuts to a medium shot showing the first and second trumpet playing their solo. The next shots alternate between medium close-ups of the conductor, shots showing the whole orchestra playing together and shots focusing on single instruments and their players. The camera normally does not move or zoom in, but the shots are partly rapidly cut, tracking changes in the music. The auditorium is only shown in the background when the conductor is seen.

As this short examination shows, the canvas of the recording is being manipulated substantially in order to support and guide viewers' access to the performance and the various resources at play during the concert. It captures specific gestures and body movements of the conductor from the perspective of the orchestra as well as the role of several instruments in the symphony, visually focusing on the players and thus highlighting their solo or other contributions. This is very different to the reception of the performance by the audience actually present: the changed focus is part of the specific communicative situation of the recorded performance which the live

---

1 Cf. https://www.digitalconcerthall.com/

2 Cf. https://www.digitalconcerthall.com/en/concert/20250

spectator cannot experience in such detail. He or she might indeed be able to point out the respective solo players in the symphony, but is not able to get as close as the medium or close up shots in the video. The use of these shots as specific resources in the respective recording canvas can thus be analysed as having the specific function of highlighting and foregrounding aspects, both musical and bodily, of the performers.

The video recording of the concert is thus clearly a new artefact. It uses the specific technologies and editing possibilities and adds particular semiotic systems to the canvas. In this analysis, we therefore have to deal with convergence and transmedia effects and address quite explicitly the specific effects the distribution of the concert has in its digital form. As described in Chapter 7, the recorded performance is thus a *depictive medium* (→ §4.2 [p. 126]) that displays the concert not only auditorily but at the same time visually and which may or may not bring its own capacities to bear. In this context, it is interesting to ask about the specific role digital media play for the combination of the various resources involved, bringing together the music of the concert with, for example, the various body movements and facial expressions by the actors and musicians. Similar research has been conducted with regard to media-based interactive artworks involving dance (Varanda 2016) or performance-installations (Oliveira 2016).

9.5 Conclusions

We have argued here that performing for the sake of performance is a very special kind of activity, demanding reflexive (and self-reflexive) awareness of the precise manner in which expressive resources are being deployed, often placing the most weight on the 'experiential' effects of manipulations of materials via aesthetically fine-tuned 'sensory commitment' (→ §2.2.1 [p. 27]). This is very different to the actions and performances found in other semi-scripted contexts such as the classroom or operating theatre, and raises several new challenges for multimodality research. Moreover, as more experimental performances continuously attempt to blur the boundaries between audience and performer, it will become increasingly necessary to draw on richer multimodal frameworks to support their analysis. There will also be connections to be made with other media, such as, for example immersive gaming or interactive narrative (→ §17 [p. 366]). Here as well, the ability to track meaning practices across canvases will be highly beneficial.

Use case area 3: spatial, static

10 Layout space

In this use case chapter, we apply the framework set out in this book to the analysis of page layout. Discussing printed pages from a multimodal perspective, Baldry and Thibault (2006: 57) observe that "in modern society the *page* is an important textual unit", which has evolved over time to support a number of communicative tasks across different media. They suggest that these tasks have become increasingly diverse following technological advances, which is reflected in everyday terms such as "index pages, glossy pages, financial pages, yellow pages, teletext pages and [...] web pages" (Baldry and Thibault 2006: 58). While this list certainly illustrates the wide range of communicative situations that involve page-like organisations, a closer examination reveals that the same term 'page' is used to refer to various kinds of semiotic and material entities, ranging from media (teletext, yellow pages) to specific genre stages (financial and index pages) and physical materiality (glossy paper pages).

As we have argued throughout this book, these are precisely the kind of multimodal phenomena that the analyst must disentangle during analysis. Moreover, these relations need to be set out clearly before addressing any textual organisation supported by page layout (→ §4.3.2 [p. 131]). That being said, what the aforementioned examples do have in common is that they all exploit the two-dimensional space provided by some canvas to combine inputs from one or more semiotic modes. In other words, regardless of whether the materiality that is being manipulated for communicative purposes is a sheet of paper or a screen, all the examples above work with a "page metaphor". This means that they are organised along two dimensions: height and width (Bateman 2008: 9).

Page metaphor

The term 'metaphor' is appropriate here, because the screen is not bound by the physical limitations determining the space available on a sheet of paper: a page presented on a screen may extend beyond its immediately visible dimensions with the help of an interface. Despite having the opportunity to extend the available layout space, while also being able to render dynamic content, many screen-based media nevertheless prefer to organise themselves along a limited space defined by two axes. These axes delineate what we hereby term *layout space*, whose theorisation has been taken furthest in the field of graphic and information design (for a comprehensive overview and terminology, see Black et al. 2017). This field is therefore the one that sets us on our course towards analysing layout from a multimodal perspective.

10.1 Perspectives from graphic design

In a comprehensive overview of layout in print and digital media, information designer and typographer Robert Waller (2012) identifies several approaches to layout in graphic design literature that may also be evaluated from the viewpoint of mul-

Different views of layout

timodality. The first approach draws on *perceptual principles* originally established in Gestalt psychology. Gestalt psychologists formulated principles related, for instance, to *proximity*, which states that elements positioned close to each other are assumed to belong together. The *similarity* principle, in turn, states that the same holds true for visually similar elements. For graphic designers, these principles act as a loose rule-book for expressing relationships between content elements in the layout space. Bateman (2008: 59), who considers these and other Gestalt principles from a multimodal perspective, observes that they are indeed likely to contribute considerably to the process of decomposing layouts into meaningful elements during visual perception (for a discussion of Gestalt principles, see Pettersson 2017). As the following discussion will show, decomposition is also a key task for the multimodal analyst.

The second approach attends to the *aesthetic* qualities of layout using terms such as rhythm, contrast, tension or balance. However, Waller (2012: 242) points out that aesthetic approaches tend to treat layout as a visual equivalent of creative language use, such as poetry and language arts. Consequently, these approaches are oriented towards criticism and commentary of individual works, which makes them less likely to contribute insights into the generic principles governing layout and its use. The third approach is to see layout as *diagrammatic*—using graphic qualities to articulate relationships amongst content elements. Twyman (1979), for instance, examines the principles of organisation behind different layouts, which he characterises along a continuum extending from truly linear to non-linear. To exemplify, the layout of this book may be considered semi-linear, because the lines of written language terminate at specific points on the horizontal axis to form paragraphs. Newspaper front pages, in contrast, are non-linear due to their compartmentalised structure, which offers multiple access points into the layout (Bateman et al. 2007).

The fourth and final perspective encompasses the *generic features* of a layout. As Waller writes:

> "Layout is the main signifying feature of many familiar document genres: for example, newspapers, magazines, textbooks, user guides, packaging, and reference books. These everyday genres owe their very being to their layout. When readers see them, they know what they are, and what to do with them." (Waller 2012: 212)

The importance of these generic features to our everyday interactions with different layouts may not always be clear, because they are highly naturalised and remain in the background unless they fail to meet their communicative goals. Without these generic features one would have to learn to recognise and interpret each layout on a case by case basis, simultaneously reformulating hypotheses about their organisation and its means of signalling discourse relations between content elements (→ §4.3.1 [p. 129]). For this reason, Waller (1987) describes these generic features as an *access structure*, which layout uses to provide the reader with clues about its organisation, thus facilitating access to the content. This view has also found support in reading research:

Cohen and Snowden (2008) go as far as to argue that the successful interpretation of any layout depends first and foremost on recognising its generic features, which precedes any consideration of the content presented on the layout!

In the light of the discussion above, it would be reasonable to assume that layout would have gained considerable attention in multimodal research. This is, however, not the case. Working with broad assumptions that have not been subjected to empirical scrutiny, such as the concept of information value zones proposed in Kress and van Leeuwen (2006), has had unfortunate consequences for research on layouts (Thomas 2014): it has shifted the attention away from a core issue that the analyst faces when engaging with any artefact or semiotic mode making use of the layout space, that is, the process of *decomposition*. This process involves taking apart the contents of a page with the goal of defining some set of analytical units. At this stage, the analyst also needs to decide on granularity, that is, how finely the layout should be decomposed to answer the research questions that are being asked. This decision must be then upheld to maintain comparability between analyses.

Decom-
position

Decomposing an entire page or a single diagram, or any other combination of modes for that matter, is very likely to require different lenses, as applying the same scheme of decomposition to both runs a high risk of failing to identify the specifics of their individual means of expression. To put it simply, the selected lens should be capable of picking out the detail required to capture any potential meanings arising from the use of layout space. Consider again, for instance, the layout of this textbook, which is mainly organised around paragraphs of written language, but occasionally integrates diagrams and other graphic elements into its expression. A broad lens reveals that the *entire page* rarely uses the layout space to convey additional meanings, simply embedding the graphic elements into the linear structure driven by written language. Yet zooming in on a *diagram* may reveal that the spatial placement of written language in relation to graphical elements can be crucial for interpreting text-image relations (→ §11 [p. 279]). Therefore, we will now begin to sketch out a viewpoint that allows multiple perspectives into the layout under analysis.

What kinds of analytical tools, then, are available for the multimodal analyst for decomposing layouts? How can we better understand their spatial, formal and generic qualities from a multimodal perspective (cf. Waller 2012)? For making sense of their spatial organisation, that is, how a 2D canvas turns into a page with a coherent layout, we may draw on the *grid*—an established design tool for structuring two-dimensional space (Williamson 1986). The grid, which is often planned in advance to support the design process, consists of intersecting horizontal and vertical lines. These lines serve as guides for positioning content in the layout space. By providing this kind of guidance, the grid imposes constraints on how the layout space afforded by the 2D canvas may be used, which needs to be accounted for during decomposition (Bateman 2008: 84). This stands in stark contrast to the very high degree of design freedom assumed for composition by Kress and van Leeuwen (2006). To sum up, a grid transforms a 2D canvas into a layout by providing structure to the space.

Grid

By reconstructing the grid during analysis, we can begin to bring the layout space under some degree of analytical control. Figure 10.1 shows the reconstructed grid of a tourist brochure published in 1995, which has been created by drawing horizontal and vertical lines along the edges of content elements. This kind of a grid, which uses both vertical and horizontal lines to position the content, is called a *modular* grid (for additional grid types, see Bateman 2008: 81–83). A modular grid uses lines to establish spaces into which the content may be poured, which are called *modules*. As a design tool, the modular grid is highly flexible: the modules do not have to be filled to the brim and the content is allowed to spill over to other modules, in which case they form larger spatial regions.

Module

For reconstructing a modular grid, defining two intersecting lines for each content element often suffices to locate the points of support that guide the positioning of a content element. This has been illustrated by marking some of these support points using black dots in Figure 10.1. Rectangular shapes, such as the paragraphs on the left- and right-hand pages can be supported by a single point in the corner of a module. Other shapes, such as photographs cropped into the shape of a diamond, may require two points of support within a single module, as exemplified on left and right.

Fig. 10.1. A reconstructed grid superimposed on a 1990s tourist brochure ©Helsinki City Tourist Office 1995

With the grid in place, we can already begin to make observations about the use of layout space, as the modules help to describe how the space is structured. In Figure 10.1, the layout is based around four major columns on two pages. Each page includes two modules for carrying the main content in written language and photographs. Moreover, as the generic features of the tourist brochure would lead us to expect, familiar document elements—headers and page numbers—are positioned consistently to provide entry points to the double-page spread. This supports the intended use of the tourist brochure, which is designed to be mainly skimmed and searched, instead of being read through from cover to cover (Hiippala 2015*b*: 75–76).

Applying the Gestalt principle of proximity to the layout suggests a relation between the paragraphs of text and the photographs positioned in adjacent columns. Yet the layout does not generally use captions or other means (such as typographic emphasis in the body text) to draw explicit relations between the written text and photographs. The layout does, however, feature a single exception in the form of a solitary caption, which is positioned in the middle and bottom half of the double-page spread, as indicated by the black triangle in Figure 10.1. This caption states that "The Suomenlinna Sea Fortress is placed on the World Heritage List"—a landmark that is described in greater detail in a paragraph on the left-hand page.

The discourse relation drawn between the caption and the photograph is arguably more explicit than the non-structural, cohesive relations that hold between the verbal and visual content placed in adjacent modules (Bateman 2014*b*: 165–166). However, because the caption is placed next to several photographs, cohesive relations must nevertheless be resolved to identify which of the photographs is actually being described. The large photograph adjacent to the caption, positioned on the right-hand page, features coastal fortifications, which makes this photograph the most likely candidate based on the cohesive ties holding between the contents of the caption and the photograph.

Now, if layout space can convey which content elements should be interpreted together, does this capability make layout a semiotic mode? As Waller (2012: 242) pointed out above, layout is one of the main signifying features of any page-based artefact. In other words, the layout has an immediate effect: firstly, for the recognition of the artefact, and secondly, for adopting an appropriate strategy for making sense of its content and structure. Waller (2012: 239) exemplifies this aptly by pointing out that "No sensible person chooses from a catalogue, sets up a DVD player, selects a hotel from a travel guide, or looks up a word in a dictionary by starting on page 1 and reading through until the end."

We must tread carefully, because every layout emerges as a result of one or more semiotic modes organising themselves on a medium. To put it simply, there is no layout without a set of semiotic modes leaving traces on some canvas. However, due to its intended immediate effect, the layout also invites instant interpretation. For the analyst, this is precisely the moment to stop and observe, which may be illustrated using a comparison of layouts in two school biology textbooks discussed by Bezemer and Kress (2008). These textbooks are now taken up for discussion as a part of our first example analysis in this chapter. This is also the moment when we switch gears and begin to complement the previous research on graphic design with concepts drawn from the framework developed in this book.

10.2 Example analysis: school textbooks

Figure 10.2 shows two pages from two different school textbooks, which are nevertheless intended to be used in the same biology class, but by two different groups of students. The example on the left is intended for students with lower performance in school, whereas the one on the right is for those who perform on a higher level. Both pages deal with the same topic—how the human digestive system works—but the level of abstraction in their subject matter differs considerably. And, as we shall see, this distinction is also manifested in how these examples make use of the 2D canvas provided by the textbook medium.

In their analysis, Bezemer and Kress call for particular attention to be paid to the use of layout, suggesting that it conveys a fundamental difference between the students:

> "In the version for the 'lower tier' layout is much more 'spaced out,' compared with the 'dense' 'higher tier' version. Spacing is here used as a signifier of ability. That is, this use of layout realizes an ideology of *simplicity of display* . . . providing less 'information' is seen as apt for those regarded to have a lesser capacity to process information." (Bezemer and Kress 2008: 190)

Put differently, Bezemer and Kress argue that low-ability students are provided with simpler layouts with increased spacing, because they are unable to process densely populated 2D canvases. Although eye-tracking studies of biology textbooks suggest that high-ability students are indeed better at integrating multimodal content positioned across the page (Hannus and Hyönä 1999), the two pages must be interrogated far more closely before making statements about any possible ideologies allegedly in play. This is, then, the central lesson of our example analysis: how to pursue more exhaustive analyses of layout by diving beneath their surface, in order to build up the critical capacity of being able pinpoint how layouts differ from each other.

To begin with layout, both pages use the same grid to structure the entire 2D canvas, as indicated by the white grid lines overlaid on top of Figure 10.2. However, the first distinction emerges already in how the layouts allocate subcanvases to different semiotic modes. The low-ability layout on the left features a large diagram, which combines written language and diagrammatic lines into labels that target a combination of a line drawing and a three-dimensional naturalistic illustration (→ §11 [p. 280]). In the high-ability layout, the same diagram can be found in somewhat smaller size in the right-hand side column: this suggests that both low- and high-ability students are expected to be able to decompose these kinds of diagrammatic representations.

However, high-ability students are also provided with three additional diagrams at the bottom of the page, which feature written language, geometric shapes and diagrammatic elements (arrows) organised into a sequence that describes digestion on the level of enzymes, which obviously cannot be represented using naturalistic drawings. As these diagrams are absent from the low-ability layout, it may be assumed that

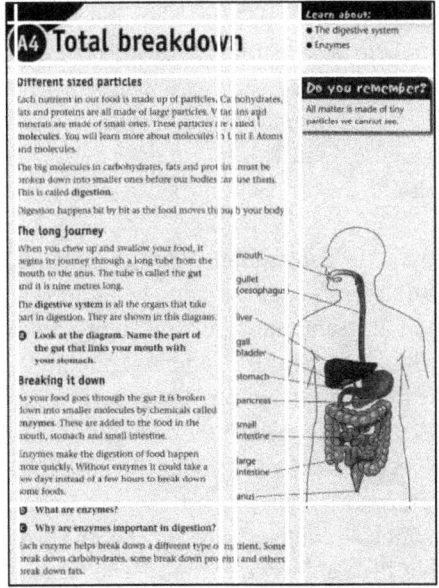

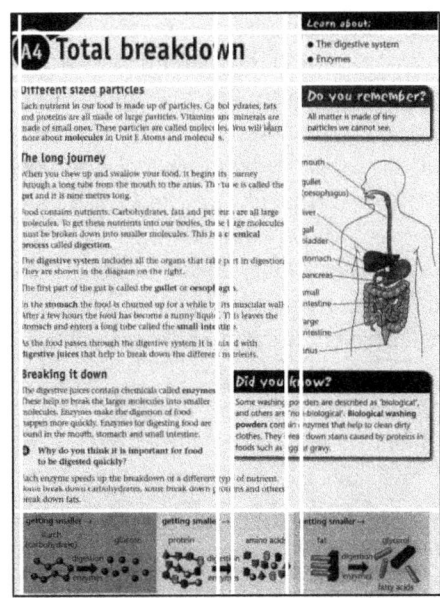

Fig. 10.2. Two school biology text book layouts for same-year students of low (left) and high (right) ability (Bezemer and Kress 2008: 191–192) ©Harcourt Education 2003

such a configuration of the diagrammatic mode is considered to be beyond the reach of low-ability students.

As said, although both textbooks are built upon the same grid, how the canvas is allocated to different semiotic modes is determined by its intended audience, as shown in Figure 10.3. Low-ability students are provided with a longer description of digestion in writing, which spans across the spatial region covering several modules from top to bottom on the left-hand size of the page. Key terms are highlighted using a bold typeface, as are several questions and instructions indented in the body text: typographic emphasis is used to instruct the student explicitly to pay attention to the diagram on the right. Moreover, the written description differs from its high-ability counterpart in terms of typographic *leading*, that is, the space between lines of text is greater.

This distinction may have lead Bezemer and Kress (2008: 190) to describe the layout for low-ability students as more "spaced out", but it should be noted that leading is a typographic feature, and hence belongs to written language, *not* layout. As such, this exemplifies some of the challenges that arise from conflating different

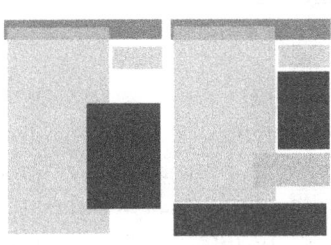

Fig. 10.3. Subcanvases on both low- (left) and high-ability (right) textbook layouts

kinds of multimodal phenomena, a situation that we attempt to avoid by identifying the canvases and the semiotic modes active on them before pursuing the analysis further. To summarise, the low-ability layout describes digestion using a long sequence of written language, while also leveraging typographic emphasis and formatting to encourage the student to inspect the diagram on the right. This layout does not, however, take further advantage of typography to highlight connections between text and the diagram.

The high-ability layout, in contrast, organises the content into separate modules, which demarcate the subcanvases assigned to different semiotic modes. In addition to the body text, these modules include two sets of diagrams and a separate box providing additional information: each module contributes towards the common communicative goal of describing digestion. To integrate the information provided, high-ability students need to break away from the linear structure of the written description and resolve any discourse relations that hold between the content positioned across the modules. Given this requirement, it is not surprising that typographic emphasis is not reserved solely for key terms in the body text, but also used to highlight certain terms that occur in the diagram labels. To put it simply, high-ability students are encouraged to integrate the content on different subcanvases, and the semiotic modes active on these canvases support this task. We will describe the means of providing this support in greater detail below.

That being said, two different kinds of organising principles appear to operate underneath the surface of each textbook, and these determine how the semiotic modes are combined in the layout space. For the low-ability layout, the principle appears linear, supporting the activity of *reading*, while the principle behind the high-ability layout seems to be spatial, encouraging the student to put the parts of the page together as a part of a process we termed *composition*. What we need to do, then, is to search our inventory of analytical concepts to build tools that can take the layouts apart and bring out their underlying principles of organisation for examination. To get started, our previous observation that the textbooks under analysis are page-based and use a grid to structure the 2D canvas provides a point of departure. With these criteria at hand, the concepts of *text-flow* and *page-flow* introduced in Bateman (2008, 2009, 2011) emerge as potential candidates for our toolkit, for the reasons we explicate below.

Bateman characterises text-flow and page-flow as semiotic modes commonly found in page-based multimodal artefacts. They are, essentially, abstractions that describe *how* written language, illustrations, diagrams and many more contributions are combined in distinct ways on canvases working with a page metaphor. The first mode, text-flow, is built around written language, whose structure unfolds in a linear manner: although diagrams, photographs and the like may be embedded into the linear structure of text-flow, the semiotic mode does not take advantage of the layout space to make additional meanings. An academic textbook such as the one you are reading, for instance, is an apt example of text-flow at work: the paragraphs of written language may be interrupted by diagrams, photographs and other graphic elements, but the underlying organisation remains linear and therefore largely driven by the discourse structure of written language (Hiippala 2016a).

In contrast, the second semiotic mode, page-flow, abandons the principle of linearity and turns towards "proximity, grouping of elements, framing and other visual perceptual resources in order to construct patterns of connections, similarity and difference" across the layout space (Bateman 2011: 26). On a more abstract level, page-flow may be characterised as a *composite* semiotic mode, which takes contributions from other semiotic modes and organises them into larger units in the layout space (Bateman 2011: 29). The individual contributions, such as instances of paragraphed text and individual diagrams, constitute the building blocks of the page—the semiotic mode of page-flow simply provides the mechanics required to interpret their combinations. As such, these mechanics bear close resemblance to what Waller (2012) has described as cues provided by a layout about its structure and interpretation.

Now, to return to the textbooks in Figure 10.2, the concepts of text-flow and page-flow can be used to explore the proposition whether there are indeed two distinct semiotic modes active in the different versions of the textbook, which are mobilised to support distinct forms of ergodic work, that is, *reading* and *composing*. Investigating this issue, however, requires zooming in from the spatial organisation of layout space to the semiotic modes present on the page, in order to consider the discourse relations that hold between them. Such a move must be supported by decomposing the layout into analytical units, which is a prerequisite for considering different kinds of multimodal structures present on the page.

So far, we have decomposed the page by uncovering the grid that imposes structure on the 2D canvas, enabling us to identify various spatial regions occupied by written language and diagrams. The next step is to establish how deep we must dig to uncover discourse relations in both textbooks, assuming that these relations hold the key to identifying their underlying principles of organisation. As Waller (2012) observes, layout can provide a considerable quantity of explicit visual cues about its *possible* internal organisation, but these candidate hypotheses must be subsequently verified. This involves checking whether discourse relations actually hold between the different modes, and if so, what kinds of relations may be identified. These relations, however, are not necessarily manifested on the level of text paragraphs and other elements commonly encountered in layouts, but require zooming in on the semiotic modes.

In terms of semiotic modes, both textbooks feature several instances of text-flow and diagrams, which we will explore in Chapter 11. For the time being, it is sufficient to note that diagrams in both low- and high-ability textbooks in Figure 10.2 feature written labels, which stand out as potential points of contact to the description of digestion realised using text-flow. These labels indicate specific parts of the digestive system represented in the diagram, such as mouth, gullet (oesophagus), liver, gall bladder, stomach, etc. With this in mind, let us consider a passage from the low-ability textbook, which retains the original typographic emphasis and indentation:

The tube is called the **gut** and it is nine metres long.
The **digestive system** is all the organs that take part in digestion. They are shown in this diagram.

> **(a) Look at the diagram. Name the part of the gut that links your mouth with your stomach.**

Now, compare the passage above to the following one in the high-ability textbook:

The tube is called the **gut** and it is nine metres long.
[…]
The **digestive system** includes all the organs that take part in digestion. They are shown in the diagram on the right.
The first part of the gut is called the **gullet** or **oesophagus**.
In the **stomach** the food is churned up for a while by its muscular walls.

To begin with, comparing the linguistic features of the two passages reveals differences in lexis (vocabulary) and grammatical mood, as the low-ability passage uses imperative mood to instruct the student. More generally, both passages obviously exploit the discourse structures inherent to written language, which act as the glue that holds pieces of unfolding text together (see, e.g., Martin and Rose 2003). These structures are naturally at the heart of text-flow as well. Both passages also make use of the typographic resources available to text-flow to highlight parts of the unfolding text. In addition, both make an explicit reference to the accompanying diagram, which takes us closer towards the issue at hand—the degree of integration between text-flow and the diagram.

Right below the reference to the diagram, the low-ability passage presents the students with a task that asks them to observe and name a specific part of the diagram. Although the sentence ("Name the part of the gut …") features lexical items (mouth, stomach) that have counterparts in the diagram labels, it is important to understand that this is not a description of the diagram, but an assignment that encourages the student to disengage from the activity of *reading* and switch to the activity of *composing* that is required to put the parts of the diagram together into a meaningful whole. This is clearly signalled using the imperative mood (look, name). The low-ability passage then moves on to another section describing the role of enzymes in the digestive system.

As shown above, the high-ability passage also makes an explicit reference to the diagram on the right, albeit using a passive and without typographic emphasis. What is worth noting here is that the two paragraphs that follow the reference do use typographic emphasis to pick out parts of the digestive system also featured in the diagram labels (gullet, oesophagus, stomach). We may either take these sentences simply as parts of the textual organisation of the entire passage consisting of text-flow, or alternatively, we may argue that this part of the passage is intended to be related to the diagram. In the latter scenario, students are encouraged to engage in *parallel* activities of reading and composing *across* different subcanvases. These activities would be then better supported by the semiotic mode of page-flow, which involves considering how the different elements in the layout space relate to each other.

Without additional examples, however, it is difficult to conclude whether the textbooks represent two alternative configurations of text-flow or whether the high-ability textbook should be considered an example of page-flow, which exploits the layout space to establish units that work together to describe digestion. In any case, this shows that the multimodal analyst must often dive beneath the visual features and delve into the use of semiotic modes to identify the organising principles behind a particular layout. Moreover, once the participating semiotic modes have been identified, it is crucial to perform any comparisons using the same set of analytical tools selected to do the job: contrasting non-structural *cohesive* ties with structural *coherence* relations quickly leads to untenable analyses (Bateman 2014*b*: 169–170).

In any case, any analysis of the participating semiotic modes must be preceded by a careful decomposition of the artefact under analysis. With well-behaved documents built upon a grid, the task may occasionally even prove somewhat trivial. However, with other external media, whose organisation is not based on a grid, the process of decomposition poses a number of analytical challenges. To illustrate these issues, in the following section we take up another example analysis, which discusses a particular genre—public service announcements—realised using the medium of a poster.

10.3 Example analysis: posters

As a medium intended primarily for public display, the consumption of posters differs from page-based media which are meant to be interacted with over longer periods of time. The poster, in contrast, is usually intended to have an immediate impact. Like any other medium, posters can carry numerous different genres that seek to achieve various communicative goals: they can persuade, instruct, explain or warn the viewer. To achieve their designated goals, the genres in the poster medium often integrate multiple semiotic modes: van Leeuwen (2005*a*: 121) aptly characterises them as multimodal communicative acts, "in which image and text blend like instruments in an orchestra". Although posters and page-based media are both founded on two-dimensional canvases, their analysis requires completely different approaches. In this example analysis, we show how to pull apart tightly integrated 'blends' of semiotic modes often featured in posters.

To begin with, some genres that have been previously studied from a multimodal perspective include movie and recruitment posters (Maiorani 2007; White 2010). Both studies build on the visual grammar proposed in Kress and van Leeuwen (2006). Whereas Maiorani (2007) provides a painstaking analysis of the three kinds of meanings of the metafunctions proposed in systemic-functional linguistics—i.e., ideational, interpersonal and textual meanings (→ §2.4.5 [p. 49])—in a series of posters advertising *The Matrix* movie trilogy, White (2010) complements the framework with the notion of modal density developed by Norris (2004*a*) to assess how much attention the posters demand from the viewer to unpack their meanings. According to White (2010), some posters are of high modal density, which means that the viewer must invest more in the interpretative process to make sense of their multimodal structure.

The process of interpretation may be illustrated by one particularly famous example, which has been discussed extensively in previous research: the so-called Kitchener poster, whose purpose was to recruit soldiers to fight for the British side in World War I, which is shown in Figure 10.4. Drawn by the graphic artist Alfred Leete, the iconic Kitchener image first appeared on the cover of *London Opinion*, a weekly magazine, in early September 1914. Due to its popularity, the design was quickly adapted into a recruitment poster (Surridge 2001).

Fig. 10.4. The Kitchener design discussed by van Leeuwen (2005*a,b*) and White (2010)

The rapid transition of the Kitchener design from the magazine to print medium has caused some confusion among researchers discussing its properties in multimodal research. What is actually the design of the magazine cover is occasionally treated as the poster, which in fact features a different text ("Wants YOU"), naturally embedded within a different medium. For current purposes, however, we continue to work with the example discussed in van Leeuwen (2005*a,b*) and White (2010), as this body of work serves to illustrate the challenges involving the analysis of posters. To avoid further confusion, we shall refer to the example in Figure 10.4—the magazine cover—simply as the Kitchener design.

Describing the multimodality of the Kitchener design, shown in Figure 10.4, van Leeuwen observes:

> "It might be thought that the text merely 'elaborates' the image in this poster. Both form an 'appeal', one with words, the other with a look and a gesture." (van Leeuwen 2005*b*: 79)

He continues:

> "They *fuse*, like elements in a chemical reaction. They become a single appeal that is at once direct
> and indirect, at once personal and official. And this fusion is effected by the visual style. There
> are no frame lines to separate image and text, and the visual style provides unity and coherence
> between the drawing and the typography. For this kind of semiotic 'reaction' process we do not
> seem to have any useful analytical tools at present." (van Leeuwen 2005*b*: 79)

Van Leeuwen attributes the 'fusion' of elements to visual style, because there are no
frame lines that separate the illustration and the accompanying written text. He also
laments the lack of analytical tools available for explaining the 'reaction' between the
two elements, which results in an appeal being made to the viewer.

The availability of tools, however, depends entirely on the viewpoint adopted in
the analysis. If one attempts to force the Kitchener design into a mould shaped for
page-based documents, expecting a clear modular grid with clearly framed text and
images, the analytical tools may indeed appear problematic. The same holds true for
abstract frameworks describing text-image relations, which define a limited set of relations that are assumed to apply across external media. Consequently, little more remains to be said than to characterise the nature of the appeal made by the Kitchener
design.

This problem persists as long as we hold on to our current set of tools and avoid
searching for viable alternatives in our toolkit. That being said, if we abandon the assumption that the Kitchener design must have some document-like qualities such as
frames (or modules), and rather approach the analytical challenge as a problem of decomposition, we are quickly put back on the right track. Once again, the analyst must
begin the process of decomposition by identifying the semiotic modes mobilised in
the Kitchener design. To this end, Figure 10.4 appears to feature contributions from
two distinct semiotic resources: a hand-drawn illustration showing a part of Lord Kitchener and an instance of written language: "Your country needs YOU".

These semiotic resources obviously stand in some sort of relation with each other,
as they appear to form a single appeal (van Leeuwen 2005*b*: 79). To arrive at this interpretation, however, the viewer must first determine which discourse semantics are
appropriate for making sense of their combination: the same applies to the analyst.
Luckily, on this occasion we are provided a rather clear cue: the presence of quotation
marks in the text invokes the discourse semantics of a specific semiotic mode—comics
(→ §12 [p. 295]). This semiotic mode provides precisely the kind of text-image relations
required for resolving the relation that holds between the illustration and the accompanying text in the Kitchener design through a mechanism known as *projection*.

For comics, projection offers the means to organise contributions from different
semiotic resources, which represent events that involve processes of saying or thinking (Bateman 2014*b*: 90). To exemplify, speech bubbles that are attributed to some
character or object using the 'tail' of the bubble, constitute one common form of projection. Essentially, comics use projection to render what natural language realises

using grammatical constructions such as "X said that Y". Recognising that the Kitchener design draws on the semiotic mode of comics, and more specifically, on mechanisms of projection, provides a candidate interpretation that effectively shuts out alternatives offered by text-image relations in documents, such as 'elaboration' (cf. van Leeuwen 2005b: 79 above). Moreover, this interpretation can be easily consolidated with the accompanying drawing, as this semiotic resource is central to comics.

What this shows is that the initial process of decomposition must be followed by a careful consideration of the semiotic modes at play. Most importantly, this process must not stop after describing the underlying medium (poster) and semiotic resources that are being drawn on (illustration, written language). Unless we take the next step, that is, determine which semiotic modes provide the discourse semantics necessary for interpreting the semiotic resources and their combinations, our analysis remains incomplete. This is precisely the kind of situation in which analytical tools may appear challenged, but such judgements should not be made before we have applied every tool in our kit. In other words, analyses must be pursued exhaustively.

The need to follow through and bring analyses to a conclusion may be exemplified by a more challenging example, a public service announcement intended to raise awareness of risks associated with obesity. The poster, shown in Figure 10.5, features a composite visual, which joins together two photographs showing a female and a male body. These photographs are connected by an additional element—a time bomb—an element that may be characterised as *diegetic*, that is, as something belonging to the 'story' conveyed by the composite visual. Put differently, the time bomb is not intended to be interpreted as a separate entity on top of the background photographs, but as a part of them. Superimposed on the subcanvas occupied by the composite visual is another subcanvas, which invokes the diagrammatic mode to introduce written labels that relate to a specific part of the underlying composite visual, namely the belly.

The main message of the poster, as conveyed by the headline, is clear: obesity kills. How obesity kills, however, is communicated less clearly. The viewer is invited to consider the labels superimposed on the composite visual to indicate various diseases associated with obesity. Yet a more careful consideration reveals that the label 'Uterine Cancer' is superimposed on the female side of the composite visual. Assuming a general knowledge of human biology, this raises an alternative interpretation: the positioning of the diagrammatic labels is meaningful, that is, they designate specific female/male parts of the composite visual. Therefore, the question is whether the composite visual should be decomposed into female and male photographs, which act as targets *only* for the labels superimposed on them, or alternatively, whether the composite visual should be taken as a single image, which is described by *all* of the labels. The central diegetic element in the composite visual, the time bomb, adds to the confusion, because it may be interpreted as a connector, which explicitly suggests that the two photographs form a single composite visual.

Fig. 10.5. A public service announcement poster from Hong Kong ©Hong Kong Department of Health

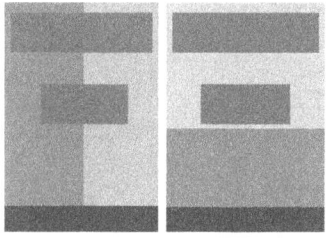

Fig. 10.6. Alternative decompositions of the poster into subcanvases

Mapping these kinds of alternative interpretations is crucial for describing the discourse semantics at play, which is in turn essential for determining the semiotic modes that are being mobilised in any medium. In the case of Figure 10.5, the conflict arises from invoking the diagrammatic mode to label the photographs, which provide a candidate interpretation that the labels refer to individual photographs, while simultaneously contradicting this interpretation by introducing a diegetic element that extends across both photographs that make up the composite visual—this is suggested graphically in terms of two alternative decompositions in Figure 10.6. At the same time, this also shows how theories of multimodality can be used for a critical evaluation of decisions made in graphic design (cf. Delin and Bateman 2002; Bateman 2017*a*).

10.4 Summary

To sum up, this chapter has attempted to show that any investigation of layout space must be preceded by an identification and description of the semiotic modes that are being mobilised and the subcanvases they occupy. We first showed how this can be supported by decomposing the layout space using a grid—a standard design tool that guides the placement of content in the layout space. As our subsequent example analysis of school biology textbooks showed, a grid can be used to pinpoint the location of subcanvases that the embedding canvas allocates to different semiotic modes.

This provided us with inroads into the semiotic modes on these subcanvases, which we then examined more closely to interrogate the difference between the two layouts. In the second example analysis, we put the grid aside, because not all layouts are founded on a grid-like organisation. Instead, we turned towards the notion of subcanvases to decompose posters—a medium that can occasionally integrate contributions from different semiotic modes very tightly. By identifying the different subcanvases and the semiotic modes active on them, we could describe how they subtly guide the viewer towards certain discourse semantic interpretations—although sometimes proposing conflicting interpretations on the way!

The lesson here is that different semiotic modes exploit the layout space to different degrees, which is why theories that assign permanent meanings to specific areas of the layout space collapse rapidly when faced with data that do not fit the model. This holds true for both page-based media with a grid-based layouts, and media that frame contributions from different semiotic modes less clearly, such as posters. Therefore, to build up the capacity to discern different semiotic modes, in the following chapter we proceed to consider the diagrammatic semiotic mode in greater detail.

11 Diagrams and infographics

This use case chapter engages with a broad range of multimodal ensembles that tightly integrate different means of expression: diagrams and infographics. Beginning with diagrams, this chapter considers the *diagrammatic mode* a particularly challenging target for multimodal analysis for two reasons. Firstly, the diagrammatic mode can realise independent, 'self-standing' diagrams, such as charts, graphs and schematic drawings, which may be readily combined with many other semiotic modes in equally many media. Some genres in page-based media that deploy the diagrammatic mode frequently include, among others, scientific articles (Lemke 1998; Hiippala 2016*a*), school textbooks (Guo 2004; Bezemer and Kress 2008) and assembly instructions (Bezemer and Kress 2016).

There is, however, more to the diagrammatic mode than its independent contribution in the form of charts, graphs and the like, which brings us to the second issue. While our toolkit already includes some of the pieces necessary for considering how diagrams work together with other semiotic modes in the *layout space* (→ §10 [p. 263]), zooming in on the diagrammatic mode will require us to sharpen our analytical tools. This requirement arises from the fact that the diagrammatic mode is highly compatible with several other semiotic modes, which enables them to form composite units. Our analytical tools must therefore be capable of detecting the diagrammatic mode at work, while also being sharp enough to cut through any composite unit to identify the participating modes and to describe their contribution.

That being said, one does not have to look much further to find examples of the diagrammatic mode working together with other semiotic modes: diagrammatic elements are often combined with illustrations, drawings, maps and photographs into what will be discussed in this chapter as *information graphics* (Holsanova et al. 2008; Dick 2015). We will address the distinction between diagrams and information graphics as the chapter proceeds, building on the analytical framework set out in this book.

In order to get started, let us consider a historical example to illustrate how the diagrammatic mode can join forces with another semiotic mode: Figure 11.1 shows reconnaissance imagery taken from a military aircraft during the Cuban Missile Crisis in 1962. Upon encountering this image in some medium that provides a two-dimensional canvas, for instance, in a history book or a Wikipedia article, it is tempting to dismiss the example simply as a photograph taken from an aircraft, which has been captioned and labelled.

However, just as when decomposing the layout space in entire page-based documents, we must stop here and determine which semiotic modes are being mobilised and on which canvases. Subjecting the image to a more diligent examination reveals that the example at hand is not just a photograph: it is a combination of multiple semiotic modes, in which aerial photography is complemented by written language in the form of a caption ("MRBM FIELD LAUNCH SITE") and labels ("MOTOR POOL, TENT AREA", etc.) that designate specific parts of the photograph using connecting lines.

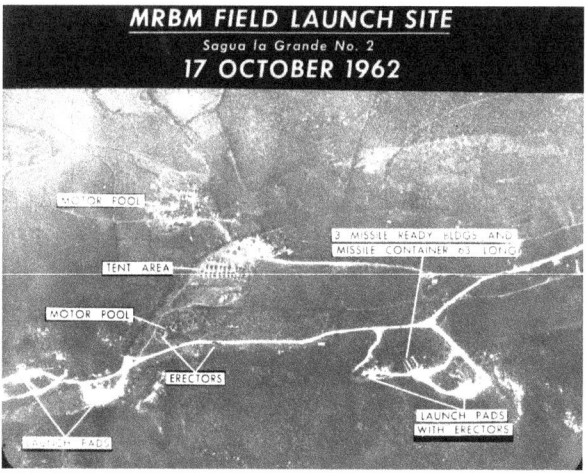

Fig. 11.1. An aerial photograph with diagrammatic elements taken during the Cuban Missile Crisis in 1962

In this communicative situation, the main burden in Figure 11.1 is obviously carried by the aerial photograph, whose interpretation requires special training, as human perception is not naturally geared towards observing scenes from a high-altitude, top-down perspective (Loschky, Ringer, Ellis and Hansen 2015). To support individuals not trained in the interpretation of aerial photography, such as politicians, students and researchers, the diagrammatic mode is mobilised to provide labels, which consist of written language superimposed on two-dimensional elements for added legibility and connecting lines that designate specific parts of the aerial photograph.

At the risk of stating what might sound like the obvious, but which actually needs to be adequately understood in terms of its multimodal import, the written labels and connecting lines are *not native to* the aerial photograph: they are contributed by the diagrammatic mode, which forms an additional, superimposed layer on top of the photograph. These layers may be seen as two subcanvases stacked on top of each other—as the following discussion will show, diagrams frequently make use of stacking to extend the space available to semiotic modes. To summarise, the aerial photograph features two different semiotic modes active on two different subcanvases, which are combined to form a composite unit.

The crucial difference between the example shown above in Figure 11.1 and 'independent' diagrams, such as those found in our case study of school biology textbooks in the previous chapter (→ §10.2 [p. 268]), lies in which capabilities of the diagrammatic mode are exploited. As we will see shortly below in connection with an example from assembly instructions, the diagrammatic mode can mobilise a wide range of semiotic resources for visual representation—line drawings, illustrations, and written language—as necessary, depending on the level of abstraction required for the task at hand.

The aerial photograph in Figure 11.1, however, does not belong to the semiotic resources directly available to the diagrammatic mode. In other words, the diagrammatic mode cannot 'generate' this kind of representation independently. However, lacking this capability obviously does not prevent the diagrammatic mode from forming a composite unit with the aerial photograph: diagrammatic elements such as labels and connecting lines appear to have no problem working together with the mode of aerial photography. For the multimodal analyst, the central question is, then, what enables these diagrammatic elements to be combined with the aerial photograph?

The short answer is that the diagrammatic mode is a full-blown semiotic mode. As pointed out above, the diagrammatic mode has several semiotic resources available for meeting its needs for visual representation: being able to draw on and manipulate a multitude of resources has enabled the diagrammatic mode to develop organisation that is not only compatible with the semiotic resources directly available to the diagrammatic mode, but also stable enough to be carried over to other semiotic modes. The labels and connecting lines in Figure 11.1 exemplify such contributions, whose presence provides the necessary cue for evoking interpretations associated with the diagrammatic mode, which viewers are able to consolidate with the mechanisms of interpretation appropriate for the aerial photograph.

At this point, it is also useful to consider the theoretical foundations of this capability: for composite units such as the annotated aerial photograph in Figure 11.1, these mechanisms arise from the respective *discourse semantics* (→ §4.1.1 [p. 116]) of the participating semiotic modes—aerial photography and the diagrammatic mode— which blend with each other to create a hybrid semiotic mode (Bateman 2011: 29–30). Acting as a site of integration, composite units such as annotated aerial photographs provide a fruitful ground for the emergence of hybrid semiotic modes, which is precisely why this chapter will not attempt to classify different types of diagrams, but rather attempt to lay down the foundation necessary for mapping the diagrammatic mode and its derivatives in further studies. This goal will be pursued by reviewing the previous research in several related fields of study, before taking on a number of example analyses by bringing the full arsenal of our theoretical concepts to bear on diagrams and information graphics.

11.1 Aspects of the diagrammatic mode

Not surprisingly, given their prominent role across many media, diagrams and graphs have attracted the attention of researchers working in various fields of study (for a recent overview, see Tversky 2017). Among others, the French cartographer and geographer Jacques Bertin (1981; 1983; 2001) and Edward Tufte (1983; 1997), an American political scientist and statistician, have made seminal contributions to theories of diagrammatic representation, formulating guidelines for designing and using diagrams to represent underlying data, objects or phenomena. Their contributions have mapped the extensive repertoire of visual expressions available to diagrams, while simultaneously evaluating their aptness across different contexts. Particularly well-known guidelines emerging from this work include Tufte's proposal that graphs, charts and other diagrams should avoid 'chart junk', such as decorations and embellishments, and commit instead to a high 'data-to-ink' ratio, that is, ensure every element of the visualisation should serve the purpose of representing data.

Although visualisations are often heralded as making complex meanings more accessible (Zambrano and Engelhardt 2008), critical perspectives into their use have also emerged in recent years (Dick 2015). Kennedy et al. (2016), for instance, argue that visualisations can privilege certain perspectives on the data and thus maintain and create power relations in society. Drawing on social semiotic theories of multimodality, the authors examine several conventions, including two-dimensional (top-down) viewpoints, geometric shapes, and 'clean' layouts.

Top-down views into maps, pie charts and scatter plots are suggested to create a sense of objectivity, because the data can be perceived 'at a glance'—which just means, more technically, that basic Gestalt principles operate rather transparently with respect to the intended meanings (→ §10.1 [p. 263]). In addition, the use of geometric shapes to represent data is argued to result in a dramatic simplification of its features, thus downplaying the individual characteristics of each instance represented in the data. Finally, clean layouts with a high data-to-ink ratio are proposed to be surrounded by an aura of simplicity, suggesting that selecting the data for representation in the visualisations is an unproblematic affair. All of these capabilities may be harnessed by those in power.

The work of Kennedy et al. (2016) underlines a particular challenge in studying diagrams, infographics and dynamic data visualisations. Connecting low-level features such as viewpoint to abstract issues like power or questioning the representational capability of geometric shapes runs the risk of sidestepping several crucial issues in the middle, which are highly necessary for understanding multimodality. As such, mapping the middle ground—media, canvases and semiotic modes—would also place any critical approach on a far stronger footing. Giving appropriate consideration to the middle ground is also essential for building up our capacity to distinguish between diagrams, information graphics and data visualisations (→ §15.2 [p. 351]).

From this perspective, one vital piece of the puzzle thus often appears to be missing: previous research has mainly focused on how diagrams and graphs represent things, events or processes, and whether they do so effectively, truthfully and in a socially responsible manner, rather than capturing the underlying principles that enable diagrammatic representations to emerge in the first place. Achieving this kind of perspective requires diving beneath the surface features of diagrams: to do so, we build on the previous work to theorise various properties of diagrams and infographics, while also relating them to the framework proposed in this book.

11.1.1 Unpacking diagrammatic structures

Engelhardt (2002) makes a major contribution towards unpacking the structure of the diagrammatic mode by treating diagrams and graphs as graphic representations, which make use of different *visual languages*. Within Engelhardt's framework, traffic signs, subway maps, bar charts and colour-coded maps are all considered to have their individual visual language. All of these visual languages are assumed to be founded on certain shared properties—a common graphic syntax, which enables them to generate meaningful expressions. According to Engelhardt (2002: 14), describing the graphic syntax of any visual language requires decomposing graphic representations into meaningful units, which then opens a view to their structure and interpretation.

The sharp-eyed reader will observe that syntax is a term borrowed from grammar—a borrowing which is entirely appropriate given the manner in which Engelhardt defines the term. Specifically, Engelhardt argues that graphic syntax has many features that resemble those generally ascribed to linguistic grammars. Just like natural language, Engelhardt proposes that a graphic syntax must have a set of basic units, which he terms graphic objects. Moreover, drawing another parallel between linguistic and graphic syntax, Engelhardt considers both to be *recursive* in terms of their structure, that is, one graphic object may embed another graphic object. To exemplify, bar charts superimposed on a map, which convey information about temperature and rainfall at specific geographical locations, are graphic objects embedded within another graphic object—the map. Moreover, the bar charts themselves include bars and lines, which are also considered to constitute their own graphic objects, adding another level of embedding to the recursive structure (Engelhardt 2002: 18–19).

Regardless of how many graphic objects are present and how deeply they are embedded as a part of a recursive organisation, all objects must occupy some *graphic space*, in which the graphic objects can stand in various relations to other graphic objects or to the graphic space they occupy. To describe these relations and how graphic objects participate in them, Engelhardt provides a broad inventory of syntactic principles and roles that are at work in the diagrammatic mode. To exemplify, Figure 11.1 can be used to exemplify one basic object-to-object relation—*superimposition*—in which a graphic object (the aerial photograph) serves as a background for several insets (labels and captions) (Engelhardt 2002: 84).

It is worth noting here that this shows once again how important and valuable it is to be very specific about the particular varieties of semiotic modes being addressed. The properties attributed to graphic syntax are very different from the kinds of properties that we set out in contrast to language in Chapter 2. We cannot, therefore, simply talk about 'visual modes' without making simplifications that will prove difficult to manage when our analytic attention becomes more focused. Dividing pictorial representations into the kinds of 'syntax'-like objects that Engelhardt sets out will generally be doomed to failure; in contrast, *not* doing this for diagrammatic depictions would hopelessly complicate any account and virtually preclude detailed analysis.

These kinds of recursive structures in graphic syntax then naturally influence the interpretation of diagrams: the viewer may need to work through several nested graphic objects to establish how they relate to each other. Naturally, the multimodal analyst is not excluded from the interpretive work expected from the viewer: attending to the layout space and the layering of content is of utmost importance for decomposing the diagram or information graphic in question. As we already saw in the case of the aerial photograph in Figure 11.1, the notion of canvas is crucial to this task, and may be seen as a generalisation of the 'graphic space' proposed by Engelhardt (2002).

However, given the wealth of diagrams and information graphics in circulation today, drawing up an inventory of graphic objects and their possible combinations in graphic syntax in the manner proposed by Engelhardt (2002) appears a daunting task. Moreover, as we have already established, any 'grammar' developed for a specific mode is rarely directly applicable to other modes. The same holds true for fitting together grammars defined for different semiotic modes, because their contributions only become integrated during the interpretative processes that operate as a discourse unfolds (→ §4.1.1 [p. 116]), *not* at a level equivalent to grammar.

Such interpretations are, in short, created dynamically and they are by nature *situated*, that is, they are dependent on their context of occurrence. In other words, these interpretations are not drawn from some pre-defined inventory, but are formed case-by-case following the principles of discourse organisation. As we will see below, this is precisely the moment when the theoretical extensions introduced in our framework begin to gain value in relation to the diagrammatic mode, as they may be used to explicate more precisely just how the diagrammatic mode guides its interpretation.

11.1.2 Reception and interpretation

Before taking up interpretation from a multimodal perspective in the case studies, it is useful to consider how the processing of diagrams and information graphics has been treated in other fields.

Experimental psychology, for instance, has a long tradition of using eye-tracking methods (→ §5.4 [p. 159]) to investigate the perception and processing of diagrams in textbooks and other media (see e.g. Hegarty and Just 1993; Hannus and Hyönä 1999). This avenue of research concerns itself with how humans construe high-level mental representations of objects, actions and phenomena with the help of diagrams. Another body of work has focused on human perceptual processes on a lower level: Ware (2012), for instance, provides a thorough discussion of information visualisation from the standpoint of visual perception, describing how humans perceive colour, texture, motion, shapes and patterns on two- and three-dimensional canvases, and how these fundamental processes may be exploited in design. Both are obviously highly relevant to multimodal research, as both high- and low-level features influence the perception of diagrams and information graphics (Holsanova 2014*a*).

Cognitive scientists have also applied eye-tracking methods to study the perception of information graphics: in a relatively recent study Holsanova et al. (2008) examine the perception of information graphics in newspapers. Drawing on theories of *multimedia learning* (Mayer 2009), Holsanova et al. (2008) build on two principles that have been suggested as beneficial for learning from multimodal artefacts. The first is the spatial contiguity principle, which states that related content should be positioned close to each other in the layout space. The second is the signalling principle, which proposes that explicit cues, such as typographic emphasis and formatting, support the interpretative process. However, as we have pointed out earlier, such broad principles can easily be overwhelmed by complex multimodal artefacts (→ §2.4.4 [p. 47]).

Holsanova et al. (2008) propose that the signalling principle may be extended by what they term the *dual scripting principle*, which states that:

> "People will read a complex message more deeply when attentional guidance is provided both through the spatial layout (supporting optimal navigation) and through a conceptual organisation of the contents (supporting optimal semantic integration)." (Holsanova et al. 2008: 1217)

To test the validity of the dual scripting principle, Holsanova et al. (2008: 1221) designed two different information graphics, whose purpose was to explain how one catches a cold. The first information graphic followed what the authors termed a 'radial' design, placing a naturalistic drawing in the centre, and surrounding this element with smaller layout elements positioned in the periphery, which provided details about what happens when someone catches a cold. This design was intended to provide the viewer with several possible entry points to the information graphic.

The second information graphic, representing a 'serial' design, carefully organised the contents into a sequence consisting of clearly demarcated panels, which describe catching a cold in a logical sequence, setting out the whys, whats and hows. Eye-tracking experiments then showed that the viewers not only followed the serial design in the anticipated logical order, but also *spent twice as much time* viewing the information graphic and performed twice as many integrative eye movements between text and images in comparison to the radial design.

Some commentators have suggested that allowing the viewer 'choice', as made possible by a variety of entry points, is a more reader-friendly alternative as it allows the viewer to select according to their interests and needs. However, such designs also force the viewer to "choose the entry point, to decide about the reading path, to find relevant pieces of information and to integrate them mentally" (Holsanova et al. 2008: 1224), which requires much more effort on the behalf of the viewer. Viewers may well select not to expend that effort at all, leading to open superficial readings of the material on offer. Designs which *help* their intended audience by guiding interpretation are not therefore to be dismissed.

For current purposes, the experiments performed by Holsanova et al. (2008) are particularly useful for highlighting a key aspect introduced in our typology of communicative situations: the *ergodic* dimension (→ §3.3.1 [p. 105]) . To reiterate, the ergodic dimension determines the kind of 'work' required from the viewer during the communicative situation. As Bucher (2011) points out, two- or more-dimensional canvases open up the ergodic dimension, forcing the viewer to take its possible effect into account. Whether this dimension is actively exploited, however, depends on the designer and the communicative situation at hand. As Holsanova et al. (2008) show, the ergodic dimension can be manipulated to a considerable degree in information graphics—up to the point where it affects whether viewers engage with them at all!

Along the ergodic dimension, communicative situations involving the interpretation of diagrams and information graphics on some canvas may be characterised as *micro-ergodic*, that is, to make sense of them, the viewer must engage in the kind of activity we termed *composition*. In other words, the viewer needs to take stock of what is being presented, usually through selective inspection, and put these parts back together. It is important to understand that composition here does not refer to active manipulation of some canvas, but to an internal process of interpretation undertaken by the viewer (cf. Bucher 2011).

As pointed out above, the extent to which the micro-ergodic capabilities are actually harnessed depends entirely on the communicative situation at hand and the realisation of the producer's intentions as explicit cues within that situation. For some purposes, such as attracting the viewer's curiosity, a maximally micro-ergodic information graphic intended to be freely explored may be an appropriate choice, whereas diagrams that are tasked with instructing or explaining, say, how to assemble a product, may choose to downplay the ergodic dimension, keeping the focus on the overarching goal of the communicative situation and the genre selected for handling the situation in question. Either way, the ergodic dimension provides potential that both the designer and the viewer must take into account, and this also applies to the analyst, as we will see with our move into the case studies below.

11.2 Example analysis: assembly instructions

Instructions provided in written language are notoriously absent from the products sold by the Swedish ready-to-assemble furniture manufacturer IKEA, much to the chagrin of customers whose diagrammatic literacy is not exercised daily. At the same time, the company, which sells its products around the globe, is likely to save substantial costs by not having to translate or localise the instructions for each market. These assembly instructions have also caught the attention of multimodal analysts: Bezemer and Kress (2016: 81–82) discuss a particular stage of the instructions for assembling a shelf, reproduced here in Figure 11.2.

Taking stock of the gains and losses of using diagrams in instruction manuals, Bezemer and Kress (2016: 81) observe that what diagrams gain in genericity, they lose in specificity: in Figure 11.2 "an 'ideal' screwdriver is shown, not one that is scratchy,

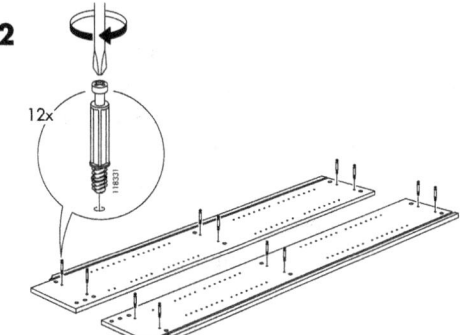

Fig. 11.2. Excerpt from IKEA 'Billy' shelf assembly instructions discussed in Bezemer and Kress (2016) ©Inter IKEA Systems B.V. 2008

used or odd in some way." They also suggest that diagrams cannot represent how "actions" actually unfold in time, as they can only refer to 'frozen' actions occurring at particular points in time: the elements in Figure 11.2 must then simply represent actions and objects at different times—moreover, the actor performing the actions and manipulating the objects appears to be deleted from the scene (Bezemer and Kress 2016: 81–82). Obviously, given the underlying *static* canvas, diagrams mobilised in a printed instruction manual cannot unfold in time, but this by no means prevents them from representing actions effectively. In fact, the human brain excels at extrapolating motion from representations on static canvases (Urgesi et al. 2006), an ability which finds very practical application throughout the history of comics and visual narrative as well (cf. Chapter 12). It is, as always, the semiotic mode that applies that determines what can be 'meant', not the material form alone.

That being said, despite all these alleged 'losses', the diagram in Figure 11.2 appears to successfully instruct the person at this particular stage of assembling the shelf. Let us therefore slow down and consider the diagram in detail. Proceeding from bottom to top, there are illustrations of the two sides of the shelf, whose sizes appear to be in somewhat correct proportion to the screws. The screws and their positions on the shelf sides are indicated using connecting lines—a diagrammatic element that we are already familiar with from our aerial photograph discussed earlier. Above and to the left, the diagram contains a large screw accompanied by the number 118331, circumscribed by what resembles a speech balloon that originates in the illustration below, labelled '12x'. Positioned right above is an illustration of the tip of a screwdriver surrounded by another diagrammatic element—a circular arrow.

How can we then describe this combination of illustrations and diagrammatic elements in terms of multimodality? As Bezemer and Kress (2016: 82) note, recognising that the segment represents different actions and objects is crucial, as this allows us to place this diagram into a context—Figure 11.2 represents a particular stage of a genre that may be tentatively called a *visual procedural instruction* (Bateman 2016). Recognising the genre provides a number of candidate hypotheses for identifying the semiotic modes at play, which narrows down possible interpretations about the relationships holding between the elements present in the diagram. For this reason, we are unlikely to treat the balloon as a speech balloon, as nothing else in the diagram tells us that the semiotic mode of comics is being drawn on. Shutting out comics for this particular diagram also removes the possibility of interpreting the speech balloon as an instance of projection—the visual equivalent of 'saying' or 'thinking' (→ §10.3 [p. 275]). The logical conclusion naturally involves the balloon acting as a container with the modifier ('12x') (Engelhardt 2002: 48), whose purpose is pick out the action that is to be performed at this stage and to numerically identify how many times that action is to be carried out.

Once the semiotic mode active in this genre stage has been identified as the diagrammatic mode, we adjust our expectations accordingly, often towards generic and abstract representations—this is why the current condition of the screwdriver represented in the diagram is irrelevant. What matters is identifying the parts and the actions they require, which requires the viewer to resolve the discourse relations between the different elements of the diagram. These relations may be inferred quite simply by applying the Gestalt principle of proximity, regardless of whether the relation holds between parts of the 'magnifying' container, or the entire container and the accompanying illustration of the shelf.

Because identifying the genre stage is crucial for interpreting the semiotic modes, the instruction manuals take care to signal the different stages of the genre very clearly. Figure 11.3 shows another stage of the genre, which could be named 'preparation'— a stage that precedes the actual assembly. This stage exploits a different mode of expression—namely that of comics—for which the interpretations that were considered invalid for the previous example appear as highly applicable. The sequence in Figure 11.3 exhibits several features that steer the interpretation towards comics: projection in the form of a thought bubble and the sequential organisation of images (→ §12 [p. 295]).

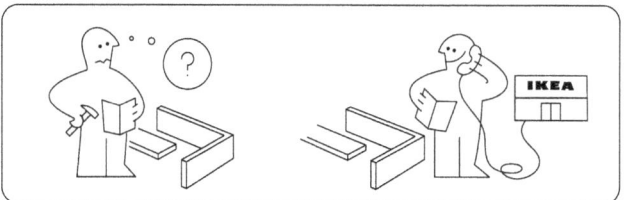

Fig. 11.3. The semiotic mode of comics deployed in the visual procedural genre ©Inter IKEA Systems B.V. 2008

The crucial issue, then, is this: before making judgements about text-image relations or any other issue commonly taken up in discussions of multimodality, we must determine which semiotic mode or collections of modes we are working with. Moreover, we need to account for the contextualising effect of genre. These issues need to be kept in mind as we now make the jump from diagrams to information graphics.

11.3 Example analysis: information graphics

In the second case study, we characterise information graphics as a composite semi-otic mode that integrates contributions from other semiotic modes (→ §4.1 [p. 112]). In other words, information graphics do not contribute any content themselves, but rather take contributions from other semiotic modes and provide the 'glue' that binds these contributions together. To explicate our view, the following discussion exam-ines multiple stacked canvases, the kinds of communicative activities performed on them, and their consequences for discourse relations across the information graphic.

Figure 11.4 shows a particular type of information graphic—a news graphic—produced by a news agency that licences them to other news outlets. Given its topic, a solar-powered aircraft completing a flight around the world, the information graphic in question might appear in different media ranging from daily newspapers to weekly magazines, popular science magazines, and on various general and special interest websites. For current purposes, we proceed with a general discussion founded on the view that these media share one fundamental characteristic: they are built upon two-dimensional canvases that provide the affordance of *layout* (→ §10 [p. 263]).

To support our analysis, Figure 11.5 provides an abstract representation of the par-ticipating semiotic modes and their placement in the information graphic. This shows that the semiotic modes generally occupy their own spatial regions, but regularly over-lap each other. In contrast to entire page-based documents, the layout space is rather densely populated, but it nevertheless retains certain document-like features, such as modularity, which involves dividing the layout into modules into which the content may be placed (→ §10.1 [p. 265]). This information graphic does not, however, estab-lish modules using a grid with horizontal and vertical lines, but instead defines loose spatial regions occupied by individual semiotic modes or their combinations.

To reiterate, let us consider for the time being that the embedding canvas is provided by a newspaper double-page, which features a grid-based layout that as-signs content into dedicated modules (cf. e.g. Bateman et al. 2007). In one module, we find a subcanvas hosting the information graphic, whose layout is not founded on a modular grid, but built on spatial regions that are placed extremely close or on top of each other. In other words, the information graphic features a number of subcanvases claimed by the contributing semiotic modes (→ §7.1.3 [p. 223]).

What all of this means for our analysis is that we cannot assume that the same principles of organisation apply all the way down from the embedding canvas (the double-page) to the specific subcanvases in the information graphic (spatial regions). Acknowledging the hierarchy of canvases is also essential for considering the kind of activities required in the communicative situation: taking stock of what is being presented on the entire canvas will undoubtedly require a different type of effort than the information graphic and its parts.

Now, to move on to the semiotic modes active in the information graphic, Figure 11.4 features several different kinds of three-dimensional illustrations. In addition to

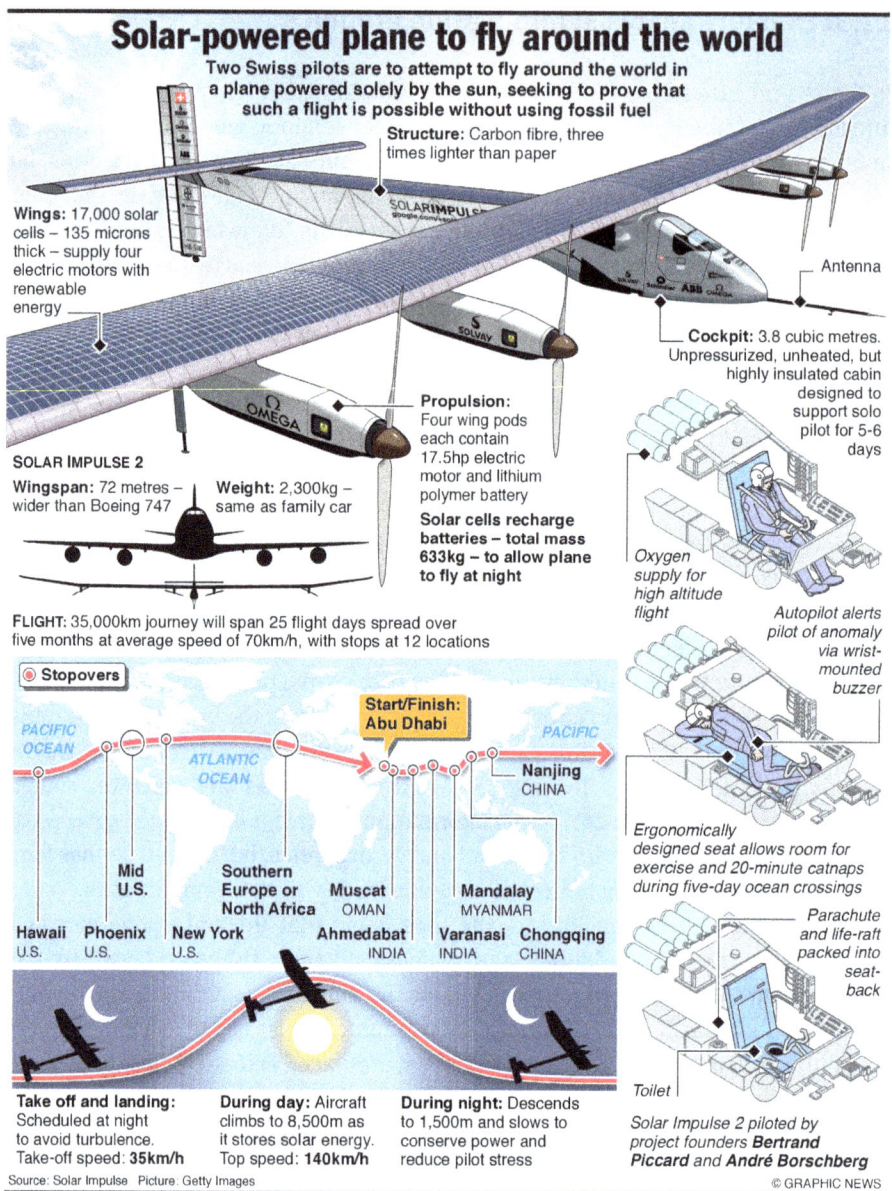

Fig. 11.4. An information graphic describing *Solar Impulse 2*, a solar-powered aircraft and its planned flight route around the world. ©Graphic News 2015, used by permission.

Fig. 11.5. Distinct contributions from different semiotic modes to the *Solar Impulse 2* information graphic: 3D illustration (blue), written language (brown), diagrammatic elements (yellow), 2D illustration (green) and map (red). Empty space is marked as white.

the *naturalistic drawing* positioned on the top, the right-hand side of the information graphic features several schematic drawings that provide *exploded views* of the cockpit. Essentially, what we have are two different kinds of illustrations complemented by the same diagrammatic elements—connecting lines and written labels. These diagrammatic elements not only describe what is being portrayed in the illustration, but also serve the broader goal of allowing the viewer to access the schematic drawings, performing a similar supportive role as in the annotated aerial photograph discussed earlier.

The information graphic features two-dimensional illustrations as well. The two silhouettes on the left-hand side represent another kind of technical drawing—a *planform*—commonly used to portray the shape of an object. Below the planforms, diagrammatic elements superimposed on the geographical map establish the flight route, whereas the diagram underneath the map illustrates distinct phases of the flight during night and day. Having identified the distinct contributions, realised using various semiotic modes, we can then begin to consider how they are combined in the information graphic and set out in relation to each other.

In order to begin resolving the discourse relations in Figure 11.4, the viewer must form working hypotheses about the relations between different contributions. Candidate interpretations are provided by the principles of spatial proximity and connectivity, which arises from the presence of diagrammatic elements (Bateman 2008: 61). This may be effectively illustrated using Figure 11.5, in which narrow yellow areas (connecting lines) frequently connect those coloured brown (text) and blue (3D illustration). Testing these hypotheses against the configurations found in the information graphic allows the viewer to zoom in and consider *what* kinds of discourse relations hold between the elements produced within the distinct semiotic modes.

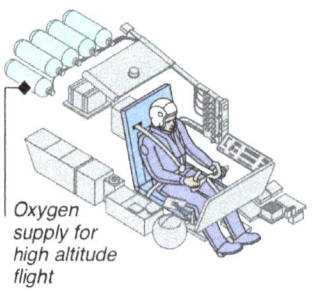

Oxygen supply for high altitude flight

Fig. 11.6. Nucleus (cockpit) and satellite (label) related to each other using IDENTIFICATION

For the analyst, this obviously leads to the question of how to select an appropriate framework for describing these discourse relations. There are several possibilities and the decision should be made with respect to how well the kinds of relations offered match the analytic goals in a particular case. Given the present target of analysis, a good candidate framework is the extension of Rhetorical Structure Theory (RST) proposed by Bateman (2008), since this has been used to describe text-image relations in several very different kinds of documents, including instructional texts (André and Rist 1993), entire newspaper spreads (Taboada and Habel 2013; Kong 2013), and diagrams in school textbooks (Bateman 2014*b*: 215–217). By employing the same analytical framework across a range of canvases, we can more readily compare what kinds of discourse relations are mobilised when presenting, for example, news stories using written language or information graphics. In this case study, however, we will limit ourselves to the analysis of the information graphic.

Before proceeding to describe specific examples of the information graphic using RST, a brief recap of the principles of this account of discourse is in order. RST was originally developed to characterise why written texts are perceived as coherent wholes rather than as simple sequences of sentences. To achieve this, RST defines a set of relations (with specific criteria) that may be considered to hold between text segments. Local text coherence is provided when a relation can be found; a text as a whole is coherent when a complete RST representation can be built relating all the text segments making up the text. Each relation defines two roles for the segments it relates: one takes on the role of the 'nucleus', any others are 'satellites'.

The nucleus carries the 'main point', to which the satellites add secondary information. Bateman (2008: 162) then introduces additional relations to describe rhetorical relations between composite units consisting of verbal and visual elements. One of these relations, which are generally intended to capture the relations holding between "fragments, individual entities and incomplete propositions", is that of IDENTIFICATION. This relation is used to assert identity in part-whole relations: we can find this relation in Figure 11.6, for example, when it uses a noun phrase ('Oxygen supply …') to identify a part of the aircraft cockpit.

Bateman (2008: 162) notes that IDENTIFICATION bears close resemblance to another relation, ELABORATION, in which the satellite provides additional information on the nucleus. For some combinations of the exploded cockpit views and their associated labels, ELABORATION may be considered a more appropriate choice than IDENTIFICATION, as the labels are not fragmented or incomplete, but consist of entire clauses. Both relations nevertheless build on the same spatial configuration of *label-connector-*

labelled, which emphasises proximity and connectivity, as described above. The spatial regions on the lower left-hand side of the graphic, however, require the viewer to formulate alternate hypotheses about the discourse structures, which need to be resolved to make sense of the information graphic. The subcanvases shown in Figure 11.5 provide a strong clue about this, as their spatial organisation differs from the previous examples.

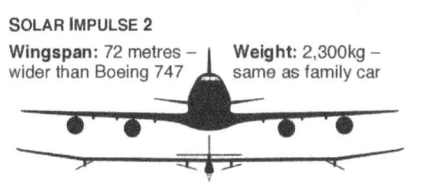

Fig. 11.7. A more complex example with CONTRAST

To draw on a single example from the left-hand side of the information graphic, the combination of planforms and written language requires additional work from the viewers to decompose increasingly complex text-image relations. In order to arrive at the correct (i.e., mostly likely intended) interpretation, the viewers must recognise that the fragments ('Wingspan ...' and 'Weight ...') do not function as labels to the planforms, but act as satellites of the header ('SOLAR IMPULSE 2'). The relation holding between the nucleus and the satellites may be described as PROPERTY-ASCRIPTION, which asserts a "relation between an object and something predicated of that object" (Bateman 2008: 162). Moreover, the other satellite is restated visually by contrasting the two planforms. This is visualised using the common notation for representing RST structures in Figure 11.8.

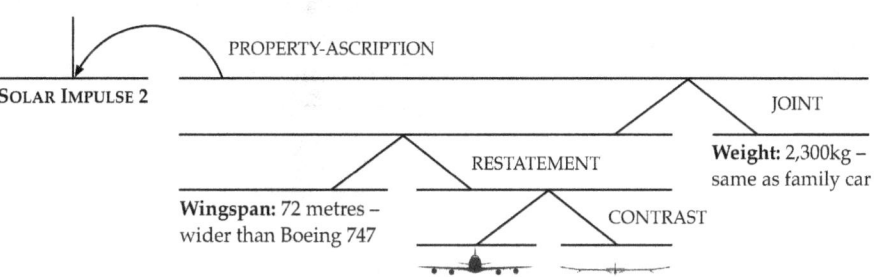

Fig. 11.8. A partial rhetorical structure in the *Solar Impulse 2* infographic. Horizontal lines designate segments of the graphic; other lines and arcs show the relations holding between those segments.

In contrast to the established spatio-visual *label-connector-labelled* configurations in Figure 11.6, it may be argued that the combination of planforms and written language in Figure 11.7 demands more ergodic work on behalf of the viewer. This claim can be supported by several points. Firstly, no connecting lines or other explicit visual cues are provided to support the interpretative process. Although the written fragments relate to Solar Impulse 2, they are positioned next to the planform of a Boeing 747. This creates a discrepancy between the visual representation and the intended rhetorical structure, which arises from the potentially conflicting interpretation based on spatial proximity. In other words, the viewer must not associate the fragments with the Boeing 747, although their placement suggests a possible relation!

Secondly, as Figure 11.8 shows, the rhetorical structure is considerably deeper, as the hierarchical RST tree embeds three levels. To underline this contrast, most IDEN-TIFICATION or ELABORATION relations designating or describing a part of the drawing do not embed further relations: their organisation could be represented on a single level. Taking these issues into account, we may argue that resolving the discourse structure from the infographic to yield the structure shown in Figure 11.8 demands more micro-ergodic work, which we described earlier using the notion of *composition*—taking stock of what is presented and putting any parts found back together.

On a more general level, the entire *Solar Impulse 2* information graphic appears to impose little control on the *order* in which the ergodic work is to be performed. This stands in contrast to the information graphics discussed in Holsanova et al. (2008), which cued a reading order by introducing panel-like sequential structures invoking the semiotic mode of comics (→ §12 [p. 295]). In any case, information graphics are clearly capable of integrating contributions from multiple semiotic modes in a manner which may or may not simultaneously exert control on the activity of *composition*. This capability arises from manipulating the layout space, which enables information graphics to set up subcanvases occupied by various semiotic modes.

When we move to dynamic data visualisations, the situation changes. Compared to information graphics on static canvases, dynamic data visualisations "are usually more interactive, allowing viewers to explore, manipulate unedited data and discover their own story" (Schwalbe 2015: 432). Engaging with dynamic data visualisations therefore entails *exploration*, ergodic work of a different kind that we will describe in connection with dynamic data visualisations in Chapter 15.

11.4 Summary

In this use case chapter, we considered the diagrammatic mode, discussing its structure, and particularly, its capability to 'attach' to contributions from other semiotic modes. As the examples showed, diagrammatic elements can be effortlessly combined with photographs, illustrations and drawings, to name just a few. Therefore, to catch the diagrammatic mode in action, the analyst must be constantly on the lookout. This became evident in our analysis of the information graphic, in which diagrammatic elements were used to support a range of rhetorical relations between contributions from an equally wide range of semiotic modes. We suggested that this also characterises information graphics more generally, as they mainly provide the 'glue' required to bind together contributions from quite diverse semiotic modes. This glue, however, is not based purely on contributions from the diagrammatic mode, but makes extensive use of the layout space as well.

12 Comics and graphic novels

Following on from the preceding chapter, our concern now will be another case of static multimodal ensembles that integrate written language and pictorial elements: comics, graphic novels, and related forms. In terms of our classification of media and communicative situations from Chapter 3, we are still then concerned with a 2D, static, observer-based and immutable micro-ergodic canvas, but this characterisation does not yet tell us about the specific ways in which meanings are being made within this canvas—for this, as we emphasise in all of our examples, we have to engage with the semiotic modes that are in play as well.

The kinds of 'meanings' that are constructed for comics, graphic novels and related forms are, in fact, very different to those of the infographics of the previous chapter, even though they draw equally on the affordances of their canvas. This offers another rich source of examples where knowing about a 'medium' is not yet sufficient, since any particular medium may be drawing on, or allowing, considerable variety in the semiotic modes that may apply. Moreover, as we shall see, we must also consider further sources of variation that contribute: first, comics and graphic novels are now used for several quite different 'communicative purposes'—they therefore exhibit *generic variation* by participating in, or realising, distinct genres; second, comics and graphic novels are also subject to considerable *cultural variation*. American superhero comics look very different (and employ different conventions) to both European comics and Japanese manga, for example. We will see where these phenomena find their place in our multimodal view as we proceed.

In comics, graphic novels, etc., one of the possibilities that we introduced in Chapter 10 for layout—that of defining framed, segmented 'modules' within the two-dimensional page space—is taken up and given some very specific communicative tasks. Considered quite generally, layout modules are defined which, on the one hand, contain tightly interwoven combinations of pictorial materials and, possibly, written text and, on the other, are themselves used as building blocks for broader sequences intending to achieve particular informational and emotional effects. This formulation in fact re-constructs a classic definition of comics offered by well known comics researcher Scott McCloud. Comics are, according to McCloud:

> "juxtaposed pictorial and other images in deliberate sequence, intended to convey information and/or produce an aesthetic response in the viewer." (McCloud 1994: 9)

Placing this against the background of our broader approach to multimodality serves as a reminder that media and their semiotic modes never exist in a vacuum! As we probe McCloud's definition further in this chapter, we will see some rather specific properties of the 'comics'-medium emerging—but we will *also* encounter places where other semiotic modes are at work as well. Moreover, we need to know the conditions that are required for the ways of making meanings we discover in comics to travel to *other* communicative situations—such as the web, films, embedded modules within other layouts, and so on. This is part of the work that a strong anchoring in canvases does for us.

Starting then with a fairly informal characterisation of comics, we will consider how to go about their analysis employing the principles we have set out so far. This will move us progressively from particular physical manifestations and publication forms that we might accept methodologically as offering a starting point, to abstract combinations of semiotic modes which may now well appear in a variety of media. This will be an important step since there is considerable discussion nowadays concerning the different places where something 'comics-like' appears, the potential relationships that may hold with neighbouring forms, such as picturebooks, as well as the possibility of defining comics at all. In fact, none of these questions can be pursued very profitably without a more robust grounding in multimodality.

Some researchers attempt to restrict what is to be understood as comics by means of the genres that they are used for. And here, the usual choice is *narrative* (cf. Pratt 2009). Comics researcher and cognitive scientist Neil Cohn, for example, is very careful to explain that his extensive empirical work is concerned not with physical artefacts that might or might not be called 'comics' but rather with an abstract account of a visually-based cognitive system, a *visual language*, whose properties appear centrally oriented towards narratives (Cohn 2013c). Philosopher Aaron Meskin also takes narratives to be important:

> "In the first place, almost all uncontroversial examples of comics that we know of are narrative in form. Superhero comic books, the comic strips that appear in daily newspapers, the underground comics of the 1960s and 1970s, and the contemporary 'serious' graphic novel—all these are narrative in form." (Meskin 2007: 371)

However, from the perspective of multimodality we might well have some doubts about the existence of a form of expression that is oriented to one and only one communicative purpose. Considering the diagram of the relationships between texts, media and modes that we set out at the end of Chapter 4 in Figure 4.4, one might expect that any bundle of semiotic modes identified operating in the medium of comics would be put to work for a range of communicative purposes, or genres.

And this does indeed seem to be the case. The trend towards using comics as informational or persuasive media—i.e., two further functions of texts and discourse— without any fictional appeal is certainly increasing. In addition, Bramlett et al. (2016) provide a list of no less than eleven different, narrative and non-narrative 'comics genres', including the typical superhero comics, western or horror comics as well as autobiographical and journalistic comics. We can see, therefore, that there a now a variety of areas in which the medium of comics is employed, including textbooks, documentaries, biographies, instructional texts and illustrations for science education. Narrative may well be a 'standard feature' of comics (cf. Meskin 2007: 371–372), but can hardly be considered an essential property that comics *must* perform. And, even for those comics which do appear to be employing the reach of their semiotic modes in order to tell stories, it would still be necessary to be rather more specific about just what 'narrative' is before coming to many conclusions; this is then also an issue that we will take up, albeit briefly, below.

For a more neutral and, at the same time, more incisive analysis from the perspective of multimodality, it is thus particularly important to ask for the specific multimodal affordances of comics in contrast to, for example, diagrams and infographics or websites in order to begin examining their patterns and regularities for meaning construction. The ability of comics to fulfil narrative functions, as well as any other functions they may be used in service of, will certainly need to be built upon the more foundational and general properties of this form of communication as such. Some of these properties may well have 'evolved' over time in order to facilitate story-telling, but that is something best revealed by empirical analysis rather than by fiat.

12.1 Comics: basic ingredients

So what goes into a comic? We mentioned above that the particular use made of identifiable segments of the composed layout appears rather central for the kind of artefacts that we are closing in on. In the context of comics, these segments are called *panels* and they indeed, as McCloud pointed out above, typically combine written text and pictorial material in an unusually tightly interwoven form. Several researchers have also pointed out, however, that the medium of comics readily incorporates not only these usual combinations that are most commonly thought of when considering comics, but may also freely use other visual data, such as photographs, paintings, maps or charts; sometimes they may even dispense with written text in order to convey meanings which are less narrative than a result of "creative experimentation and innovation" (cf. Pedri 2015: 2). And, conversely, some (e.g., Cook 2011) have even asked whether pictorial material is strictly necessary as well!

One quite common method for dealing with less than clear situations such as that painted in the previous paragraph is to propose a *prototype* approach—an idea suggested by the psychologist Eleanor Rosch (1978) and going back to Wittgenstein's famous notion of *family resemblances*. Essentially the idea here is that a category may be defined by a collection of properties; some of these properties might be more criterial than others and no property is essential. Thus an object that has many of the strong properties is a better exemplar than one that has fewer or weaker properties. The better candidates are 'closer' to the prototype.

To see this applied, lets take a look at a quite famous and already rather old example of a visual narrative: the Bayeux Tapestry, which was created in the 11th century to tell the story of the decisive battle between Normans and Saxons at Hastings in 1066. Interestingly and despite its rather exceptional status as a tapestry with its sheer quantity of cloth material, the Bayeux Tapestry is occasionally claimed as one of the first comics—for example by Scott McCloud (McCloud 1994: 12–13). Figure 12.1 shows an extract from the tapestry in which Harold, the Earl of Wessex and later King of England, sets off for France across the English Channel.

Fig. 12.1. Extract from the Bayeux Tapestry, showing Scene 4: 'Here Harold set to sea' (*Hic Harold mare navigavit*) ©John McLinden, Flickr Creative Commons Licence

Just as frequently, however, the status of the Bayeux Tapestry as a comic is called into considerable doubt: how might we decide? Certainly, from the information expressed and the form of that information, there are many similarities to be drawn with traditional notions of comics. For example, the event of sailing is displayed in this extract by several visual depictions placed next to each other, each accompanied by a textual addition above the images describing the event, such as: "Hic Harold mare navigavit" – "Here Harold set to sea". And, if we take a closer look at the details of the depictions, we also see other events, such as, for example, people eating and drinking (on the left side of the image), characters bringing animals to the boat and others preparing the departure or looking out to sea. All these events tell us parts of the story depicted in the Bayeux Tapestry—which is therefore clearly a narrative. Thus, along the dimensions of narrative, segmented story elements, combinations of text and image, use of visual materials, its use of a static medium to display dynamic events, etc., the Bayeux Tapestry would evidently score rather highly.

However, in its material form as a tapestry, it certainly exhibits some rather less common properties. Furthermore, the perception and interpretation of the events it portrays and their overall story does not happen in the same way as typically found in comics. Given the fact that the tapestry is over 70 metres long, recipients have to walk along the material to read and view all the texts and depictions. The story thus has to be read in sequences of the verbal text and individual pictures that happen to be in view of a 'travelling window' as one moves along its length. On dimensions of form of material, distribution, methods of access and so on, therefore, the Bayeux Tapestry scores rather poorly and so can certainly be seen as a less prototypical case.

Moreover, some argue that it is obviously the case that the Bayeux Tapestry is not a comic because comics only came into existence gradually from the 18th century onwards; the simple fact of finding stories that are told in sequential segments, which actually goes back considerably further than the Bayeux Tapestry at least to Roman, Greek and Egyptian times, is not considered sufficient. Meskin, for example, considers the hypothetical case that one should find a piece of music from the 17th century that exhibits properties reminiscent of jazz and pointedly notes:

> "Nothing could have counted as jazz in the seventeenth century, and any theory that implied that there was an instance of jazz 350 years ago or more would be anachronistic. Think of how the incorporation of the Bayeux Tapestry into the category of comics would reshape our appreciation of it. 'Where are the speech-balloons?' we might be led to ask. 'How radical to embroider a comic!' I take it that these would be distortions." (Meskin 2007: 374)

In this case, then, the sociohistorically situated medial properties of 'comics' are made strongly criterial—and this appears to be the most sensible choice.

While it is true, and we will see some of this in more detail when we set out characteristics of the semiotic modes deployed in comics, that there are some similar meaning practices involved in comics and stories told with sequential images of other kinds, 'comics' is primarily a term that labels a medium and, for *media*, issues of their manner of distribution, publication models, embedding in society and so on must be given priority (→ §4.2 [p. 123]). This also means, as a necessary corollary, that before studying such artefacts in any detail, one should familiarise oneself with the history of the medium as far as possible (cf. e.g., for comics, Lefèvre 2009; Duncan and Smith 2009; Heer and Worcester 2009; Goggin and Hassler-Forest 2010*b*; Stein and Thon 2013; Baetens 2015), so as to avoid the equivalent of postulating 17th century jazz!

The term *graphic novel* is also much discussed and questioned and is today commonly used more by the publishing industry to promote the genre as 'serious literature' (particularly after *Maus* by Art Spiegelman won the Pulitzer Prize). In general, the distinction is based on formal aspects. On the one hand, the average page number for graphic novels is normally higher than that typical for comicbooks (e.g., 32 pages). On the other hand, researchers underline that the "change in format from the center-stapled pamphlet to the larger, bound graphic novel" (Jacobs 2013: 205) is also accompanied by a focus more on single, coherent narratives, whereas comicbooks more often exhibit characteristics of 'seriality', that is, smaller, self-contained stories leaving the status quo unchanged from story to story. Moreover, as Humphrey (2014: 72) observes: "the term 'graphic novel' has become established solely as a way of attributing cultural authority to comics. It has been successfully applied to many books that are neither particularly graphic nor novelistic, leading some to argue the term simply refers to 'comics with a square binding'".

What, then, does the medium of comics—and several other closely related medial forms, such as graphic novels, comic strips and so on—use in order to make its meanings?

First, we must orient ourselves using our multimodal methodology, following the multimodal navigator from Chapter 7 in order to begin to subdivide the communicative situation involved in the use of any artefact being analysed. For the type of canvas at hand, therefore, we can imagine communicative situations with and around 'comics-like' artefacts in which people are reading a printed book (i.e., a 2D, static, observer-based and immutable ergodic canvas). From this vantage point, we can ask about the material details of the comic, picturebook, graphic novel, etc. This would involve its design and the effects this design and material have on readers—for example: how do they have to turn the pages? Is the book read from left to right or from right to left?—which is, in our Western culture, for example the case for many Manga books which follow the traditional Japanese order from top to bottom and from right to left even when translated into English or German.

As we explained in Chapter 7, *any* of these aspects of the broader communicative situation may be sites of multimodal interest. Consider, for example, the further 'consumption'-oriented aspects of communicative situations involving 'comics' in which readers have to follow certain instructions in order to be able to access the structure of the artefact at all. This can be seen in the case for books such as *Toutes les mers part temps calme* by Alex Chauvel or *La véridique* by Olivier Philipponneau and Alain Enjary (for discussion, see Bachmann 2016). In order to analyse situations of these kinds, we also have to access the broader canvas in which people approach the book in question, for example in a reading group or in a comicbook store, as well as the supporting context of talking about the books or just experiencing them in isolation; here ethnographic methods of various kinds might be appropriate (→ §5.2.2 [p. 144]).

If we narrow our focus to just the 'printed artefact' canvas, we still have some preliminary work to do to set out potentially different types of information as well as their respective canvases and sub-canvases and their own typical deployment of semiotic resources. Here, most analysts of comics arrive rather quickly at the sub-canvas of individual pages, which we must see as one individual type of subcanvas within the context of a whole comicbook. This subcanvas is itself usually quite complex: its subcanvas may deliver several further subcanvases that we might deal with in our analysis, such as the overall compositional form of the page, its smaller units, and its selection of further elements such as photographs, diagrams or the use of colour, for example—these all add individually and collectively to the meaning-making process.

Just as was the case in our discussion in Chapter 7, whereas this kind of division may appear at first glance to be complex, it is in fact essential for making sense of 'comics', and of particular pages within comicbooks, *no matter how complex they get*. For example, a comic may well employ an infographic, in which case one can usefully pass the analysis of that subcanvas on to the kinds of abstractions and categories we introduced in the previous chapter: there is little sense in attempting to reconstruct a theory of infographics as a component of the theory of comics. More interesting then is where there are genuine intermedial effects—such as, for example, when the organisation of an infographic is used not locally but, perhaps, for the composition of entire comicbook pages. The analysis of designs of this kind requires that we bring together *both* knowledge of comics *and* knowledge of the workings of infographics; in other words, such comics will in all likelihood be requiring a higher degree of *media literacy* in order to be appreciated.

In addition, also as stated in Chapter 7 and following on from basic semiotic principles of the kind we introduced in Chapter 2, even for a single page we must consider the interpreter who interacts with the specific communicative situation of reading that page and engaging in its interpretation. The interpreter might be an experienced trained comics reader who knows of comics genres, time periods and their conventions and can use this to understand the page s/he has in hand. Alternatively, that reader might be a novice in reading comics, not knowing much about their attention-directing elements or how to separate and order relevant units from each other. As we shall return to below, it has now been established that the level of comics literacy makes substantial differences in how a reader goes about interacting with and extracting information from any comics material. When describing interpretations, therefore, we always have to keep in mind such details about a recipient's contextual and discourse knowledge in order to say more about his or her likely inferences and meaning-making processes.

Finally we begin to arrive at the kinds of units and elements that are usually talked of when introducing comics and comics analysis—such as grid-like panel compositions, speech bubbles, captions, and other strategically placed written text. For many, Neil Cohn included, it is thought bubbles and speech balloons that are "the most recognizable morphemes from comics", serving generally (but not necessarily!) as carriers of textual information (Cohn 2013c: 35). In addition, sound effects, such as 'bang!', 'craaash!', etc., belong here as well; these are often labelled 'onomatopoeia' because the sound made by reading the word is meant to be reminiscent (i.e., iconic) of the sound intended.

Discussions of cases of onomatopoeia continue to be quite popular across the comics research literature. This is because of their explicit combination of a linguistic reading (i.e., one has to know how to read the characters in order to obtain the sound) and a visual reading caused by suggestive typography or spacing. This phenomenon is rather less noteworthy considered from the perspective of multimodality as we are exploring it here, however, because all written text has a visual side which may then serve as a material medium for many visually-driven distinctions in its own right—nevertheless this is taken quite far in the case of comics and there are interesting cultural differences (Pratha et al. 2016). Manga, for example, regularly appeals to a very broad range of 'onomatopoeic' conventions, some of which not even involving sound at all! Although several writers describe this as a 'tension' between two modes of expression, using this to point out one of the primary features of comics and graphic novels, there is actually rarely tension in any multimodal sense—far more common is that the various technical features available, spanning across visual and linguistic semiotic modes of various kinds, work together in service of the meanings made.

A number of good introductions and examples of these kinds of elements are available and, when beginning with comics research, it is a useful to consider pages in terms of such 'technical details' in order to group the visually available material on a page into established functional classes. Cohn (2013c) sets out a broad characterisa-

Table 12.1. Standardly discussed 'technical details' of comics

Technical feature	Brief description
panels	basic unit of comics page composition
frames	manner in which a panel may be marked off from others or from its environment on the page
panel scale	the amount of content within a panel, referred to by Cohn (2013c) in terms of 'attention framing' ranging over macro (containing two or more characters), mono (single character), micro (detail) and amorphic (no characters)
captions	small marked off visual units containing text, often with distinctive colour or frame and giving background temporal information
thought bubbles	small marked off visual units connected to an 'origin' or 'thinker' by a broken or dashed graphical trace
speech balloons	small marked off visual units connected to a 'sayer' by an unbroken graphical trace or path
inset panels	small panels inside other panels
speed/motion lines	lines added to or around some object indicating either trajectories of motion or intrinsic motion (e.g., trembling)
emotional state markers	lines added around or near a character designated a mental state such as surprise, anger, etc.
text and typography	written language appearing will generally also make use of typography for additional subjective meanings, separating characters, authorial voice, etc.
onomatopoeia	typographically distinct text floating near to a source of sound or situation designating a sound
characters	narrative characters
postures	expressive postures or positions taken up by characters
dress/clothes	clothing of characters, also generally serving narrative functions

tion of the various forms and functions in some detail, showing some cultural variation as well (Cohn and Ehly 2016); while Forceville (2010, 2011) focuses on categorising and classifying some of the additional stylistic features of balloons of various kinds and diverse visual indicators of emotional states and the like—properties that Forceville refers to as 'pictorial runes' and Cohn as 'upfixes' building on linguistic morphology. Forceville also distinguishes 'pictograms', which are visual elements that already have a standardised and 'stand-alone' meaning, such as 'hearts' for love and dollar signs for money. Cohn places most of these within his category of upfixes, as they all 'modify' some information given elsewhere in the panel, regardless of whether or not they have their own meaning; and indeed, the degree of conventionalisation of such additional material is in any case constantly developing within specific genres and cultures. For ease of reference we summarise the most common technical features used in the comics literature in Table 12.1; a gentle overview with examples can be found in Forceville et al. (2014).

Although comics may be described by picking out these kinds of elements—and we will refer to them in our examples below— multimodally we are often actually more interested in how particular combinations of selections from these options function *together* in the service of meaning-making. And for this, as we have described in more detail elsewhere, we need to distinguish these elements at various scales of description:

> "there are very small units, i.e., individual identifiable visual elements contributing to a visual depiction, as well as larger units, such as regular geometric patterns of panels or other identifiable visual configurations spread over an entire page or spread." (Bateman and Wildfeuer 2014*b*: 376–377)

Each of the elements found has its function in the meaning-making process and the communicative situation of the canvas being analysed and it will generally be beneficial to consider pages at all scales; there may even be structures within structures—as in collections of smaller panels within larger panels, speech balloons within thought balloons, and so on. The primary multimodal question then becomes what those functions are and how we can find out. There are many questions open concerning the relation of any elements to other resources in the overall interplay of semiotic elements and so there is a need for substantially more systematic research.

There are also further interesting categories that arise from the particular use of page composition and layout within the medium. For example, the arrangement of panels on a page gives rise to the much discussed notion of the *gutter*, the interpretative gap left between panels that readers need to fill inferentially to provide coherence to a sequence as a whole (cf., e.g., Barnes 2009; Goggin and Hassler-Forest 2010*a*; Postema 2013; Cohn 2013*b*; Wildfeuer and Bateman 2014).

Comics not only blend image and text, but are inherently multimodal in the way they also convey further layers of meaning through the use of such things as colour, fonts, balloon shape, panel composition, and further visual cues that can present nuanced meaning to the whole. We process these visual aspects simultaneously (as with a piece of art), and therefore greater meaning is created through the resonance of the different modes than any single mode could achieve alone.

— Sousanis (2012)

To describe how analysis can proceed, and the questions that arise when we look at real cases, we turn now to an example and characterise how to approach it according to our multimodal methodology.

12.1.1 Example analysis: visual narrative canvases and analytical dimensions

Figure 12.2 shows the first page of Neil Gaiman's *The Sandman* (1988), a very well received work within the 'canon' of graphic literature. Following several introductory

Fig. 12.2. First page of the narrative in the comicbook *The Sandman: Volume 1. Preludes & Nocturnes*. From: *The Sandman #1* ©1988 DC Comics. Written by Neil Gaiman and illustrated by Sam Kieth. Courtesy of DC Comics.

pages, illustrations and a dedication page, this page is the first one with the typical grid-like layout which it shares with almost all other story-related pages in the book. The page introduces the story and some characters using several different semiotic resources and modes and is a reasonably complex, but still quite representative, example of contemporary comics design practice.

Our question is how to go about revealing both the effects that such an information offering has on readers and how those effects are brought about by the material distinctions exercised on the page. We need to undertake this in as systematic a fashion as possible, so as to avoid simply 'describing' what is going on—as with all interesting multimodal artefacts, there is generally *far too much* going on to just describe it. Such a description is generally what one has when applying and cataloguing the various kinds of comics elements as listed in Table 12.1. In this one page, there are interesting uses of colours, of drawing styles, of shapes of panels, of captions and speech balloons (themselves of different shapes and colours), of framing and angles of depiction within panels and much more besides.

More specifically, therefore, and to focus our attention on the page: we have what is generally termed in comics descriptions a 5 tier grid (i.e., there are 5 horizontally organised 'rows' within the 2D grid); the top, fourth and fifth tiers are singe panels running the width of the page, while the other two tiers contain two and three panels respectively. The first panel in the first tier contains a speech balloon, whose 'tail' points to a speaker in the panel who we cannot actually see. The panel also contains a caption, distinguished by a different colour (yellow) and a somewhat serated edge, iconically suggesting parchment or torn scroll; as often the case, the caption gives us circumstantial information for locating the narrative, and the placement of this information on a torn fragment may perhaps suggest historicity. Other speech balloons follow; in the last, thin panel in the third tier, we see the outline of the balloon being used for narrative effect—its spikiness probably suggesting some aspect of difference in voice quality. In the last panel of the second tier, we also see motion lines as the passenger gets out of the car. We can also see that the panels are not simply arrayed in a grid but have various 'overlaps' with others, adding to the overall composition of the page.

When performing analysis, one should certainly be in the habit of routinely *seeing* these options and any other options that are taken up in any page being studied. But we also need to go further. We mentioned that some readings of what is occurring were 'suggested'—we need to consider more precisely how an artefact makes such suggestions and how particular readings rather than others tend to follow from the task of finding overall coherence among the elements as a complete communicative offering. Thus, in particular, in order not to get lost in the detail, we must always apply principles of multimodal analysis to guide our encounter with the artefact so as to draw out generalisations that might subsequently be subjected to empirical investigation.

To show this, we will focus now on something that the listing of elements does *not* effectively reveal. As discussed in Chapter 10, the *layout* of a page also clearly plays an

Composition

important role in shaping communicative engagement with an artefact. It is, in fact, this facet of a visual artefact's form that constitutes a reader's first point of access: as eye-tracking and other attention studies reveal, information is gathered in 'gist' form at least for the overall organisation, the size and colour of panels that are available for viewing and so on. The precise manner of working of this process is not yet well understood, however: there is much here to research as we will return to at the end of this chapter.

Thus, for the vast majority of readers, and certainly for all members of the readership intended for the current case, it will already be clear that we are dealing with a comicbook or graphic novel (rather than an infographic) and so the application of semiotic modes appropriate for this kind of artefact will be drawn on rather than those of diagrams, graphics, paintings and so on. This in turn suggests that the composition of panels on the page be subjected to a particular kind of reading in the sense of using a particular set of strategies to impose coherence: this is the work that semiotic modes always undertake for their users. On the one hand, then, we have panels that will be assumed to provide episodes or fragments of information relevant for constructing a coherent interpretation of the whole; on the other hand, we have an overall composition that we will engage with in order to draw out indications of just how the panels are to be related to one another.

Z-path

This latter step is particularly interesting as we also still know relatively little about how shapes and arrangements of panels, those panels' graphic design and their portrayed content interact to guide consumption of the material on offer. Neil Cohn has performed experiments with systematically varied panel configurations (without content) in order to see effects on the order that readers assume for the page (Cohn 2013*b*). His results show that, although readers would tend to read a regular grid-like organisation on the page similarly to lines of text, i.e., in a typical 'Z-path', zigzagging from top left to bottom right, certain page layouts push readers away from this strategy.

Staggering

To conduct the experiments, it was necessary, as we explained in Chapter 6 above, to *operationalise* (→ §6.3 [p. 178]) the compositional decisions under study. Cohn achieved this by classifying particular layout configurations so that they could be reliably recognised in any page so that readers' responses to those configurations could be tested in a number of *systematically* varied layouts. The dependent measure was then deviations from the Z-path. One of these compositional configuration is when a border is forced around intervening panels because of different panel shapes, a situation termed 'staggering' by Cohn. As an example, imagine following a border down the side of a panel and you then run into a panel, stopping you from going further. If you can find another border to follow down a little to the left or right, then this is a case of staggering. We can see an example of this in *The Sandman* page if we attempt to following the border between the two panels on the second row downwards: we then bump into the panel below showing the hand on the ornamental door knocker and so have to 'move' either left or right to find a path continuing downward. The same situation can also occur when following a border horizontally.

Another configuration investigated, termed *blockage* by Cohn, is like an extreme case of staggering. Here a border being followed is completely blocked—i.e., there is no border to find either left or right (when moving vertically) *or* up or down (when moving horizontally) that is available for continuing. An example of this in *The Sandman* page can be seen between the first and second rows: we see that, moving upwards, there is nowhere to go from the single border separating the two panels of the second row. The border is therefore 'blocked' at this point. Particularly when a border is blocked *horizontally*, i.e., a number of panels are stacked vertically next to a single panel, reading paths are strongly effected because it is unclear just how a reading path is to be continued (cf. Cohn and Campbell 2015).

In the present case of *The Sandman* page, however, these cases do not cause any particular problems for the reading path, since they do not interfere with progressive scanning of the tiers down the page, following panels left-to-right. Cohn's characterisation would show both that this is the case and explain how the page composition supports this. Research is now beginning to examine the consequences of more varied, more complex page layouts and we can expect that such results will be refined considerably in the near future by eye-tracking studies (→ §5.4 [p. 159]); these can reveal how the eyes of the reader are guided by salient elements on the page in addition to the panel composition alone.

This is also necessary because, even though there is no particular doubt about the *order* of reading in the case of *The Sandman* page, there is still considerable organisation to be teased out and examined for its communicative effect. For example, we might explore how features such as the yellow caption at the top of the page, on the left of the first panel, which stands out against the rather dark colours of the rest of the panel, or the close-up panel depicting an eye at the end of the middle tier, etc., might systematically influence not only the paths that readers explore but the lengths of time they assign to particular features, which is itself often an indication of further cognitive processing, or 'regressive' *returns* to points already 'read'. All of these indicators offer traces of the readers' work in assigning coherence to the page. We can generally assume that such design features are deployed by comics' creators, perhaps not consciously, to increase the aesthetic effects of comics for their readers—but *how* this works precisely and *what* it does is still largely unknown.

The question of operationalisation raises here with considerable force. All investigations, no matter what techniques are employed, require that we find ways of characterising reoccurring 'situations' that can be identified in a range of contexts so that how readers respond to those situations can be examined across a broader range of data. Collecting and annotating data in this fashion allows one to explore whether the situation one has characterised has reliable effects of its own or not. This can be done for any level of information that we may be interested in within the comics we are examining. And it is only when we pursue an empirical investigation of this kind that we can move away from the description of individual, unique artefacts and discover

Blockage

properties and mechanisms that are deployed in comics in general. It is, therefore, a crucial step.

We might, for example, focus on how characters and other objects are picked up across panels, thus giving us information about continuity of various kinds. Organisational strategies of this kind fall under the general heading of *cohesion*. Multimodal studies of cohesion by Chiao-I Tseng have demonstrated that cohesion, traditionally a linguistic notion, is equally important 'cross-modally' in media artefacts as it help readers or viewers connect elements regardless of their semiotic modalities (Tseng 2013). Cohesive connections are called 'ties' and come in various forms. In *The Sandman* page, for example, a cohesive tie may be established between the visual representation of the main character in the second and third row, continued further by depictions only of his hand; simultaneously, there are also cross-modal ties between the character and verbal mentions of the same character in the speech balloons. The analysis of cohesion thus helps construct and maintain reoccurring elements, such as the identity of characters, across many levels of representation, showing how the 'text' holds itself together. Mapping out the chains and interconnections of cohesive ties is consequently a strong indicator of the *texture* of any artefact under study.

One can also ask about the kinds of 'logical' connections exhibited between panels since it is evident that this is also information-bearing. Once the reader has decided to attempt one particular reading path rather than another such relations become very relevant for any attempt to construct the coherence of the whole. In *The Sandman* page, these relations between panels turn out generally to be temporal succession— i.e., first an event in one panel happens, then the event in the next panels happens, then the next, and so on. For example, since the event of *arriving* in the first panel is directly followed by the response depicted in the second panel, the two events follow each other in a temporal (and also spatial) sequence. As we shall return to below, this provides supporting evidence for reading the page as a narrative at all.

Coherence relations Assuming *coherence relations* to hold allows many further interpretative inferences to be drawn. For example, the details revealed in the second panel can also be inferred as giving more information about the characters and the overall setting and thus *elaborate* on the whole situation. Relations of this kind have been described in several accounts of both verbal (cf., among many others, Hobbs 1990; Asher and Lascarides 2003) and multimodal (cf., e.g., Wildfeuer 2014; Bateman and Wildfeuer 2014*a*) discourse; we also saw the use of RST for this purpose in Chapter 10 above. These frameworks constitute an important resource for characterising in detail the *discourse semantics* (→ §4.1.1 [p. 116]) of the semiotic modes involved.

Alternatively (or rather, in addition), we might take a look at possible classifications of the *perspectives* adopted in successive panels: for example, the middle sequence of panels begins with one character, then offers a 'close-up' view of (presumably) that person using the door knocker, followed by a similar 'close-up' of an eye (presumably) behind the door. 'Presumably' is written here to emphasise the fact that these are again all matters of inference and interpretation: the design of the page leads the reader to make these assumptions but they are not explicitly expressed. They are therefore again contributions from the discourse semantics of the semiotic modes employed.

Finally, returning to the page as a whole composition, we can ask whether there are generalisations to be pulled out at this level as well. And, here again, in order to work methodologically towards broader empirical investigation, we would need to define general classification schemes that can be applied to many cases. We saw Cohn's approach to this above when he characterised the reoccurring situations of 'blockage' and 'staggering' and then examined the effects that these situations had on readers' selection of reading paths across a broad selection of data. Comics scholars have also suggested that there are more 'Gestalt' characteristics operative on comics pages that result from the 'entire composition' of a page (cf. Groensteen 2007 [1999]). Characterising reoccurring situations at this level is challenging as we do not yet know just what descriptions of entire pages will be effective; there are, however, some classification schemes, or *transcriptions* (→ §5.2.3 [p. 146]), now emerging that suggest possible lines of investigation (Bateman, Veloso, Wildfeuer, Cheung and Guo 2016; Pederson and Cohn 2016).

We will often need to compare contrasting characterisations of data in order to see how they may differ with respect to the kinds of analysis they support. In the end, one may well then need several distinct classification schemes in order to address different questions. As an example, let us consider how we might characterise the composition in the middle of *The Sandman* page in order to draw out its possible effects and possible motivations as questions we might investigate empirically. We noted above how the page looks like a regular 'grid' of the kind most commonly associated with comics and that its particular panel composition would not give rise to deviations from a basic Z-path for reading: nevertheless, a closer look does show considerable subtlety. The third row, for example, is not really a 'row' at all—what we have is a 'gap' in a row that lets a background image show through. That background image in fact extends up to the top of the page, where it has become a darker blue and we see some leaves and sky. This is then why the gutters between panels at the top of the page are coloured blue—because they are placed against the sky. The background image also runs all the way down to the left-hand edge of the page as well, but does not cover the entire page—it is only behind a very narrow strip of the lower two rows of panels on the left of the page, for example.

How then might we characterise the compositional choice here in a manner that would let us probe its possible motivations? Empirically this requires that we can characterise the situation sufficiently clearly as to be able to pick out 'similar' cases from a broad range of comics so that we can explore similarities in effect or communicative purpose. One approach would be to adopt a description where the large panel in the background which shows through the gap and colours the borders of the panels in the top half of the page is actually a *separate* layer of design organised in 'depth', i.e., along an imaginary axis orthogonal to the plane of the page. This would explain why we can still see it 'behind' the panels at the top of the page. Another approach, employing a different kind of description, would be to see the page as 'in principle' flat, i.e., as exhibiting a grid-like organisation of the kind usual in comics, but with one of the panels 'relaxing' its boundaries so that its content can 'spread out' and *bleed* behind the other panels.

The former characterisation would be a more production-oriented approach—creating separate layers is certainly a straightforward way of actually producing the page with graphic design software; the latter approach attends more to the design practices of comics and graphic novels where the technique of letting content 'bleed' off the page or behind other panels is a common effect in its own right, regardless of how it might be produced technically. The purpose of any transcription is to create 'equivalence classes' for situations; these are the situations that are then probed when we work empirically. If our transcription suggests that two situations, A and B, should receive the same classification, then we want to be able to find similarities in responses to those situations. The approach of having separating layers might then suggest that compositional choices of the kind we see here are similar to other cases where we have an effect of different layers or planes stacked up on a page; the approach with a gap and bleeding suggests instead that the current page should be grouped alongside regular grids. A choice between these is then made an *empirical* issue as well as any other properties it may have. The two alternatives are depicted graphically in Figure 12.3; the 'bleeding' alternative is the one favoured in the scheme defined in Bateman, Veloso, Wildfeuer, Cheung and Guo (2016), one of the most detailed schemes for comics page composition to date.

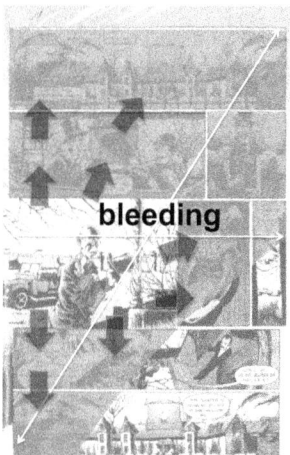

Fig. 12.3. Comparison of descriptions of the page from *The Sandman*: the left-hand image suggests a classification in terms of layering in depth, the right-hand image suggests that the central panel's content runs out ('bleeds') to form a colouring effect for the area indicated

Ideally, then, we want different *predictions* to follow from any classifications we pursue—predictions, for example, about where readers might have problems or deviate from otherwise normal reading paths or concerning different interpretations of how the events depicted might be related. Collecting a body of data containing page compositions that are classified in the same way and seeing how readers actually respond then offers the basis for empirical multimodal investigations of the kind suggested in Chapter 6 above. Concretely: we might see first of all if the use of the panel that appears to spread out behind the others actually leads to any disruptions in reading. If it tended to call for different interpretations or created uncertainty in readers as to how it relates to the remaining panels, then this might be evidence in favour of a classification that places it on its own layer. This assumes that a useful classification of comics pages should correlate with how readers deal with those pages and go about interpreting them. If, conversely, we found no evidence that this panel was read in a way that is different from a panel that behaved like all the others on the page, then evidence for placing it off on its own layer is reduced. Going further, one might then ask questions concerning potential influences on narrative interpretation: is, perhaps, the panel that bleeds behind the others seen as more 'central' to the narrative unfolding on the page? And so on; there are many such questions to pursue.

At this point, however, because we have not done these experiments, we cannot state which way these studies would go. This is one of the exciting things about empirical research: in many areas we simply do not know the answer beforehand and so new knowledge about what we are examining can readily result. Our hypothesis would be that there *is no* difference *in reading path* between that panel that bleeds behind the others and an equivalent version that has strict boundaries like the other panels on the page. The basis of this hypothesis is that the page composition as a whole still 'reads' like a complete grid and, when such a grid can be found, reading strategies of the kind that Cohn explores apply relatively straightforwardly. With the information given in Chapter 6 you could design your own study to explore this. This then leaves open the question of just what this compositional choice *is* doing.

Whereas the choice might appear to be a case of the designers of the comic making one-off, individual creative choices that could, in principle, do anything, a more methodologically-focused characterisation in terms of general classification schemes illustrates how we can take analysis further. As a basic methodological principle, we need to be cautious with assumptions of free variation because this is not in general how semiotic modes work. Instead, making the assumption that differences have meanings is a powerful methodological tool that pushes you to look more closely. Sometimes one might not find differences but most often one does—therefore the search is always worthwhile.

Employment of a semiotic mode naturally draws on an entire history of creative choices made previously. Choices among possibilities are then used to make meanings, sometimes bending or extending what has been done before but always relating to that history in some respect. If this were not the case, readers would not be able to make sense of what they are seeing. The present case appears in fact to be like this as well. Adopting our classification of situations in terms of panels that 'bleed' behind other panels on a page allows us to collect many pages where a similar compositional choice is made and lets us ask quite concretely what range of meanings the compositional choice supports. The answers that we receive are quite specific to the medium of comics, showing again how we can use more robust methods to work out both general principles of layout *and* media-specific meaning-making strategies.

Moreover, this also pushes us further in the direction of specifically *narrative* interpretations. Note that we have so far said rather little about *The Sandman* being a narrative—that is, we have not assumed that it is a narrative before doing our analysis. On the contrary, we can now use our analysis both to *gather evidence* that a narrative interpretation is worthwhile and to show what follows from this interpretative assumption. As we suggested above, this is a natural tendency when examining comics and comics-like artefacts but not a *necessary* situation.

In the present case, however, we can see that the design decisions made on the page are in fact strongly supportive of a narrative reading. First, the cohesive ties construct characters that persist across the panels and participate in actions, the events shown appear to be organisable logically into patterns that 'make sense' as recognisable activity sequences, and so on. An assumption of narrative also helps us interpret aspects of the representation that are not explicitly given. Consider the first panel, for example. Here we have a speech bubble *without* a depiction of who is speaking. The semantics of this expressive resource within the semiotic modes of comics certainly informs us, however, that *someone* is speaking. We thus have the beginnings of an unresolved chain of cohesive ties. The next panel resolves this by providing two figures. One of these says something ("Already? I must have dozed off") that clearly extends the utterance in the first balloon as a response in a dialogue. We can now, therefore, extend the chain of cohesive ties backwards and know (i.e., infer) who was talking before.

Note that this is also a direct illustration of the importance of 'interdependence' between expressive resources that we have mentioned at several points in the book so far. In the present case, we can only interpret aspects of the image when the dialogue contributions are given; and often we will find the reverse situation holding as well. The visual depiction in the second panel, for example, reveals further details of the characters involved (i.e., the white beard of the character on the left and his top hat, which might lead to the inference that he is a person of some social standing) and also gives us a closer look at the estate and its entrance. As we suggested at the outset in Chapter 1, analysing these verbal and textual expressive resources separately and subsequently trying to combine them will often raise more problems than it solves (→ §1.2 [p. 17]): we need, particularly for semiotic modes that tightly interweave distinct resources such as comics, to place these together at a very early stage in our analyses.

On the basis of this reasoning, then, we are gathering evidence that a discourse hypothesis of 'narrative' is going to be well motivated for the present case. This gives us a further possible hold on the remaining question we have left open until now concerning the compositional choice of a bleeding panel in the third row. Considered narratively, the arrival of this character at the mansion shown as a setting in the first panel is a major event, one which is no doubt meant to open up a host of narratively relevant questions: who is this person? what is his business? what is his relationship to the occupants of the house (evidently he is not one of them)? The use of the bleed may then well help signify that this event of arrival is in some sense not just another step in a sequence of equal events but is itself in some sense 'focal' or 'topic-worthy'. If we were to describe narratively the entire page, it could well be with a phrase such as 'arrival at the house'. The composition of the page expresses this with impressive clarity, using the resources of layout modified in the service of comics to express both a momentary event *and* narrative prominence. Given this, we then have a range of further hypotheses: Is this kind of bleeding strategy always used for such moments of narrative significance? Are there other forms?

Note what this achieves for us. We have constructed a bridge between details of composition that can be recognised largely independently of interpretation and potential narrative uses. Finding out about the use of this particular expressive design choice is thereby made accessible to empirical study. This lies at the heart of multimodal empirical research and is a clear example of how this can move us forward in exploring how multimodal communication operates, both in general and for quite specific media choices as considered here.

There is a considerable amount of research to be undertaken. For a comprehensive analysis, it is not satisfactory to look, for example, at image-text relations only; we must also delve deeper into the details of the depiction and the use of 'smaller' resources such as colour, specific panel perspectives, and so on. This general complexity and multidimensionality of comics has only been addressed sketchily in the literature so far. Müller (2012), for example, focuses on graphical details such as body posture and facial expressions as indices for stereotyped social interaction and thus includes smaller semiotic resources in his description of multimodal interaction. Forceville (2010, 2011) has characterised both particular forms of speech balloons on a local scale as well as broader 'embodied metaphors' (→ §8.2 [p. 244]) that may be used to structure the visual depictions employed. Lim (2007) also highlights the need for an integrative model of all semiotic resources involved in the meaning-making process but, nevertheless, mainly focuses on the more general visual and verbal level without naming specific details such as colours, the specific form of balloons or other visual elements, for example. Such studies will need to go considerably further in involving ever more of the details of the multimodal comicbook page, its semiotic resources, and how these interact to construct meanings.

12.2 An aside on the notion of 'narrative'

As mentioned above, the use of comics as visual narratives telling a story is one specific type of communicative purpose which has characterised comicbooks over the years. 'Visual narrative', 'graphic literature', 'graphic novel' or 'visual storytelling'—these and similar terms are currently topical issues, not only in several disciplines and research areas, but also in the print and media industry. The booming commercial interest in both static and dynamic media that represent their content mostly with visual depictions has led to a similarly booming focus on how to analyse and examine those media with regard to their possibilities for storytelling. In fact, several prominent researchers in the field have stated that narrative, as a fundamental human capacity for making sense of the world, is not essentially tied to language and that, similarly, visual and multimodal artefacts are endowed with the ability to tell stories without any or with only little involvement of verbal text.

Attention to this possible line of exploration has only been taken further relatively recently. Several substantial publications are now available focusing on the phenomenon of *transmedial narratology* (Ryan 2004*b*; Wolf 2005; Ryan and Thon 2014; Thon 2016). The general position adopted here is the following:

> "[Story] is independent of the techniques that bear it along. It may be transposed from one to another medium without losing its essential properties: the subject of a story may serve as argument for a ballet, that of a novel can be transposed to stage or screen, one can recount in words a film to someone who has not seen it. These are words we read, images we see, gestures we decipher, but through them it is a story that we follow; and it could be the same story." (Bremond 1964: 4 as translated by Chatman 1978: 20)

It is noteworthy that the discussion in the field of narratology always talks of 'trans*medial*' narrative; there is rarely any explicit orientation to multimodality or explicitly visual narratives as such (Herman 2010 offering one exception).

Nevertheless, although raising important questions of its own, many of the practical tasks of tracking narratives 'across media' also involve issues central to multimodality. On the one hand, many forms of narrative these days are inherently highly multimodal—film offering perhaps the clearest example; and, on the other hand, typical transpositions across media, such as from novel to film, also necessarily involve changes in semiotic modes. It will then be unavoidable to consider multimodality more explicitly when probing issues of transmedia narratology more closely.

It is useful when considering such investigations, however, to have a more articulated notion of just what a 'narrative' might be and there are certainly many attempts to offer definitions in the literature. As suggested in our analysis above, we take a particular perspective on this issue as well—one that is specifically situated within our account of multimodality. We did not state in our analysis that the page or comicbook analysed 'was' a narrative, we instead described how one can use cues in the material to offer an explanation of the page's coherence in terms of a narrative *interpretation*.

This follows several suggestions in narratology, for example by Werner Wolf, that one can explore artefacts and performances for 'narrative indicators' that support a narrative reading to a greater or lesser degree (Wolf 2003). An approach of this kind is also compatible with accounts that see narrative as a cognitive achievement of an interpreter rather than as an inherent property of something analysed (cf. Herman 2003; Ryan 2003).

Taking this position on the status of narrative generally supports tracking narrative and its traces across diverse modalities more effectively than accounts that attempt to anchor narrative in specific media or materials. This then places 'narrative interpretation' alongside any other kind of *discourse interpretation* that is supported by a semiotic mode or by combinations of semiotic modes. In short, and relating back again to our diagram relating genres, media and texts in Figure 4.4, narrative is seen as one possible organisational strategy that a genre may select to coordinate its use of the discourse semantics provided by semiotic modes in and across media. This implies both that narrative can be used for a variety of communicative purposes—such as explaining, arguing and so on—and that various modes may be mobilised for (and by) narrative—such as diagrams, image sequences, films, dance and much more.

12.3 Beyond narrative: comics as non-fiction and metacomics

We emphasised above that 'narrative' may be just one of a variety of communicative purposes that can be pursued with the particular canvas at issue in this chapter. Bramlett et al.'s (2016) list of comics genres already offered several further alternatives (see above). A particularly interesting case is the *metacomic*, a use of the comics medium which is "'about' comics in one sense or another" (Cook 2016: 257). Metacomics pick up and thematise a property often considered constitutive of comics as a medium: *self-referentiality* or *reflexivity* (cf. Cook 2012). This property refers to the tendency of many kinds of comics, from all 'genres', to explicitly direct attention to their own expressive resources via use of those very same resources. When this is done within a fictional work, the term 'metafiction' from literary studies may also be used, although, as we shall see below, in the case of comics this is by no means limited to works of fiction. The traditional literary purpose of metafiction of drawing attention to distance been reality and fiction is thus not so prominent.

The prevalence of 'self-reflexivity' with comics is often seen as a logical consequence of the medium's reliance on combinations of words and pictures. Cook (2016: 262–263) suggests, for example, that it is this 'unique' combination—i.e., comics' particular flavour of *multimodality*—that makes the affordances available almost ideally suited for generating the specific effects of a metacomic or metafiction. Referring to Atkinson (2010) and Baetens (2004), who both stress the particular working of multimodality in comics, Cook considers comics as generally:

"foreground[ing] the constructed nature of comic art. [A]s a result, readers of comics are already cognizant, at some level, of the mechanics of the medium and are thus 'primed' for metafictional excursions in ways that, for examples, readers of novels are not." (Cook 2016: 263)

In this section, we will explore this area in more detail, presenting a detailed analysis of a non-fiction work to illustrate how a closer multimodal analysis can effectively show reflexivity in operation.

12.3.1 Previous studies

One of the most famous metacomics is clearly Scott McCloud's *Understanding Comics. The Invisible Art*, that we have already cited above. This work is defined by the author himself as a "sort of comicbook about comics" (McCloud 1994: 1), which is what 'meta' usually means. McCloud's comicbook is still primarily presented as a narrative, however, telling the story of McCloud talking to friends, colleagues and the reader about definitions of comics, their history and analysis. As Kuhn-Osius (2016) states in a footnote, McCloud's book can be described well as a visual essay in which "the implied author (or even the real author) addresses readers using the pictures as sample material to be commented on".

A further recent and popular example of a metacomic is Nick Sousanis' *Unflattening* from 2015, a dissertation entirely written and drawn as a comic which has been awarded both the American Publishers Award for Professional and Scholarly Excellence as well as the Lyn Ward Prize for the best Graphic Novel in 2015. In this case a narrative reading is far less likely. Moreover, as both a graphic novel and a scholarly and professional academic monograph, the work shows that graphic novels today "address an array of subject matters" (Connors 2012: 31) and are certainly not to be limited to 'plot-driven' narrative. Several academic papers have now appeared in comics form (e.g., Humphrey 2015*a,b*; Helms 2015; Kashtan 2015)—an overview can be found in Humphrey (2014).

Metacomics publications have increased significantly in recent years, with contributions covering topics ranging from historical events and personalities, scientific theories and philosophical trends to social and political issues or instruction manuals for technical issues (cf. Blank 2010). Echoing McCloud and, similarly, other comics scholars such as Will Eisner (1992, 1996), the rise of metacomics, both fictional and non-fictional, follows rising acceptance of comics as a medium suitable for a range of audiences, including children, adolescents and adults (Connors 2012: 31) and for a variety of purposes.

Researchers in the area also naturally address the question of defining and categorising the kinds of metacomics that exist. Cook (2012), for example, describes several types of mostly fictional metacomics, while Gonzalez (2014) differentiates between narrative and discursive metacomics—the latter are those comics in which a 'metafictional break', i.e., an explicit pointer to the artificiality or fictional nature of the work, already occurs at the level of its material and textual constitution, i.e., its 'discourse', rather than in the content. We will see this illustrated in the example analysis that follows. It is also worth pointing out before we start that analysing meta-comics is often considered difficult or problematic—when we apply a multimodal analysis, however, many of the mechanisms are shown with particular clarity.

12.3.2 Example analysis: Metacomics and their canvases

Our example analysis here will focus on a page from Humphrey (2015*a*) reproduced in Figure 12.4. We begin, as always, with a demarcation of the canvas that will form our locus of analytic attention. For present purposes, we will take the page as present in the visual field of any reader, i.e., as a static, flat, two-dimensional and micro-ergodic canvas as with our previous example above. The first stage of analysis is then again to consider possible divisions of the page into different parts which might then present further sub-canvases. Within media relying on visual materials for their form, we generally consider this decision to be best pursued *visually* in order to get started (cf. Bateman 2008: 104–106).

Even though we do not have a straightforward grid at hand here, we can still see several suggestive 'blocks' on the page and these offer the first steps into analysis. The first appears as a written paragraph, then there is a block of four comics-like panels of more traditional form, then a larger drawn pictorial image, unframed and with larger speech balloons, and finally what might be a borderless caption and a further single image (with four speech balloons). In this case, these can all be subordinated to the general expressive resources available in the comics medium—in general this might not be the case, for example if photographs or diagrams had appeared here also.

Next, we make use of the unframed written language that appears at the top of the page as a flowing text that helps tie the subsequent contributions on the page together. We can take the verbal text with its formatting as one sub-canvas for our analysis. It employs some of the standard expressive resources of comics that are employed for written language: a specific 'handwritten' font type, which is used throughout the whole comic, as well as a specific size of the font, which is different in the first line of the text on this page, as well as a decoration of the initial letter. The emphasis of this line of the text suggests its function of starting a new paragraph and introducing the discussion that follows, similar to text-structuring resources in written texts. Underlining as well as several punctuation characters are used throughout the text to highlight words and expressions in the manner long established for comics generally. The text itself introduces the focus of the discussion to be presented on the page: i.e., the concept of written language which "is often linked to speech".

Following this verbal text are four panels perceptually forming a 'block' to be considered for their combined input to the whole. The first three of these panels are organ-

Fig. 12.4. Extract from Humphrey (2015a): page 3 of the metacomic representing a scientific article ©Aaron Scott Humphrey; used by kind permission of the author

ised in a small strip and the fourth is placed underneath; readers will assume that this grid-like module is making use of the expressive resources of comics and is to be read left-to-right, top-to-bottom in the standard Z-path. This hypothesis is then supported when the textual contributions presented in the speech balloons are also seen to build a single coherent (i.e., in this case, grammatical) contribution: "Speech is … here and

... then it's gone". The panels also depict a character, who is maintained across most of the page with visual cohesive ties of 'repetition'.

Already in this first mini-grid we see some of the power of comics with respect to self-reflexivity at work. The speech balloons directly exemplify one of the ways in which written language in comics is linked to speech as stated in the introduction above the panels. The text within each of the first three balloon is fragmented, forcing a 'performance' of the words on the reader. The fourth panel then repeats the previous panels in miniature and slightly overlapping in the distance, together with the main speech balloon of the panel—shown much larger and even escaping the borders of the frame: "But text sticks around!".

With this illustration, the comic emphasises the differences between written and spoken text (again, as said in the introductory paragraph) in that it displays speech first as a rather volatile and fleeting entity, whereas written language or verbal text "sticks around" and can be repeated, as done in the fourth panel by repeating the speech bubbles. The comic thus uses the material qualities of its resources to exemplify these qualities and make them visible to the recipient. By working with typical features of the comic, such as the speech bubble as well as the different panels, it not only explains how comics can emphasise the differences between written language and speech, it *performs* those differences.

The second part of the page then supports this further by focusing on writing in contrast to speech within a larger module, or subcanvas, of its own. This unframed panel takes the self-reflexivity of the page to new heights, which, again, we can bring out particularly clearly with multimodal analysis. First, the speech balloon making the point of the panel is itself 'replicated' four times, shown as overlapping speech balloons. The figure is therefore *saying* that text can be replicated and the comic actually *does this*, replicating the speech balloons and their texts—the visual replication of the speech bubble thus exactly represents the replication mentioned on the level of the written language. Moreover, the character is *physically carrying* his speech balloon from the previous panel, while the text in new speech balloon says that text has a 'physical presence' and is 'transportable'.

This use by the comic of the material qualities of the written language and the different elements of visual representation is constitutive of its style of making meaning. The links between visual and textual elements may be described in analysis as cohesive ties across levels of representation that would normally be quite separate—according to Abbott (1986), for example, one would usually see the naturalistic visual depictions and the text and speech balloons as occupying different planes: when we analyse the current page, we have cohesive ties going across these two planes, clearly showing a metacomic at work. Thus the self-reflexivity about its own ways of expressing and creating meaning is constructed in a strikingly clear fashion.

The last part of the page then first summarises the contrasting positions represented in the first two parts by stating in written language:"Writing and speaking are clearly very different...". Resulting from this, it then uses written language to pose a question: "So why … when we read … do we construct a voice … in our heads?". The text is again broken into several parts in four different speech bubbles all emerging from a book lying on a table. The drawing thus visualizes the voice constructed when reading the text as an example of reading a page in a book; since speech balloons generally entail a speaker.

With a view to the overall page layout, the different parts discussed in detail are clearly combined with each other mainly to demonstrate the contrast between writing and speech. We can thus identify several relationships between the different parts of the page: Whereas the textual part on top of the page introduces the example discussion (as a continuation of the overall paper's argumentation from the pages before), the second part with the four panels elaborates on this example by visually *exemplifying* the written text. This exemplification of speech then stands in *contrast* to the exemplification and representation of writing in the third part of the page. A summary of these contrasts is given in the second textual part underneath the drawings. And the last visualisation represents a *resulting question* based on the summary of the contrast. We illustrate these discourse relationships arising from a multimodal analysis in Figure 12.5.

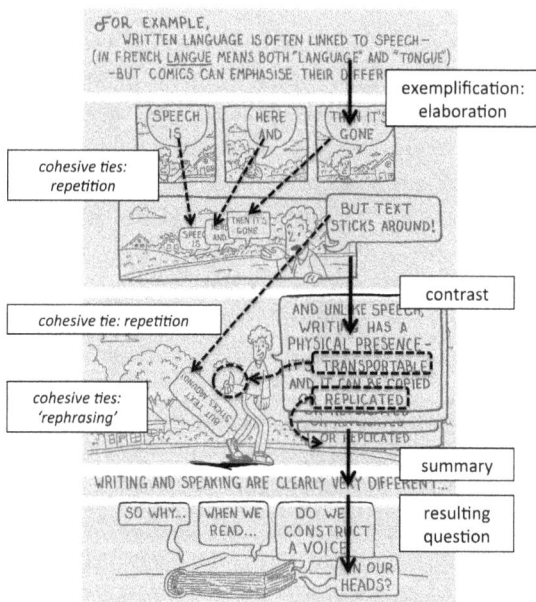

Fig. 12.5. Page analysis of Humphrey (2015a: 4), showing several of the various relationships between the individual parts of the canvas

The analysis shows that this particular page gives rather different cues for its interpretation than the more strictly narrative page of our previous example, which used more temporal relationships. We can, of course, isolate a similar rather basic narrative that would be motivated by the passage of time (by virtue of the unfolding speech) and the visual construction of distinct spatial settings, settings that are similar but sufficiently different to offer cues of movement. Nevertheless, far more of the relevant content is provided by the relationships picked out in Figure 12.5, and these are by no means 'normal' narrative relations. The recipient of this page needs to look for cues concerning the coherence, or discourse, relations that apply as is always the case when performing analysis. This does not, however, need to lead inexorably to a *narrative* interpretation. Other things can, and often are, done with discourse semantics. In the present case, finding the exemplification-elaboration relationship, for example, points more directly to an interpretation in terms of argumentation.

On the basis of the various conventions at play in this medium, therefore, a recipient is led to interpretations where the article not only verbally describes the specific qualities of written language in comics, but also its realization on the visual level. The page uses the different ways of constructing meaning in comics to talk about this meaning construction and thus introduces its own metalanguage working precisely with those semiotic resources native to comics. Questions of narrative, metacomics, metafiction, argumentation and so on are thus properly placed with respect to one another as different kinds of *discourse hypotheses* that an interpreter is more or less licensed to pursue given the material distinctions drawn.

Comics' design is thus a combination of exactly the modes we have now described in our analysis: written language, page layout, the visual system of drawing, etc. For these modes, there are different conventions available which can be used by each designer to construct the comicbook page. In the case of our example analysis, the designer used the specific conventions of speech bubbles, and the use of text within these, as well as the opportunity to reduplicate verbal text in written form.

12.4 Issues of literacy

Both of the examples discussed in this chapter have shown that, as with *all* media, competence in 'reading' the cues that semiotic modes deploy for guiding interpretation is presupposed. If we had never seen a comicbook before, then we might be hard-pressed to see how the composition of the page was being used. These compositional cues and conventions can be of considerably subtlety and there is now growing awareness that comics may provide valuable training for engaging more actively with rich visual materials. This runs in a similar direction to the goals generally promoted in *multiliteracy* research (→ §2.4.4 [p. 45]).

As a consequence, comicbooks and graphic novels are beginning to find more application both *in* educational contexts and *for* explicitly educational aims. Comics can not only encourage a deeper engagement with their presented material (Eisner 1992: 153) but also encourage more self-reflexive and aware engagement with *the medium itself* (Humphrey 2014). We thus find ever more comics on the market concerning themselves with scientific, historical and other broadly 'factual' themes.

Discussion of the effects of such comics on learning and literacy is then also increasing, often with explicit consideration of the issues of multimodality. Jacobs, for example, underlines the need for a framework "that views literacy as occurring in multiple modes" (Jacobs 2013: 7). Understanding just what benefits or drawbacks the comics-form may have, and for whom, naturally raises many issues—both theoretically and practically. To address such concerns, detailed characterisation of the interplay of semiotic resources will be critical. This then constitutes a further potential area of application for the accounts sketched in this chapter.

12.5 Moving onwards: empirical multimodal comics research

There are several areas of empirical research that are currently gaining momentum in studies of comics and these are certain both to give rise to a host of interesting results while also establishing further, more refined questions. We briefly characterise some of these here, as they are all constitute directions where further applications of multimodal methods will be highly beneficial both for understanding comics and graphic novels and for revealing interesting insights concerning multimodal integration and communication.

Text-image relations To begin, although comics are commonly described as a medium in which text and images combine almost by definition, there is still considerable need for more precise characterisations of just what relationships between these modalities are possible and the communicative purposes they achieve for the medium. One classification is proposed by McCloud (1994: 153–155), who sets out seven categories of text-image relationships on the basis of whether the text 'repeats', 'contradicts', or 'reworks' the visual material; this is similar to a broad range of text-image relationships discussed in the literature.

Cohn has also set about describing these interrelationships at work within comics. In Cohn (2013*a*), for example, four types of 'multimodal interfaces' are proposed between text and image—'inherent', 'emergent', 'adjoined' or 'independent'. Inherent is where the 'world' depicted may also include text (as in shop signs, etc.) or mixtures of text and pictorial materials (e.g., "I ♡ NY"); emergent is where text and image are closely interfaced as in, for example, speech balloons; adjacent is the situation in captions; and independent is where there is no visual interaction at all, although there may be cross-references (as in "see Figure X"). Cohn offers many illustrations of these relationships, suggesting how they can be recognised when they occur. This framework brings together both small-scale visual and larger-scale textual material, such as sound effects and speech balloons, within a single coherent account.

Considered more generally, some authors propose text-image relationships as general constructs that hold whenever language and visuals combine—for example, Martinec and Salway (2005)—others focus more on specific media, as in instruction manuals as we saw in earlier chapters, or in picturebooks (cf. Nikolajeva and Scott

2001; Colomer et al. 2010), which are similar but also strikingly different to comics and graphic novels. A detailed introduction to text-image relationships across a range of media and artefacts is given in Bateman (2014*b*). From the multimodal perspective we will need to proceed cautiously and examine just what combinations of texts and other forms of visual semiotic modes are achieving for the artefacts in which they appear. Establishing such relationships is itself, therefore, an empirical question.

Knowing the kinds of text-image relationships at issue is important for comics research because it can have significant influences on other aspects of a comic's reception. For example, we mentioned eye-tracking at several points above and the increasing accuracy and decreasing price of eye-tracking equipment is already leading to a considerable increase in efforts applying cognitive measures and eye-tracking experimental paradigms of various kinds to comics comprehension as well. Consequently, when text and image are working together, it will be likely that eye-tracking results will also pick up on this—perhaps showing numerous 'integrative' saccades as readers bring together different sources of information. Eye-tracking will also be an essential method for coming more to terms with the effects and consequences of larger-scale page composition choices. As we discussed above with respect to *The Sandman*, quite subtle composition effects may be functioning to signal upcoming important events or to establish hierarchies of salience among the panels, characters and narrative episodes involved.

Another important variable that needs to be controlled for whenever empirical research on comics is undertaken is the degree of familiarity of individual experimental participants with the comics medium. Cohn has again been particularly influential in showing how a variety of reception phenomena exhibit statistically significant interactions with measures of comics fluency. Therefore, leaving this information out is quite likely to make any data collected 'noisy' since it includes very different kinds of behaviours. To achieve more control of this variable, Cohn has developed what he terms the *visual language fluency index* (VLFI: Cohn et al. 2012). Prior to any experiment, participants are asked to complete a questionnaire concerning their exposure to, and consumption of, visual narratives, whether they also draw visual narratives themselves, which kind of visual narratives they are familiar with, and so on. From their responses, a 'fluency score' is calculated which can then be used as an additional independent factor when evaluating any experimental resources.

VLFI

The ability to subject comics and graphic novels to more in-depth analysis, showing how combinations of technical devices support particular paths of interpretation rather than others, also opens up new angles on the more traditional study of narrative that we introduced above. There is considerable potential for applying constructs developed within the field of *narratology* to comics, for example. Narratology has proposed important notions for explaining the workings of narrative such as the *narrator* ('who tells'), *focalisation* ('who knows, perceives or feels') and more—but it has been difficult to show these originally *textual* constructs reliably in visual media. Narrative notions are essentially examples of *discourse* interpretations and so demand attention

Narrative

to precisely the kinds of combinations of technical details we have introduced here; this needs to be taken considerably further.

Finally, there are also interesting distinctions and similarities to be drawn out more systematically with *other* media. Sometimes other media 'quote' the medium of comics directly, using established comics conventions and techniques for quite different purposes. Conversely, comics are increasingly experimenting with other forms—for example, building a narrative in a form more reminiscent of an information graphic or a technical flow diagram. These intermedial references are made possible by the shared properties of the canvases involved. This is also a classic environment for the emergence and development of new conventions, building from the historical trajectories of previously separate forms of depiction. Approaching this process multimodally will be crucial when moving on to follow developments among webcomics or combinations of media as exhibited in comics, film and animation. There will certainly be much to do.

12.6 Summary

To conclude the chapter, it is only fitting to rely on the affordances of the comics medium to both tell and show the relationships involved from the perspective of multimodal analysis. That is, as performed in Figure 12.6: "writing with comics allows us to manipulate those modalities in interesting ways! They give us a way of showing what we mean rather than just telling it!" (Humphrey 2015a: 13).

Fig. 12.6. Conclusion of the scientific article in comics form from Humphrey (2015a: 13). ©Aaron Scott Humphrey; used by kind permission of the author.

Use case area 4: spatial, dynamic

13 Film and the moving (audio-)visual image

The study of film, movies and cinema constitutes an enormous field in its own right that lies largely orthogonal to the work of many engaging with multimodality research. This makes it all the more important when considering film from the perspective of empirical multimodal analysis that one draws fully on the breadth of experience already accumulated in various branches of film studies. This is doubly necessary because the medium of film is immensely challenging when seen from the perspective of multimodality. As a medium with very strong depictive capabilities for any visual and audial information in other media and in the real world, the opportunities that film offers for tightly synchronised *integration* of modalities are some of the most sophisticated and most extensively developed in any medium currently available.

When we in addition consider that *everything* that is seen *and* heard on screen may have been planned and designed, we can begin to imagine just how highly complex the resulting product might become. Everything from someone blinking to someone else walking across a scene in the background to a bird chirping off-screen *may* have been designed into the film precisely when seen or heard. It is this property that led some theorists of film and the visual in the past to suggest that film's medium is actually 'reality itself' (Pudovkin 1926; Panofsky 2004 [1934])—i.e., the real world is taken and shaped to perform as the manipulated material of the medium.

Such a characterisation is both too broad and too narrow: it is too broad in that there are certainly (still) aspects of our engagement with reality that are not captured or manipulable, although the entire history of the technological development of film has been one of attempting to overcome this boundary—as we can currently see with explorations of immersive virtual reality as a potential medium for film. But it is also too narrow in that there is much that happens in film, or rather in how film is structured and organised as a communicative artefact, that we do *not* find in reality—and it is precisely these aspects which primarily contribute to the power of film as a communicative medium. For example, we do not, in reality, suddenly leap to an extreme close-up of someone's face to receive apparently unmediated emotional cues while at the same time deafening orchestral music raises the intensity of the moment still further. Neither in reality do we find our perception apparently being drawn along parallel or even alternative strands as competing or contrasting events play out before us. All of these possibilities and many more are 'business as usual' as far as film is concerned. The 'canvas' of film therefore has some very specific features of its own that are constitutive of film being able to do (with us) what it does.

Film has been, and continues to be, explored from many standpoints, ranging from the psychoanalytic readings that were prevalent in the 1960s and 1970s, through to many broader cultural, aesthetic and political readings commonly practised today. There are many excellent introductions to the broader practice of film studies available (with more appearing every year) and so working through some selection from these should certainly be done before taking on any multimodal film analysis of your own (cf., e.g., Lacey 2004; Prince 2007; Monaco 2009). Classics of the field that go more deeply into the area include Bordwell (1985) and Bordwell and Thompson (2013); and these need to be engaged with before, during and after doing any film analysis of a more serious nature.

In this chapter we will not provide a general introduction to how to do film analysis but will instead, as in the rest of the book, focus particularly on what we see as the *primary multimodal challenge* presented by film. As already hinted, that challenge lies in the extreme degree of orchestration and integration of very diverse forms of expression in order to achieve a coherent audiovisual artefact. It is precisely the fact that film manipulates not only a rich variety of visual cues (themselves ranging over naturalistic images, animation, written language) but also an almost similarly rich variety of *audial* cues (ranging over sound, music and spoken language) in an integrated fashion that makes it so powerful.

The resulting audiovisual artefact may also serve a broad variety of purposes, ranging from the relatively straightforward narratives of the Hollywood blockbuster, to critical documentaries targeting changes of behaviour and attitudes amongst its viewers, to experimental art films challenging perception, notions of narrative, notions of truth and more besides. Film is therefore far more than the 'feature film'. Extensive discussions of all these aspects can be found in the literature; Nichols (2010) provides a useful introduction to the study of documentary film, and Bordwell and Thompson (2013), again, gives an excellent starting point for other types of film as well. *All* of these distinct uses of film may equally utilise the full expressive capabilities of the medium.

It is thus important not to see, for example, 'documentary' films as necessarily more restricted in their styles or approaches—non-fiction films increasingly use techniques established for effective fictional narrative, while fiction films increasingly use techniques known for their documentary effects. This emphasises the point that to engage with film we need to address the basic mechanisms involved. These mechanisms are *essentially multimodal* and so multimodal analysis is relevant and necessary no matter what kind of filmic artefact one is addressing.

Some of the orienting questions raised in the other use case chapters motivated by our multimodal navigator are, nevertheless, of equal relevance here and need to be mentioned before proceeding. Just as with all 'media', we have particular, often institutionalised, circumstances of reception. When we adopt a broad focus on the canvases within which film is operative, this leads to considerations of recipient and audience studies and the effects of viewing (often emotionally engaging) material together with others—there is here, as with most areas concerned with film, considerable research and this should be consulted if adopting a focus of this kind. There are also accounts of the film making process, also with its own considerable literature, and so on. In each area, quite different communicative situations are involved and so different multimodal methods may be relevant.

We focus in now on the filmic artefact itself as a micro-ergodic, immutable, 2D, audial canvas capable of depicting any visual material, and primarily including 'photographic' or otherwise pictorial renderings of some selected subject matter.

13.1 The technical details of the virtual canvas of film

In order to engage in empirical multimodal analysis we must be able to 'fix' certain properties of our object of analysis in the manner we explained in our chapter on methods above (Chapter 5). Many of the categories necessary for describing very fine-grained properties of any film have already been set out extensively in film studies. The basic unit of analysis that is most commonly adopted at this level is that of the *shot*, most simply described in production terms as a continuous 'take', i.e., what is recorded between starting the camera and stopping the camera.

The 'shot' is, however, more accurately seen as a *perceptual* unit—i.e., an interval of film that appears to be continuous (regardless of whether it actually was produced in that way and even whether a camera was used at all). This difference is increasingly important when we consider the now standard use of computer generated effects to mix natural and photorealistic but artificially-generated scenes freely and to blend together different production 'shots' to form single, perceptually continuous units. For multimodality, it is generally advisable to work *only* with perceptual units since it is only these that provide the material distinctions that a semiotic mode has to operate with. The shot

Once we have our shots, we may then need further to distinguish the 'scale' of what is being shown—i.e., is it a close-up, at a medium distance, or at a longer distance? We may also need to determine the height and angle of the 'camera's view', as well as properties of lighting, colour balance, sound quality, and more. In parallel there may be ambient sounds, music, spoken dialogue and so on. Although extensive, most of these properties can be anchored temporally into the unfolding film and so can be characterised with the kinds of corpus tools we introduced in Chapter 5; a broader overview of such tools for film can be found in Bateman (2014c).

More challenging is what happens when any of these properties *change* during a shot. This is particularly an issue with more recent films as technology has made it ever easier to have highly dynamic film compositions with camera movements that in earlier times would have been far too expensive or difficult to achieve—this now also includes 'impossible' camera movements created with digital effects. Camera movement is as a consequence a much researched topic in its own right because of the apparent increase in 'immersion' that it brings about. Even relatively small movements during an otherwise static scene will raise a sense of 'being there'. In many films nowadays, however, the camera is hardly ever still and the physiological, perceptual and interpretative effects of such constant motion require considerable and very interesting further research work.

There is then also the question of how such units are pieced together, or edited, into more extensive structural sequences. Here we have considerations of the kinds of 'cuts' or transitions that may be effected between shots and the relations (temporal succession, simultaneity, flashback and so on) that the film may cue for the viewer. These relationships provide a backbone for constructing ever more complex structural units, ranging from the simplest of shot/reverse-shot sequences used for dialogues and chases up to entire films with complex narrative and other organisations. These larger organisations are then themselves the subject of extensive investigation across films, historical periods, and cultures. Questions of filmic genres (cf. Neale 2000; Grant 2012), 'national' cinemas (cf. Shohat and Stam 1994), and historical 'poetics' (cf. Bordwell et al. 1985; Bordwell 2007) are ongoing topics of research interest.

Film analysis at the technical level begins by deciding on the kinds of technical features that are going to be transcribed or annotated, and then applying those analytic categories to a selected body of material. The technical features decided upon should be selected with respect to the research questions posed and the material under study. That material should, as always, be selected on the basis of some systematic 'sampling scheme' (→ §5.2.1 [p. 142]). It is increasingly possible to use computational methods here, for example, for automatic divisions into shots, ascertaining camera distance, movement and so on. Such techniques can be very worthwhile in moving beyond the basic details of the filmic material efficiently. For ease of reference, we set out in Table 13.1 a (non-exhaustive!) list of the technical terms commonly used for filmic technical features in the literature; analyses will often draw on these when describing particular properties or design details of the selected material.

The result of such transcription and annotation can very quickly pose considerable challenges for further analysis—the sheer quantity of material requires methods, both practical and methodological, for focusing attention. Here a multimodal perspective can be particularly useful: when we adopt the position, as we have done throughout this book, that any expressive resources being used are deployed *together* in order to achieve communicative effects of various kinds, this can offer a way of determining the likely *relevance* of the regularities present in the artefact.

For film, this can be especially crucial because, as several scholars have made clear (e.g., Branigan 1984: 29), it is *not* the case that particular technical features are always used for particular communicative effects. The relationship between material regularities and their intended interpretations is far more flexible and can even extend over sets of very different expressive resources. That is: what in one scene may be achieved by camera distance may in others be achieved by music or sound; or what in one scene may be achieved by a combination of cuts and music, may in another be achieved by split screens and camera movement. This naturally establishes film as an exceedingly interesting 'object' for detailed multimodal analysis!

It is also the reason why many of the rules of thumb commonly voiced in visual analysis—such as 'angling the camera up gives power to the character looked at',

Table 13.1. Selection of standardly used 'technical details' of films

Area	Technical feature	Brief description
Shot Scales	Extreme long shot (ELS)	a shot taken from a considerable distance giving an overall sense of the setting (*establishing shot*)
	Long Shot (LS)	a shot showing an entire action or entire character, with some background visible
	Medium Shot (MS)	a shot typically framing a character from the waist up, showing the character's face and gestures
	Close-Up (CU)	a shot showing just the face of a character, drawing attention to expression and emotion
	Extreme Close-Up (ECU)	a shot showing a fine detail or part of a face, focusing specifically on what is shown
Camera Angles	Tilt	Camera is rotated on an horizontal axis parallel to the action to look either up or down on the action
	Roll	Camera is rotated on a horizontal axis orthogonal to the action to tip the entire picture over (*slanting*)
	Pan	Camera is rotated on a vertical axis, for example when following a movement
	Low angle	A tilt looking up at a figure or object
	High angle	A tilt looking down at a figure or object
	Bird's Eye	A shot looking down on action from a height above
Camera Movements	Tracking	A smooth camera movement parallel to the action (typically along a pre-laid track)
	Dolly	A smooth camera movement towards or away from the action
	Crane	A smooth large-scale camera movement supported by a piece of equipment holding the camera
	Steadicam	Camera is attached to the camera operator for smooth, complex movements
Shot transitions	Cut	One shot followed without visual break by the next
	Fade in/out	The image of one shot fades out (typically to black, but may be to other colours) or fade in from an empty image
	Dissolve	The ending of one shot is gradually replaced visually by the next, showing both images for a brief period
	Wipe	The image of the following shot moves across and replaces the image of the preceding shot
Sound	Diegetic	Sounds that are 'naturalistic' with respect to the action portrayed
	Non-diegetic	Sounds added that are not 'caused by' any actions or events being depicted
	Music	Musical compositions, which may also be either diegetic (played in the action of the film) or non-diegetic (part of the sound track or accompanying film music)

'loose framing creates uneasiness and danger' and many more—are inadequate and often wrong. Film theory knows this and so adds jokingly that rules are made to be broken, or that 'there are no rules' (apart from that whatever you do has to 'work'!). For multimodal analysis this is not so helpful: our concern must be to find out just how some *combinations* of expressive resources have the effects they do in specific contexts of use—that is, when something works (or not), we want to be able to explain why.

13.2 Multimodal film analysis: an example

To illustrate both something of the complexity of film and how multimodal analysis of the kind we suggest allows us to approach that complexity and make it manageable, we briefly work through the opening minutes of the 1995 Hollywood film *Die Hard with a Vengeance*, directed by John McTiernan, written by Jonathan Hensleigh and Roderick Thorpe and starring Bruce Willis, Jeremy Irons and Samuel L. Jackson.[1] The opening sequence is a highly-crafted multimodal composition that demonstrates the levels of narratively-motivated formal complexity that audiences take for granted—it combines a variety of transitions between shots, non-diegetic music imposing both a general atmosphere and rhythm on the cutting, changes in sound levels for specific effects, variations of shot scales and much more besides.

Discussing film is consequently a long established technical problem in its own right: how does one 'fix' the ongoing dynamic experience of a film for presentation in a static document such as this book? Although it is becoming increasingly common to use electronic publication to include film clips, more commonly one will want to refer to the film from a written text. This task can be handled more straightforwardly when one remembers that the point of an analysis is not to 're-create' the object analysed but to give sufficient cues to a reader of the analysis of what is going on. It is usually far better practice to watch any segment being analysed in its original form as well.

Film tran-
scription

One standard presentation style for film is to select representative frames and present these in sequence visually. The frames may represent individual shots or particular events or technical framings within shots as necessary for the discussion. There is no point in attempting to show the film frame-by-frame because this is never perceived by any viewer in any case. Figure 13.1 sets out the frames that our analysis will need to reference; some shots are omitted due to space constraints. Whenever issues of movement are at issue—either of characters or of the camera/viewpoint—the limitations of such static depictions are, however, quickly evident. The frames lower right

---

1 For further production information on films of all kinds, the website www.imdb.com is an essential research resource.

in Figure 13.1 represent this by frame depictions that are to be read as following each other in rapid succession as will be discussed below.

Another considerable limitation of such representations is their restriction to the visual. The film opens in image (1) with a loud and lively non-diegetic (cf. Table 13.1) rendition of "Summer in the City" performed by the Lovin' Spoonful. The colour balance of the shot, together with the shots following presenting various scenes of city life, show summer activities and give an impression of heat (2-7)—although not in any particularly oppressive fashion. The beginning seconds of the film are thus quite up-beat; the diegetic sounds of the city, buses passing, street noise and so on, gradually come in under the music, constructing a conventionalised film opening where the viewer is gradually moved into the storyworld. Image (8) gives the first more defin-ite information, although as part of the previous sequence this can also be interpreted as a further temporal detail: "Back to School Sale!". The music continues over the city noises.

In image (9) we have the first shot that clearly picks out a detail that we have seen before, i.e, the entrance to the department store already seen in image (8). In the terms we met in Chapter 12, therefore, we have an example of a cohesive tie (→ §12.1.1 [p. 308]) binding the shots together and suggesting that there may be something of thematic interest for the film. The music continues.

Image (10) shows the final fraction of a second of the shot begun in image (9) in which the front of the department store is blown out by a loud explosion. The mu-sic is stopped equally abruptly by the explosion increasing the surprise effect of this completely unexpected event. In images (11-16) we are presented with a very fast selec-tion of scenes of the explosion from different camera angles, including single actions (such as a van being blown through the air) requiring quite precise timing. The sound of the explosion runs continuously even though some images show the same action from different angles. This multimodally constructs both continuity and an intensified sense of action via the fast cutting and multiple explosive elements. The explosion se-quence then ends with a dramatic reduction in the sound level, relative quiet and a long-distance view down the street (image 17). Thus, in just over 60s seconds the audi-ence has been thrown into the narrative and has already had a rewarding surprise.

Images (18–21) then pick out frames from the scene immediately following which continues both the pace and the themes unfolding in the first minute of the film. Whereas a scene in a busy police office could be made dull, in this case what is shown is a single continuous movement of the camera past the various police officers—thereby visually introducing all the major characters in one shot—to end with a phone ringing and the female police officer in image (21) answering it. Although she is in the background and seen behind one of the other characters, her salience is clearly con-structed by leaving space around her, both visually and aurally as it becomes clear that she is receiving a call that is going to be important for what happens next. This is, therefore, again very effective and succinct visual storytelling of the kind Hollywood has developed to a high art.

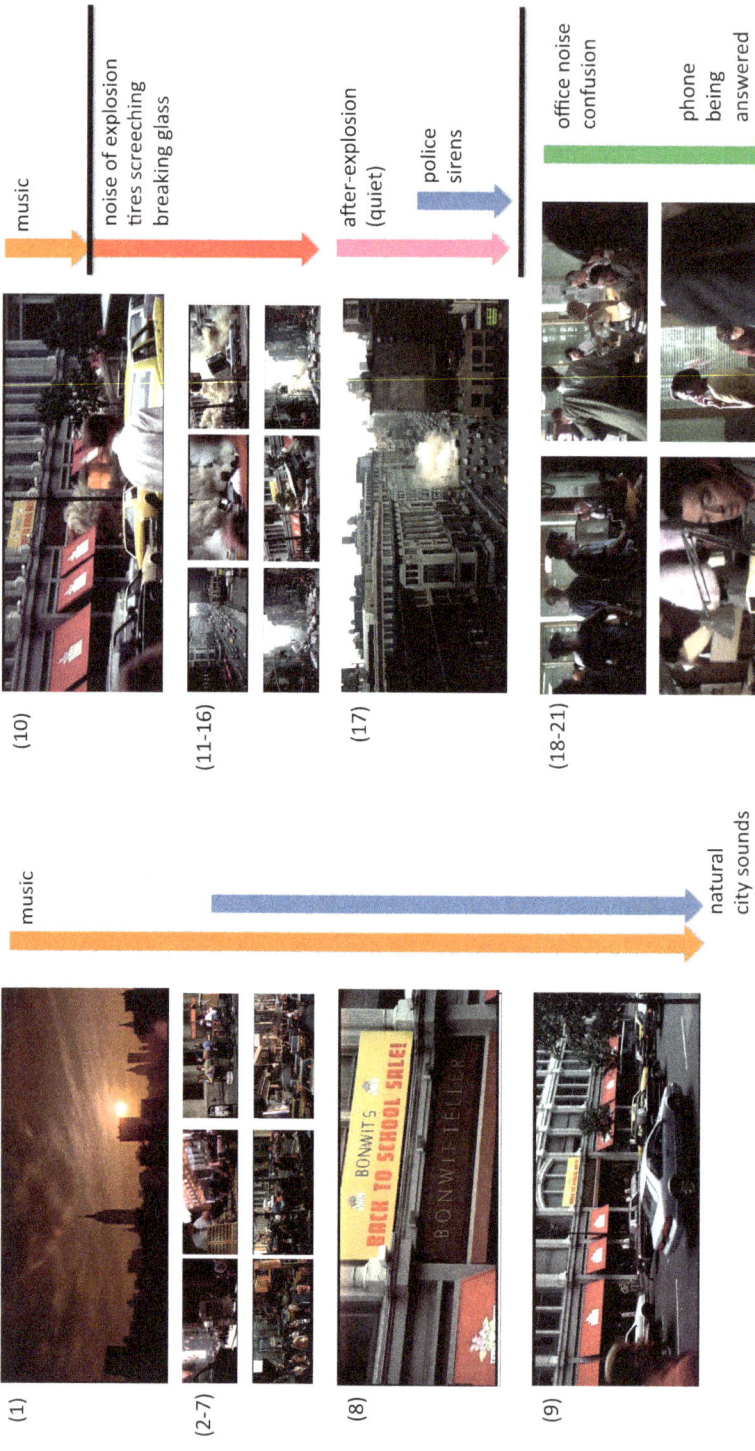

Fig. 13.1. A visual sequence protocol of the multimodal composition of the opening of *Die Hard with a Vengeance* (0:00:10–0:02:00) © Cinergi Pictures Entertainment, Twentieth Century Fox Film Corporation, 1995

We can then begin to derive from the way in which information is presented through these various technical devices what Bateman and Schmidt (2012) and Wildfeuer (2014) characterise as a *filmic discourse structure*. This constructs an unfolding and dynamic representation of a viewer's possible interpretation of the scene's coherence. As with all discourse structures, this is an *abductive* process (→ §2.5.2 [p. 61]) which may need revision as more information is obtained.

The shots of images (1-8) are then topic establishing within this structure; showing activities with no causal or strong temporal relationships is a standard technique for such scene setting—the film semiotician Christian Metz termed such sequences *descriptive syntagma* (Metz 1966). The musical accompaniment operates to bind these together structurally more strongly. The explosion then offers the first actual narrative event, again bound together by multiple angles, fast cutting and continuous sound. Fading sound and increased distance then bring this episode to an end, leaving the narrative question for the audience of what happened and why. The images in (18-21) then also operate as a single unit as described above, again offering a kind of setting before the main events of the narrative unfold. The discourse structure up until this point in the film is constructed straightforwardly, opening up the most likely successive narrative events to be dealt with.

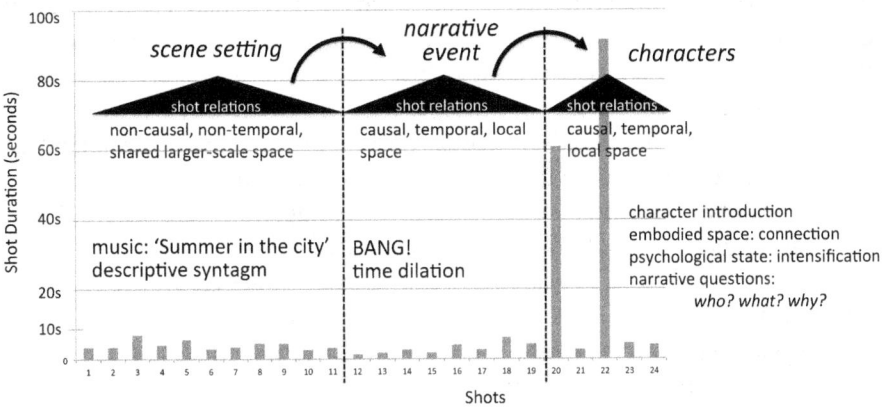

Fig. 13.2. The filmic discourse structure constructed by the opening shots of *Die Hard with a Vengeance* (above) and shot durations (below)

This filmic discourse structure is suggested graphically in Figure 13.2. This graph combines a bar plot of the durations (in seconds) of all the actual shots in the opening sequence, together with the units formed by the spatiotemporal and causal relationships holding between those shots. Segmentation of the film into intelligible discourse units is generally driven by the task of recognising which relationships hold and where those relationships run up against boundaries between the events being depicted. The graph also shows how variation in cutting rhythm can be a good indication of boundaries being drawn as well—most of the opening sequence contains shots of between 2 and 5 seconds length, punctuated by very short shots (e.g., when the department store explodes) and very long shots (e.g., when we switch to the inside of the police department).

Much more could be extracted even from this example of barely two minutes of film, but we will have to restrict our main point to the following: all of the resources being deployed in this segment, cutting, camera movement, sound track, natural sounds, framing of characters and camera distance have been combined in a way that *supersedes description of them individually*. It is only when we see all of the resources as expressing a single, developing and coherent strand of discourse, that we are guided both to particular aspects of the film to analyse and to see those aspects in appropriate combination with others.

13.3 Films and comics: adaptation and convergence

A further area of relevance to much current work on multimodality involves explicit 'transfers' or re-constructions *across* diverse media. Investigations of this kind move directly into contact with broad discussions in 'adaptation studies' and, more generally, *transmediality*. We noted in Chapter 3 how work focusing on media raises interesting issues of a multimodal nature but (with a few notable exceptions) has not itself theorised the phenomena of multimodality that make transmediality possible.

We end this brief sketch of film analysis, therefore, with an area of 'adaption' that is currently extremely popular: the relationship between (usually) narrative film and comics and graphic novels. As we can see when we compare our characterisations of film in this chapter with that of comics in Chapter 12, these media exhibit very different properties. Nevertheless, the affordances of their respective canvases also overlap in interesting ways. This makes it both interesting and challenging to explore more closely what happens when effects created with the semiotic modes available in a static medium—which allows readers to proceed at their own pace and to address entire pages of information at once—are re-created in a dynamic medium where such control is, in normal viewing situations, not available.

Although there have always been film versions of material originally found in comics, the adaptation of comicbooks to extremely large budget blockbuster films that we see today is relatively recent. Hollywood studios now appear to be producing one superhero movie after the other. Described by Burke (2015) as a "post-2000 trend, which continues to dominate Hollywood cinema", it is not only Hollywood that has contributed to this trend. There are also adaptations produced by other international filmmakers such as *Persepolis* or *Oldboy*, as well as adaptations to TV series such as *Smallville, The Walking Dead, Agents of S.H.I.E.L.D* or *Jessica Jones*.

Several studies have been published on the analysis of these adaptations, starting already with some discussions in the 1970s (cf. Lacassin 1972; Barthes 1977), and ranging from narratological and philosophical to translation and cultural studies perspectives (cf., e.g. Christiansen 2000; Wandtke 2007; Gordon et al. 2010; Cook 2012; Pratt 2012; Sina 2014). From a multimodal perspective, these adaptations present a further interesting analytical aspect with regard to the transfer of meanings from the semiotic resources available in one medium to those of another as well as to the far more general question of transmediality proper of whether different media share similar meaning-making mechanisms, principles or even narrative structures. The discussions and examples offered by Ludwig (2015) are very useful in this regard, as he addresses film and comics versions of selected scenes from a variety of works, bringing out particularly clearly what changes and what does not.

Moreover, as Bateman and Veloso (2013) highlight, it is not only interesting to ask for convergence patterns, such as the developments of comics into more 'cinematic' structures as often suggested in the literature, but also to explore which means of expressions from one medium find correlates or alternative forms in another medium. For this research question, the particular capabilities of the media being compared need to be considered very closely, and this is only possible with a sufficiently well articulated account of multimodality as a foundation.

Both McCloud (1994) as well as Groensteen (2007 [1999]) highlight that the media themselves provide different communicative situations in which materialities support distinct forms and manners of representation:

> "Each successive frame of a movie is projected on exactly the same space—the screen—while each frame of comics must occupy a different space. Space does for comics what time does for film!'" (McCloud 1994: 7)

> "The linkage of shots in a film, which is properly the work of editing, carries itself out in a single linear dimension, that of time, while the panels of a comic are articulated at once in time and in space." (Groensteen 2007 [1999]: 101)

Nevertheless, with an appropriate level of abstraction—such as that offered by multimodality—we can begin to examine such cases of media cross-over in a far more differentiating fashion. Comics and film indeed share several affordances, being both immutable and providing (still in most cases) a two-dimensional visual field for the observer and a sophisticated collection of broadly pictorial semiotic modes. By pulling apart the particular ways in which the respective canvases can support *aspects* of the canvas of the other, interesting transmedial generalisations can be explored. We offer two brief examples to conclude.

First, strong parallels can be drawn between, on the one hand, the visual-spatial layout on the comics page made up of collections of framed or unframed panels and, on the other, the use of split screens in film adaptations. One of the most flamboyant film treatments of panels is that used in Ang Lee's film *Hulk*; Bateman and Veloso

(2013) have examined the extent of this transmedial transfer in some detail, showing that the possibilities of film (motion) and those of comics (framed panels) can be merged to produce new narrative effects.

Second, it is then interesting to consider how the use of different narrative structures in comicbooks, as analysed above, might need to change when transferred to the medium of film and to what extent (and where) the perception and reception by recipients may similarly change. Far more empirical reception studies (→ §5.1 [p. 140]) will be necessary to probe these aspects further. Similarly, a more detailed analysis of film reception in terms of eye-tracking would then also help further compare the recipient's hypotheses and inferences while understanding a sequence. But in order to place such studies on a firm basis, analyses of the kinds illustrated throughout this chapter and those preceding will be essential.

13.4 Summary

In this chapter we have suggested just how interdependent the expressive resources involved with the medium of film are—even for fairly straightforward and 'mainstream' productions. We have not been able within the limited space here to proceed to the next steps of detail and pick out how precisely one might compile and perform detailed analyses on the basis of the transcriptions of expressive resources provided. That this is now a very necessary and worthwhile step is shown by the broad range of more empirical studies now emerging with respect to film (cf., e.g., Reinhard and Olson 2016; Wildfeuer and Bateman 2017).

Empirical studies

We mentioned the applicability of eye-tracking studies in the previous section: there is now a growing body of work looking closely at how films and attention interact using methods of this kind. The results so far are fascinating; on the one hand, eye-tracking shows that film is extremely effective, which means that the creators involved in the fine-grained details of editing and so on are extremely effective, in directing viewers' attention precisely to support uptake of the intended narratives (cf. Smith 2012; Loschky, Larson, Magliano and Smith 2015).

Empirical studies go even further. Brain imaging studies of people watching different kinds of films have shown a remarkable degree of 'inter-subject consistency' in brain activity (cf. Hasson et al. 2008). This shows again that films are very effective in guiding quite diverse kinds of responses to film, going far beyond purely visual responses. Interesting regularities have also been recognised with other kinds of behavioural responses, such as heart rate, skin conductivity and facial expressions (cf. Suckfüll 2010). Studies of these kinds open up a range of further questions concerning just what, precisely, in the physical medium is serving to trigger effects of one kind or another. And here, more detailed analytic decompositions of what is occurring in films are essential for making precise predictions that can then be subjected to analysis. Such fine-grained characterisations can only be pursued with sufficiently detailed multimodal frameworks.

Empirical analyses can also be pursued at many other levels of abstraction. Even characterisations of narrative effects, of approaches to guiding or misguiding the interpretative hypotheses that viewers will follow, can be carried out with more likelihood of success if multimodal characterisations are employed. Thus, for any film, picking out the cohesive ties (→ §12.1.1 [p. 308]) that offer cues for just what elements are going to be useful for constructing a coherent narrative (Tseng 2013) will offer testable proposals for just what elements in the film are guiding interpretations. Using discourse relations between shots and events will also provide useful characterisations of the ways in which larger-scale coherent structures can be hypothesised by a viewer for a film (Wildfeuer 2014). Tying these elements into a coherent overall structure has also been subjected to detailed investigation (Bateman and Schmidt 2012).

Once one has regularities of these kinds accurately annotated for bodies of film, many opportunities are opened up for the empirical investigation of characterisations of narrative strategies, genres, changes across time and much more. We then also have a firmer basis for turning the results back on multimodal studies: can we characterise adequately just how different expressive forms are being orchestrated? Is there now sufficient evidence for a *semiotic mode of film*? We have deliberately not used this phrase in this chapter—until a lot more basic research has been done pulling out just how expressive resources work together, this will remain an empirical issue.

Finally, we can now also consider the multimodal 'added value' of approaching phenomena across different media. Whereas the view from within one medium might still leave a range of potential descriptions or explanations open, contrasting and comparing *across* media can offer an additional standpoint that allows each medium, and its solutions for particular communicative challenges, to be seen in the light of another. This methodological position is termed *triangulation* in analogy to finding distances or angles by employing a combination of measurements from different positions.

Triangulation

For example, imagine that our empirical studies have shown that a certain point within a film scene gives rise to a particularly strong reaction in viewers. We can then ask whether a similar 'peak' occurs at the corresponding point in readers when reading a comics-based version of the same scene. Using our detailed analyses we can attempt to pinpoint just what brings about the effect in the two cases. Alternatively, we may find that the peak effect occurs somewhere completely differently—in which case we have the equally interesting research question of just which multimodal resources make that happen. This is certain to lead to new insights in the workings *both* of multimodality *and* of the media concerned.

14 Audiovisual presentations

In this sketch of a use case, we consider software such as Microsoft PowerPoint, Apple Keynote or newer systems such as Prezi for presentations of all kinds. Such software is coming increasingly under the spotlight of multimodality and related research (cf. Tufte 2006; Yates and Orlikowski 2007; Schnettler and Knoblauch 2007; Coy and Pias 2009; Bucher et al. 2010; Djonov and van Leeuwen 2011; Bucher and Niemann 2012; Djonov and van Leeuwen 2013; van Leeuwen et al. 2013; Zhao et al. 2014). Here, as with our other use case scenarios, we discuss briefly how to go about analysing these kinds of communicative situations with an explicitly multimodal orientation.

14.1 Characterising the medium

The first decision as always is to characterise the medium involved. Many studies look at a 'PowerPoint' presentation from the perspective of the individual slides, frames or 'overheads' that are designed with the software. This is, however, insufficient—imagine looking through a deck of PowerPoint slides without the accompanying comments from the authors. Unless *the authors have put particular effort into making the slides comprehensible as a stand-alone communicative artefact*, the slides will usually be rather difficult to follow. This shows that looking at the slides alone and experiencing their use in context point to very different kinds of communicative situations. In one, information is placed on the slides in a manner largely similar to the self-sufficient depictions one sees in information graphics; in the other, most of the work of 'leading' a viewer/hearer through the information on the slides is actually performed by the presenter—this needs to be considered as well therefore and raises highly suggesting connections with other medial forms more explicitly focused on performance as such (→ §9.3 [p. 255]).

The canvas for an audiovisual presentation proper needs to be seen as the entire communicative situation in which the presentation takes place. This goes beyond the slides included in the presentation and includes transitions between slides, the spoken discourse of the presenter or presenters, and accompanying gestures as well. In fact, gestures often play a crucial role in synchronising the attention of the presenter and the audience. Hand gestures (e.g., pointing to a particular part of the screen) and verbal references (e.g., "in the graph on the lower left we see…") are equally important here. The resources then function together as a single orchestrated multimodal ensemble (cf. Figure 14.1).

If the array of possibilities in this representation looks complex, remember that our goal in the book as a whole is not to 'hide' complexity, or to pretend it does not exist until it is too late. Instead we seek to make the complexity approachable, bringing it under control. Examining the different areas of multimodality in Figure 14.1, we can

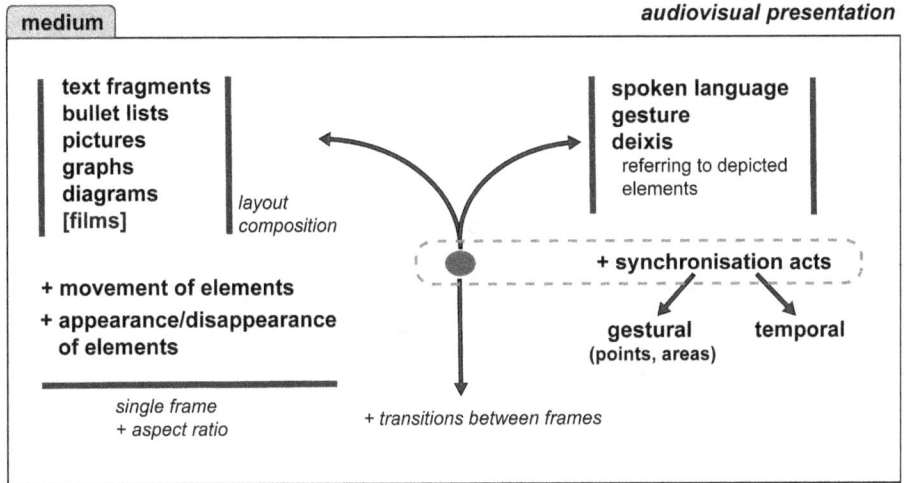

Fig. 14.1. A visualisation of the complex canvas of audiovisual presentations

see that most of them have already been addressed in other chapters. Everything that we have learned concerning layout, information graphics, spoken face-to-face interaction, gesture and performance is both immediately relevant and applicable.

Setting out the canvas to this degree of explicitness makes it apparent just what kinds of descriptions need to be drawn upon. This also provides a good basis for searching the literature for particular treatments of phenomena that may be similar to the 'slices' or canvases that are being addressed. For example, there are, on the one hand, studies of the presentation of text in digital environments and the kinds of variations (typically involving movement) that this supports (Gendolla and Schäfer 2007; Simanowski et al. 2010) as well as, on the other, the beginnings of treatments of *kinetic typography* directly, tracing its history from film titles through to the more diverse media possibilities readily available for use today (cf. van Leeuwen and Djonov 2014; Harrison et al. 2011).

The characterisation also allows us to pinpoint what then *is* new or in demand of treatment. Here these are the *synchronising acts* that hold the verbal discourse and the visual presentation together (Bucher and Niemann 2012). Such synchronising acts constitute an interesting range of meaning-making practices: they are all essentially temporal, because they rely upon two or more pieces of information being perceptually available to an audience at the same time. Thus, a picture on a slide may be referred to explicitly in words; alternatively, the area on the slide may be referred indexically with a hand movement. Alternatively, a particular point on the slide may be picked out by pointing. In addition, specific actions performed on the slide—such as making elements appear or disappear—may be synchronised with particular spoken descriptions that are then taken to relate to the visually present information by virtue of their explicit co-presentation. These details might actually be specific sub-canvases for an analysis if we want to take a concrete look at the modes and resources involved. We will draw on this further below.

Providing a detailed characterisation of these possibilities and the forms they take in actual cases of presentations would therefore constitute a valuable multimodal contribution, which is also likely to be applicable subsequently to other, more or less closely related practices. Examples here include cases such as those of the teacher indicating positions on a map or a blackboard, as suggested in Chapter 7, or a product launch presentation by a company executive. In every case, would have a further example of being more precise concerning just what modalities may be involved in any communicative situation so that information about these may be re-used and adapted for other situations.

14.2 Exploring the canvases

In the previous section, we characterised the audiovisual presentation as a medium. In particular, we emphasised that this medium does not only consist of a presentation shown on a screen, but encompasses the entire communicative situation, complete with the presenter and the audience. In this section, we proceed to briefly sketch how this medium can be sliced into distinct canvases, following the procedures set out in Chapter 7. Once we have established our targets, we illuminate these canvases by drawing on what we already know from previous studies.

Let us first set our sights on the most prominent canvases made available by the medium and their properties. To begin, we have the presenter, who can draw on the bundle of linear canvases generally available for spoken embodied communication. Fleeting instances of spoken language, gesturing, posture, gaze and other embodied modes of expression are constantly used for engaging with the audience and for synchronisation acts between canvases. Although this essentially constitutes a mutable, fully ergodic communicative situation, the constraints arising from *genre* impose further limitations on the choices available (Yates and Orlikowski 2007).

In contrast, the 2D canvas hosting the actual presentation, provided by the screen, allows access to modes that differ in terms of transience (→ Fig. 3.9 [p. 109]). Put differently, the fleeting moments of spoken embodied communication are supported by more permanent forms of expression on the 2D canvas. It is precisely this combination that provides considerable potential for achieving different communicative goals. As Kjeldsen (2013) has shown, this potential is actively exploited in visual and multimodal argumentation, in order to convince, persuade or achieve other effects on the audience (see also Kjeldsen 2015).

That being said, we know from the work of Bucher and Niemann (2012), who used eye-tracking to study the reception of scientific audiovisual presentations in real-life settings, that the audience divides their attention between the presentation on the 2D canvas and the presenter. They observe that the recipients actively integrate and combine information presented on both canvases: however, perceptual patterns are affected by the semiotic modes that are put to work on those canvases.

To focus on the 2D canvas, the semiotic modes mobilised on a slide elicit different eye movement patterns. Static slides consisting mainly of written language are typically read through before attending to the presenter, whereas dynamic slides that reveal the content gradually cause the viewer's gaze to jump frequently between the slide and the presenter. A similar back-and-forth pattern is observed with slides primarily containing images (Bucher and Niemann 2012: 294–295).

Positioned within our typology of communicative situations, it may be suggested that the audience undertakes different forms of ergodic work while engaging with an audiovisual presentation. In addition to *listening* to the presenter, the audience constantly switches between *viewing* the presenter and *reading* or *composing* the slides. With composition, we naturally come across the particular affordances of the 2D canvas, i.e., layout (→ §10 [p. 263]).

Layout, and more generally, its production using the software provided by Power-Point, has been studied extensively (Djonov and van Leeuwen 2013; van Leeuwen et al. 2013; Zhao et al. 2014). In our view, software provides the means to manipulate the semiotic modes made available by the 2D canvas. The result of this process is the concrete artefact in the form of a presentation. The analyst must therefore be clear about what is being studied: the final artefact, or the process and tools used for designing that artefact.

Finally, we can return to the synchronisation acts between spoken embodied communication and the presentation on a 2D canvas, crucial for bridging the gap between the two canvases (Knoblauch 2008). Bucher and Niemann (2012: 297) identify several fundamental features of successful synchronisation acts, such as their timing, sequence and precision among others, but also observe that external devices such as laser pointers can have a considerable effect on guiding the audience's visual attention.

14.3 Summary

To sum up, in this brief use case we explored audiovisual presentations as a medium. Although presentations unfold temporally across multiple canvases covering not only the presentation but the presenter as well—both of which are synchronised with each other—we showed how we can nevertheless impose analytical control. By slicing the medium into smaller canvases, we can examine these canvases in the light of previous research, instead of attempting to build an analysis from scratch. However, the prerequisite for doing this was a sufficiently rich—if perhaps at first glance somewhat complex—description of the medium under analysis. Understanding the medium lays the foundation for all further analysis. This will become increasingly evident, as we now move ahead and add interactivity to the mix in our final area of use cases.

Use case area 5: spatiotemporal, interactive: 'media that bite back'

15 Webpages and dynamic visualisations

The study of webpages is, of course, well established in areas ranging from students' theses to professional consumer research (e.g., Nielsen and Loranger 2006). Webpages are one of the most obvious manifestations of the growth of multimodal communication and the co-deployment of information in any form that the depicting medium can support, which is a very broad spectrum indeed. There has consequently begun to be a steady stream of theoretical treatments and practical proposals for how to go about analysing webpages from a multimodal perspective (cf. Djonov and Knox 2014). From the perspective of what we have learned so far about multimodality and methods, however, the position of webpages can often be seen to be confused. This is largely because webpages themselves offer a moving target. To talk of 'webpages' at all is to identify a notional version of what one thinks webpages might do, even though many existing webpages do very little with the possibilities that are (theoretically) on offer.

From our characterisation of multimodality, we should, in fact, begin to see that 'webpage' is not actually an appropriate level of abstraction—at least with respect to many of the questions that researchers working on multimodality want to address. This is analogous to a researcher or practitioner who wishes to analyse some of the areas that we have seen so far, such as infographics, and deciding instead to analyse the physical properties of sheets of paper. The sheets of paper are important, because they have certain affordances rather than others, but this is then only indirectly related to infographics. We need, in a nutshell, to make sure that we enforce the separation of media, modes and genres that we have illustrated in previous chapters (→ §4.4 [p. 135]).

Webpages are, at best, an abbreviation for a range of media. They cannot be elevated to the status of media proper due to their huge range of distinct uses, which take advantage of their affordances in different ways. They are also evolving as the technology matures and more can be done with something presented as a 'webpage'. This does not always mean that one has radically new artefacts to consider—some of this movement has involved the re-creation of previously existing forms of communication. One such example is the medium of online newspapers, which has now come to resemble print versions of newspapers, whereas early online newspapers looked rather different because the medium did not yet support sophisticated layout (cf. Bateman et al. 2007; Hiippala 2016*b*). This development can be seen across the board as older, rather 'clunky' looking pages are replaced by pages that are very similar in many respects to the conventions of print media.

This complex relationship between 'old media' and the development of new and old uses of webpages has been addressed in the literature on *cybergenres*. One model put forward by Shepherd and Watters (1998) saw new 'genres' either being replicated or evolving from old ones or spontaneously appearing; this model is introduced and discussed in some detail, for example, in Bateman (2008: 209). The notion of

Cybergenre

'genre' appearing here is problematic in several ways, primarily because of the confusion of properties of a medium and what is done with that medium. Thus, when we take webpages of a particular kind— 'homepages' are a common example selected for analysis—it is still often unclear just how many of the properties of those pages are due to some genre of 'homepages', or to the properties of webpages as such, or to conventions that have arisen in the use of webpages that are totally independent of the functions of 'homepages'—e.g., search bars and social media tags—or are consequences of limitations arising from the technology used (cf. Bateman 2014a).

Separating out these aspects is important when planning any research on how information presented across the web is organised on 2D canvases. Thus, we can make considerable use of what we have seen so far in other use cases—for example, as an example of a *page-based medium*, everything that we have said so far about layout and composition can be applied, at least at the outset, to the consideration of multimodal webpages (→ §10 [p. 263]). Whether the effects and practices of such layouts are the same for recipients of webpages and for those of other media and devices is still largely a matter of empirical investigation—so there is certainly a lot left to research.

When approached from this angle, any claims for distinctive properties for the 'new media' become questionable—or at least appropriately relativised. The information offerings of the 'new media' are not automatically different from what has gone before; the ways those media are organised to support different canvases and different genres via the available semiotic modes is what counts. This can move in new and interesting directions—the rise of social media is one example of this and so we address this in a separate use case of its own (→ §16 [p. 355]), but this needs to be assessed individually and always taking into consideration the potential media, modes and genres that are already established elsewhere. The extent to which any webpage has moved beyond technological dependence to support multimodally interesting meaning-making possibilities is an open question, and certainly a question that is not addressed adequately simply by assuming it to be the case.

15.1 Challenges and difficulties: determining objects of analysis

The points mentioned in the previous section combine to motivate the adoption of genre as one foundation stone in many methodologies for investigating communicative artefacts; this then applies equally to webpages. Selecting a genre for investigation is motivated on the one hand by the assumption of social relevance and coherence—artefacts to achieve particular communicative purposes in society are being considered—and, on the other hand, by the accompanying assumption that the artefacts at issue will exhibit a useful degree of homogeneity with respect to their formal properties and internal organisation.

A further essential foundation stone is then provided by the characterisations of the medium itself. We mentioned earlier how the field of human-computer interaction has had a long history of engaging with the 'digital medium' (→ §2.4.2 [p. 41]). Murray (2012: 87–103) takes this further and argues for explicit considerations of the affordances of that medium: in particular, she sets out the four affordances of spatiality, encyclopedic knowledge, procedurality—by which what is depicted may be the result of computation, and participatory involvement of the 'users' of the medium. Murray then uses these dimensions to classify both existing web offerings as well as to drive consideration of the design of potentially *new* kinds of artefacts. A slightly extended and composite view of some of the offerings available on the web classified according to Murray's scheme is shown in Figure 15.1.

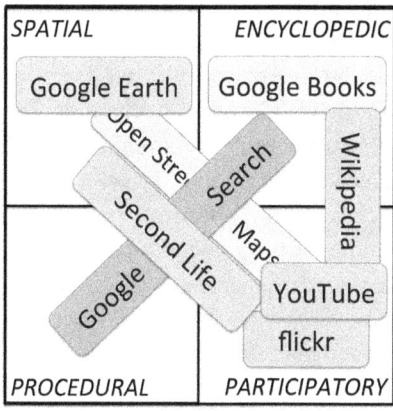

Fig. 15.1. Composite view of various web services classified according to Murray's (2012) four affordances of the computational medium

This kind of classification goes much further in relating what is actually done within any piece of web communication than superficially more straightforward divisions of analogue/digital or old/new media. Several existing media can also be placed within the diagram, therefore, and we can relate it directly to our characterisation of distinct canvases (→ Fig. 3.9 [p. 109]). Mutability, for example, requires either the affordance of procedurality, if the canvas is to be changed by the system being used, or that of participation, if the users are to effect that change. Spatiality requires at least a two-dimensional extent to deploy. 'Encyclopedic' extends the considerations involved still further, moving into what we have characterised here and elsewhere as the 'virtual canvas': there is much more to consider here at both theoretical levels and in terms of practical description.

A classification of this kind also allows us to pick apart the various modes that may be occurring in any specific webpage and to track these in their usage across genres regardless of whether those genres are web-based or not. This stretches both back in the direction of 'old' media and forwards into the future of immersive media. A virtual reality gaming situation would clearly be spatial, procedural and participatory—and perhaps even encyclopedic if enough information is made accessible within the game! These affordances in turn require particular medial properties and can stand as the material basis for a variety of modes, some of which have been studied, and some of which are certainly still to be explored. For the purposes of the present use case sketch, therefore, let us turn to a kind of web offering that is not just a reworking of an existing mode but which draws specifically on properties of the medium: dynamic data visualisations.

15.2 Example analysis: dynamic data visualisations

Compared to our earlier treatments of diagrams and static information graphics in Chapter 11, dynamic data visualisations take a step further along the ergodic dimension into a zone that may be described as properly ergodic, yet immutable. To interact with this kind of canvas, the viewer must engage in the kind of work we have described as *exploration* (→ Fig. 3.9 [p. 109]). However, introducing additional work in the form of exploration does *not* free the viewer from the micro-ergodic work required to unpack a dynamic data visualisation at some specific configuration or stage. Although visualisations may be rendered dynamically again and again, certain stages of the exploration process are very likely to involve work that may be more aptly characterised as *composition*: putting together the elements presented on the two-dimensional canvas to make sense of the visualisation.

Because data visualisations render their content dynamically, thus taking advantage of this particular capability of the underlying digital medium, they must also invest in a user interface to support their use (Weber 2017). Weber and Wenzel (2013: 10) identify several different levels of interactivity in dynamic data visualisations: the lowest level may require the viewer to navigate from one view to another or zoom in on the visualisation. The middle level may involve manipulating the visualisation through sliders or menus, which render the intended changes visible. Finally, on the highest level of interactivity, the user may choose which parts of the available data are rendered, often through filters or some other means of selection. The degree of interactivity may be influenced by which semiotic modes are mobilised and for what purpose. As Lima et al. (2014) have shown, for example, information graphics adapted from print media invest relatively little in interactivity, using the dynamic canvas mainly to distribute their content into different views accessible through a user interface.

Regardless of the level of interactivity, the data selected for the visualisation is typically immutable, that is, the producer has chosen the data for the consumer. For this reason, an increased degree of interactivity does not mean that the data visualisation in question should be moved up to the category of mutable and ergodic communicative situations. The analyst must therefore take care to distinguish between mutable and immutable situations and the canvases on which they occur, while also accounting for the possibility that a canvas may allocate a subcanvas to a user interface. An interface is, after all, a common feature of both dynamic data visualisations and computer games, which fall into immutable and mutable categories respectively (→ §17.1 [p. 368]). These issues, among others, will be taken up for a more detailed discussion in an example analysis below, which explores the use of dynamic data visualisation in journalism. The purpose of this example analysis is to show how we can bring dynamic visualisations under analytical control using our proposed framework.

Bounegru et al. (2016) explore how 'network visualisations' can be used for journalistic storytelling, proposing that these kinds of diagrammatic representations on dynamic canvases also possess narrative potential. Examining visualisations in various journalistic genres, such as investigative reports and feature articles, the authors identify several ways of using network visualisations to create narrative. One such way is the so-called ego network, which depicts relationships focusing around a single node (or actor). Bounegru et al. (2016) propose that this kind of network is particularly apt for exploring associations between different actors as well.

Connected China, a website published by Reuters in 2013, uses this kind of visualisation to map the social and institutional connections and career paths among China's political elite. Within Murray's (2012) classification, the site could therefore be placed in the encyclopedic category (→ Fig. 15.1 [p. 349]). Three screenshots of the website are provided on the left-hand side of Figure 15.2 to illustrate what Bounegru et al. (2016) describe as an instance of the dynamic diagrammatic mode, which can elicit different narrative readings. The screenshots are accompanied by an abstraction showing their subcanvases and the semiotic modes mobilised on the right-hand side.

The question, therefore, is: how can a narrative emerge from a wealth of nodes and edges stored in a database, of which only a small part may be rendered on the canvas at a time? To better understand how these data visualisations, which Bounegru et al. (2016) term 'network stories', are able to weave together a story from a set of nodes and their connections, we can evaluate them in the light of our proposed typology of multimodal communicative situations. Having already identified the ergodic dimension as a key feature for extending and contracting the narrative capabilities of diagrams and information graphics, we now proceed to unpack how a dynamic data visualisation such as *Connected China* construes a narrative. We will limit our analysis to a part of the website examining social power, that is, how family relations and personal connections shape power relations in Chinese politics.

To begin with, the principles behind diagrams used in *Connected China* have been studied extensively in graph theory, a subfield of mathematics whose applications range from finding the shortest route home using GPS navigation devices to structuring information in databases (Wilson 2010). We will not dive into graph theory here, but simply note that graphs are known to possess considerable representational capability. By storing information about individual nodes and the connections between them, graphs, which can take various forms, are inherently geared towards handling the complexity of the information presented in *Connected China*. As the following discussion will show, we can indeed identify various types of graphs put to work in this

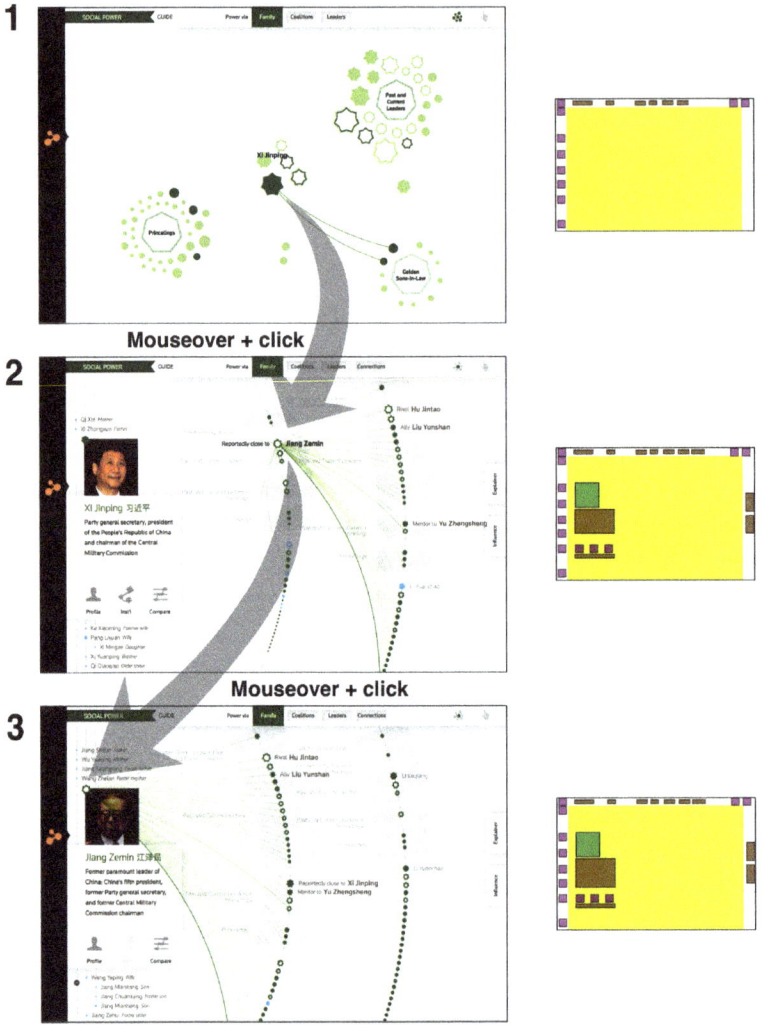

Fig. 15.2. Three screenshots from *Connected China* ©Reuters 2013

data visualisation. We will adopt the term *diagram* to describe these graphs as instances of the diagrammatic mode mobilised in the data visualisation.

The first screen in Figure 15.2 presents a diagram with multiple actors grouped into clusters. This kind of a *clustering diagram* is used to represent how the individual actors can be grouped together using different criteria, which include 'Family', 'Coalitions' and 'Leaders'. By selecting a particular criterion in the interface, the nodes move around to form new clusters. New nodes may also be introduced into the diagrams, while redundant ones disappear from view. The selected criteria is highlighted in the interface: in Figure 15.2 the active criterion is 'Family', as indicated by the green box at the top of the screen. It is worth noting that most of the 'real estate'—i.e., the page area taken up—on the canvas is dedicated to the dynamic diagrammatic mode, which also affords a degree of interactivity. Moving the mouse over a node lights up its connections to other nodes in the diagram, either by drawing connecting lines or changing their shape. As Adami (2015: 138) points out, changes in appearance provide a strong cue about interactivity, painting the element as a target of potential further action. This principle applies across the subcanvases, from the user interface to the diagrammatic mode, and is also well known in human-computer interface design more generally (→ §2.4.2 [p. 41]).

The second screen in Figure 15.2 shows another kind of diagram, which we may term, following Bounegru et al. (2016), a proper network diagram. What graph theory would describe as an undirected graph consisting of nodes (members of China's political elite) and edges (connections between them), uses labels, such as 'Ally', 'Mentor to', 'Promoted' or 'Rival', to lend specific meanings to individual edges and nodes. Individual nodes are also grouped into different subgraphs such as 'Golden Sons-in-Law', 'Past and Current Leaders' and many more, indicated using grey shading under the nodes and an accompanying label.

These labels and subgraphs are crucial to the representation of social power in *Connected China*, as they enable the data visualisation to *attribute* particular characteristics to nodes and their groupings in the subgraphs. As we already know, based on our previous use cases focusing on diagrams and static information graphics, labelling is a powerful expressive resource that becomes available by invoking the diagrammatic mode (see Chapter 11). Not surprisingly, *Connected China* uses both dynamic labels, which emerge as results of the user's actions (moving the mouse over a node), and static labels, which designate groups of individual nodes.

In the network diagram, the viewer may select individual nodes for closer examination by clicking them. Although graphs can store a wealth of information, presenting their content in a concise manner requires drawing on other semiotic modes besides the diagrammatic mode. Bateman et al. (2001) show how information stored in graphs may be used to generate coherent diagrams and short biographies combining written language and photographs. This is precisely what we encounter in Figure 15.2 as well: selecting a specific node for investigation generates a biography of the actor represented by this node in the database. What we have here, then, are two separate tasks allocated to semiotic modes that are appropriate for completing them: whereas the diagrammatic mode is the most appropriate choice for rendering the underlying graph visible, combinations of photography and written language are much more effective for describing the individual nodes.

Bringing this combination of semiotic modes together, however, is not a straightforward task, but naturally requires the contribution of *layout* to organise the semiotic modes. It is important to understand that layout supports the integration of these semiotic modes to the same extent as in static information graphics (→ §11.3 [p. 289]).

Differences arise, however, in the kind of support the layout provides: whereas static information graphics use layout to organise their content into an ensemble that the viewer can take in at a glance, dynamic data visualisations seek to control the flow of content in order to allow the viewer to keep up. This results from the fact that graphs have a high capacity for storing information, which sets certain constraints for presenting their content. This may be exemplified by a simple thought experiment: imagine if all nodes in the graph and the information they contain would be rendered simultaneously on the screen. Even if the diagram would remain legible due to the high number of nodes and edges, the user would be overwhelmed by the information.

For this reason, the dynamic diagrammatic mode in *Connected China* takes advantage of the underlying medium by rendering only a part of the graph at a time. As Figure 15.2 shows, most of the layout is dedicated to the dynamic 'viewport' into the graph (marked in yellow), which may be explored either by interacting with the diagram or using the interface positioned around the viewport. As such, the allocation of subcanvases resembles those found in games, in which the flow of information from the central viewport is aggregated into and/or controlled using a user interface. In this case, however, rendering the entire graph is obviously not feasible, as the layout space would quickly become congested. This is crucial, as the data visualisation relies heavily on grouping and labelling the data to make sense of its connections.

Given that the node under focus determines the attributes associated with all other nodes shown on the canvas at the time, Bounegru et al.'s (2016) characterisation of *Connected China* as an example of an ego network appears highly appropriate. Extracting a narrative from the data visualisation, however, requires the viewer to alternate between the activities of exploration and composition as required. Exploration involves interacting with the diagram using mouseovers and clicks, as shown in Figure 15.2, selecting parts of the underlying graph for examination. Composition is required to infer the relations holding between nodes at a given moment in the diagrammatic representation rendered in the viewport, while also making sense of the combinations of photographs and written text that make up the short biographies. Any narrative arises from moving back and forth along the ergodic dimension, as the viewer alternates between exploration and composition.

15.3 Summary

In this use case sketch, we considered some of the challenges that the analyst faces when analysing webpages. Having identified the term 'webpages' as too broad to be relied on as an analytical category, we proposed that a more careful approach should build on the central concepts of media, modes and genres. Additionally, we considered certain affordances of digital media, as defined by Murray (2012), before putting our framework into use in an analysis of a website centred around a dynamic data visualisation, picking out modes and their contributions on different canvases.

16 Social media

In this use case, we proceed to consider the multimodality of social media, a phenomenon which has become an integral part of our lives. Cotter (2015: 796) observes that "YouTube, Facebook, and Twitter are now prime players on the cultural stage—as well as sources of news for many. The medium, to adapt McLuhan's aphorism, is 'social'."

Social media platforms have indeed become 'prime players on the cultural stage', but nevertheless warrant closer examination, as social media is not merely consumed but created and generated by the producer at the same time. Coupled with technology—such as smartphones, tablets and laptops, which enable the production of multimodal content with little effort—social media platforms that allow sharing user-generated content therefore bring the roles of producer and consumer much closer together (cf. Manovich 2009).

As opposed to communicative situations involving printed or broadcast news, in which the roles of producers and consumers are traditionally distanced from each other, social media allows individuals to take on both roles. This has a profound effect on the communicative situations and our participation in them.

Consumer and producer proximity

One prominent example of this development is the video that captured a deeply disturbing moment during the 2015 terrorist attack on the satirical magazine Charlie Hebdo in Paris. The video, which showed the fatal shooting of a Parisian police officer, was first uploaded on Facebook by a private user who witnessed and recorded the event from a rooftop and subsequently decided to share the video on the social platform. The video was immediately picked up by several online news agencies and websites which shared it without providing further commentary, as typically expected of news discourse.

During the day, the video not only spread over several online platforms, but also to TV broadcasts which used the video to document the events around the attack. Although several immediate reactions on Facebook and Twitter mentioned the cruelty and immediacy of the video, more critical comments about its distribution and use only arose a few days later. By then the discussion shifted to another perspective, asking—for instance—why the video had not (always) been pixelated or why news agencies found it necessary to show it several times and with zooms on the victim. The person who captured the video later apologised for uploading it directly on Facebook without thinking about the consequences.

What this extreme case underlines is the communicative reach that the combination of technology such as camera-equipped smartphones and social media platforms lends to the user as a *producer*. However, these producers might not always be aware what kinds of communicative situations and user-generated content may unfold as a consequence of their actions. The producer may have originally intended to mediate the experience of being a witness to such a horrific event (cf. Villi 2015), but failed to

Communicative reach of social media

estimate the emotional reaction of the audience and how far the video would spread. The same applies to the news agencies, which featured the video without taking the necessary steps to integrate it properly into the communicative situation of reporting.

However, the activities we undertake on social media platforms are not limited to conveying experiences of both everyday and extraordinary events. We enact social relationships by messaging friends and commenting on the content they create, while simultaneously constructing our own identities by creating and sharing our own content. Moreover, we not only engage with other individuals on these platforms, but also interact with brands that provide products and services, expressing our disappointment and approval of them.

All these activities naturally involve a wealth of different communicative situations and roles, which are inherently multimodal. Moreover, new resources for making and exchanging meanings have also emerged to support these activities—emoji constitute one very prominent example. We will explore these and other affordances of several social media platforms below, mapping their affordances, canvases and semiotic modes.

16.1 Previous studies

Not surprisingly, social media has also gained currency in various fields of study. Within the field closest to our interests, multimodal analyses of social media have often focused on specific platforms, elaborating on their respective affordances and the semiotic modes they provide to the user (Zappavigna 2012; Eisenlauer 2013; Siever 2016: cf., e.g.), while also considering the kind of 'digital' literacies required for engaging in communicative situations on these platforms (Jones and Hafner 2012). More detailed analyses have paid attention to, for example, image-text relations in social media (Siever 2015) and the framing of news events across various platforms (Pentzold et al. 2016). Methodological developments in studying social media have leveraged computational techniques ($\rightarrow$ §5.5 [p. 162]) to cope with the wealth of data available on these platforms (O'Halloran, Chua and Podlasov 2014; Cao and O'Halloran 2015), while social media design naturally overlaps closely with more general work on human-computer interaction and interaction design (cf. Benyon 2014: 341–362).

Facebook & Twitter

Major social media platforms such as Facebook and Twitter have naturally gained the most attention in research. Eisenlauer (2014), for instance, explores the shifting roles of participants as producers and consumers of content on Facebook. These communicative situations do not only involve individual users taking on the aforementioned roles, but *the platform itself* as a 'third author', which generates content based on the users' actions. These include content such as "X commented on Y's status" which do not constitute genuine contributions by the users, but originate from the platform—Eisenlauer (2013) provides a foundational discussion on the different ways in which Facebook intervenes in communicative situations.

The 'microblogging' platform Twitter, as it is often called due to limiting the content to 140 characters, has been approached by building on the tradition of social semiotics. Working with a 100-million-word corpus collected from Twitter, Zappavigna (2012) explores different aspects of communicative situations on the platform, such as how Twitter users express evaluative meanings—for instance, positive or negative feelings—and affiliate with other users on the platform by using specific #hashtags (see also Zappavigna 2015; Scott 2015). In addition, Zappavigna (2012) takes on the elusive cultural phenomena referred to as memes (Shifman 2013), albeit mainly from a linguistic perspective.

To move forward, let us now focus on how social media platforms enable their users to initiate and engage in a wealth of communicative situations. First of all, most social media platforms make use of a 2D canvas, which is partially given over to fully-ergodic communicative situations that unfold between users. We will return to these situations and their canvases shortly below, but for the time being it is sufficient to note that these communicative situations rarely involve written language alone: many platforms allow embedding other media, such as static and animated photographs and illustrations, videos, etc. (→ §4.2 [p. 127]). This opens up a wealth of opportunities for combinations of semiotic modes on different media, which the users frequently put to creative use in communicative situations.

Naturally, the use of social media does not take place in a vacuum, but is often embedded within other communicative situations. One example of how social media changes established activities such as *watching* a television show is provided by Klemm and Michel (2014) and Michel (2015), who highlight the use of Twitter as a platform for *interaction* during a broadcast. Acting as a 'second screen', Twitter enables the previously passive viewers to take on the role of an active interactant, participating in simultaneous discussions about the television show in social media. In situations such as this, hashtags provide a crucial navigational aid for engaging with other users and thus bridging the gap between the two communicative situations (Zappavigna 2015).

Embedded social media

Hashtags and other 'metadata' available in social media also benefit researchers by structuring the data. To provide access to this data, many social media platforms provide an application programming interface (API), which allows querying and retrieving the content available on the platform using a programming language such as Python (→ §6.6 [p. 194]). Social media content is currently being harvested for research data in different fields—one prominent area where such data is consumed in very large amounts is the automatic analysis of evaluative meanings, or sentiment analysis (Davidov et al. 2010; Pak and Paroubek 2010). McCreadie et al. (2012), in turn, describe some of the current challenges in data collection, particularly from the perspectives of user privacy and Twitter's terms of service. However, moving from linguistic to multimodal corpora in the study of social media raises considerable further challenges. In order to develop a quantitative approach to analysing social media content, O'Halloran, Chua and Podlasov (2014) argue that:

Automatic analyses

"[a] multimodal digital humanities approach, where vast quantities of different social media data can be amassed, clustered, analysed and converted into interactive visualisations, provides rich resources for mapping socio-cultural trends." (O'Halloran, Chua and Podlasov 2014: 565)

The use of different resources such as written language, emojis, photographs, illustrations, videos or animated GIFs challenge the existing annotation schemes and methods of analysis of large volumes of data.

Moreover, visualisations that support analysis constitute only one aspect of the issues that have played a key role in previous studies of social media data, particularly within the so-called 'cultural analytics' community (Manovich 2012; Hochman and Manovich 2013). To pursue more in-depth, multimodally-focused analyses of social media, we need to zoom in on the communicative situations and activities taking place on these platforms. We show how this may be approached in the section following.

16.2 Communicative situations in social media

To get started, consider Figure 16.1, which shows a visualisation of social media platforms and their general functions developed by Brian Solis. Last updated in 2013, the visualisation shows 122 platforms divided into 26 categories, ranging from social networks to live broadcasting, and from marketplaces to wikis.

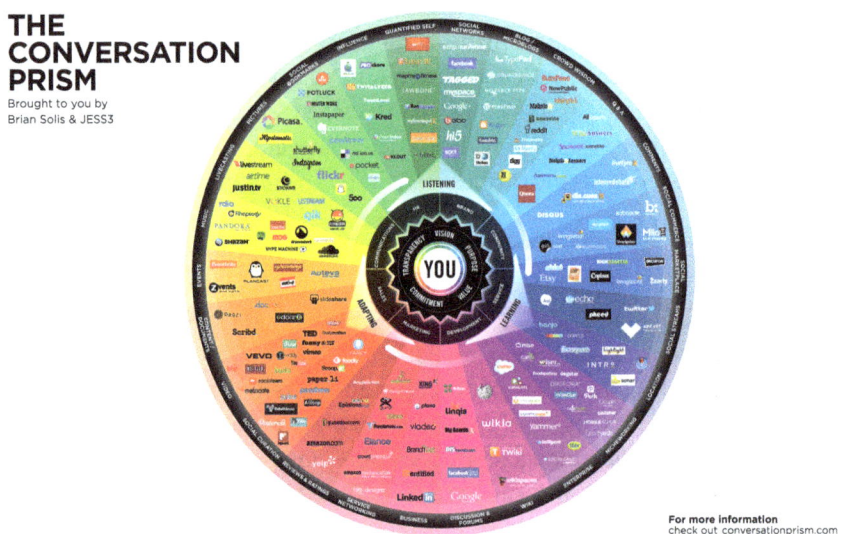

Fig. 16.1. The Conversation Prism ©Brian Solis and JESS3, https://conversationprism.com

Although the landscape of social media is continuously changing and constantly giving rise to new platforms and networks, one central issue is likely to remain the same: a wide range of different communicative situations unfolds on these platforms. Quite appropriately, Page et al. (2014) suggest that 'social media' seems to serve as an umbrella term for various kinds of communicative situations, whose individual properties can vary considerably. Drawing attention to the differences between blogs, discussion forums and platforms such as Facebook and Weibo, Page et al. (2014) consequently describe the current social media landscape in the following terms:

> "[a] diverse range of forms, with different genres [...], and social media sites and services which realise these genres in specific ways [...] and a diverse range of communicative channels and text types, some of which can be integrated within the same site." (Page et al. 2014: 5–6)

While their observation is undoubtedly correct, without a sufficiently sharp analytical approach, the analyst is quickly overpowered by the terminological jungle. In many ways, this situation resembles the one identified for 'webpages' in Chapter 15—the concept is simply too vague for pursuing effective multimodal analyses. For this reason, we must again make sure we enforce the clear separation of media, modes and genres to make headway into understanding communicative situations in social media.

To achieve a working definition of social media within the framework proposed in this book, let us first focus on the users and their possible roles. As opposed to limiting themselves to a passive role, many users choose to actively contribute content to social media platforms. Because social media platforms are not based on the one-to-many principle that defines mass media broadcasting and publishing, but instead work with a logic founded on sharing content via networks, any contributor can potentially reach a very large audience (Page et al. 2014: 5). <!-- marginnote --> User roles

Hoping to reach the largest possible audience may naturally not be the prime motivation for participation. Kaplan and Henlein (2009: 60) identify different categories of social media, including collaborative projects, blogs, content communities, social networking sites, virtual game worlds, and virtual social worlds, all of which motivate their users to participate in different ways. These categories can easily be related to Murray's (2012) classification of affordances discussed in the previous chapter (→ Fig. 15.1 [p. 349]). What is also noteworthy is that these media are not only accessed using desktop or laptop computers, but increasingly using mobile devices such as smartphones and tablets. In fact, some forms of social media under Kaplan and Henlein's (2009) proposed classification, such as games, are exclusively designed for mobile devices—a prime example being the augmented reality game Pokémon Go (→ §2.6.2 [p. 69]).

For multimodal analysis, the underlying materiality and its qualities constitute an issue that needs to be accounted for when pursuing analyses of social media, or any other phenomenon for that matter. Consider, for instance, the difference between accessing the Facebook website on a laptop and on a mobile device using the application. Although both work with a 2D canvas, the underlying materiality of the screen—and in particular, its size—determines how the canvas is allocated to subcanvases on which the communicative situations may unfold.

Having a dynamic 2D canvas at hand naturally brings along the affordance of layout, which presents one possible avenue for a further exploration of the device-specific differences between platforms (→ §10 [p. 263]). Although many platforms choose to present their content using 2D 'page-like' organisations, those drawing on augmented or virtual reality may employ the 2D canvas to render a 3D representation. This may also entail changes to other aspects of communicative situation, such as necessarily moving from an observer to an active participant, so the analyst must remain constantly vigilant!

The social There are, however, aspects of social media that may be considered more stable—one crucial aspect that Page et al. (2014) emphasise is the social itself. Investigating communicative situations always involves the question of how individuals—as social actors—choose to draw on the available semiotic modes to produce something meaningful, thus simultaneously displaying, managing and entering into relationships with others. In analysing social media, this does not only include accounting for the various roles assumed by the individuals when enacting social relations, but paying attention to the fact that these kinds of social behaviour and performances are *mediated* and consumed using a computer or a mobile device.

Self-presentation Kaplan and Henlein (2009) also underline that the concept of self-presentation dominates all types of social interaction, not only presenting one's own identity, but also when influencing others, which can be achieved, for instance, through persuasion or entertainment. The social presence of the actors, however, depends on how extensively the platform mediates the interaction. On some platforms, such as Facebook, the degree of mediation may not be as high as in collaborative projects or blogs, but at the same time, not as low as in virtual games or worlds. Whereas the latter "try to replicate all dimensions of face-to-face interaction in a virtual environment", social media engage recipients in more or less direct interactions of dialogic nature (Kaplan and Henlein 2009: 64).

To identify and describe the various canvases at play, the analyst must consider the communicative situations taking place on them. Some canvases, such as those occupied by the user interface of the platform, may be micro-ergodic and immutable. Other canvases, in contrast, may host communicative situations that are fully ergodic and mutable. These include, for instance, the comments under some post, which allow the interactants to deploy the full arsenal of semiotic modes available in a way that best supports their communicative goals. For example, Twitter or Facebook posts generally offer (if not prevented by the producer's settings) an option to share their text in terms of a re-tweet or a shared post on a(nother) Facebook wall, which not only technically enables users to produce new texts but also implicitly motivates them to do so. Video-conferencing tools such as Skype or video chats invite people to re-create an as-if face-to-face situation to talk to each other. This is precisely the terrain for new combinations of semiotic modes. As Page et al. (2014: 18) observe, this continuous and dynamic interaction exerts "important influences on the choices people make when communicating with each other."

These choices, however, are naturally constrained by the broader activities performed, which we can describe using the notion of genre (→ §4.3.1 [p. 129]). In relation to the genre- and medium-specific affordances of social media, Page et al. highlight that:

> "the affordances [...] are not exclusive properties of a single genre or set of genres. They are to do with how people actually make use of them. [...] More often, the possibilities of social media genres exaggerate the characteristics found in antecedent forms of communication." (Page et al. 2014: 18)

The imitation of the communicative situation is thus necessarily accomplished by the actual use by, and actions of, people involved in this situation. By responding to the respective affordances by, for example, clicking the re-tweet button, quoting the tweet, and adding their own statement, recipients make use of the possibilities available and become active producers themselves.

Similarly, by commenting on a post and producing new text which dynamically adds to the conversation, the consumer likewise is transformed into a producer. This is particularly the case for contexts in which a direct exchange with others is explicitly motivated, i.e. in communicative situations in which consumers are intentionally stimulated to become active. Corporate marketing accounts on Facebook and Twitter, for example, ask consumers for replies and comments, thus following a pattern of customer relationship and impression management, while enacting certain social roles on both sides (Zhao et al. 2013).

The social media environment is therefore generally a convergent situation of both production and consumption of the texts and performances evolving from the users' activities (cf., e.g., Sindoni 2013; Benson 2017). Both production and consumption activities are nevertheless constrained by the genres and discursive contexts given in each communicative situation in social media. This creativity results not only in various material forms of texts and performances, but also allows certain new practices and the semiotic modes used to realise them to come to the fore. The increasing use of *selfies* in social media, for example, is today seen as "a way of speaking and an object to which actors (both human and non-human) respond", thus forming a dominant social practice in our daily life (Senft and Baym 2015: 1589).

Messaging services such as WhatsApp or Threema allow the use of semiotic modes such as animated or static photographs and illustrations, memes and emojis, and written language, to communicate to other users in private or group chats. In contrast to traditional text messages (SMS), the meaning potential available to such mediated communicative situations has increased immensely and is no longer primarily based

Social media convergence

on written language (cf., e.g., Dürscheid 2014). In other words, these platforms are moving towards supporting increasingly multimodal dialogue in communicative situations that are fully mutable and ergodic.

Finally, the online dating platform Tinder provides a certain *swipe logic* of navigating through other users' profiles, which David and Cambre (2016) suggest as redefining and resituating perceptions of intimacy. The platform naturally emphasises self-presentation (Kaplan and Henlein 2009), which is also reflected in its design. Instead of organising the content on a timeline, Tinder is practically built around two separate 2D canvases: the first consists of an immutable canvas for photographic self-presentation, while the second features a mutable and ergodic canvas for dialogue. What this shows is that social media platforms are designed to support their intended communicative functions, which we can evaluate from the perspective of multimodality using the framework presented in this book, as the following case study shows.

16.3 Social media analyses and possible methods: Instagram

For a detailed analysis of communicative situations in social media, the first step is to identify the canvases that the social media platform provides to its users. In the following case study, we focus on Instagram, a photo-sharing service commonly used on mobile devices. As in many cases, which canvases of the platform may be manipulated are defined by the user's chosen role. By identifying the canvases that can be manipulated, we can also begin to outline the semiotic modes made available to the user. Figure 16.2 illustrates this by mapping the canvases on Instagram for the roles of producers and consumers.

User interfaces

To begin with, the screen is allocated to several subcanvases. Many of these subcanvases are dedicated to user interfaces, either for the operating system or the platform, which makes these canvases generally immutable and either minimally micro-ergodic or purely linear, depending on the extent they make use of the layout space. For instance, the two subcanvases on top of the screen (1) display the status of the device, the user's name and, if activated, a geographical location associated with the uploaded content.

Below the content, two further subcanvases are occupied by interfaces: the first features buttons for interacting with the content, that is, for liking, commenting and sharing (3), while the second displays statistics on the content, i.e. the number of likes (4). Note that although these interfaces allow the user to perform common actions within the platform, they are immutable, that is, the user cannot manipulate them! Nevertheless, the activation of a like for a certain posting by a consumer, for example, indeed changes the layout of the pictogram indicating this liking: the heart symbol fills with red colour. With this, the textual structure of the interface changes or, as Aarseth (1997: 62) (→ §3 [p. 74]) points out, the content of the so-called scriptons, i.e., the strings of signs or resources—in this case the pictogram of the heart, empty or filled

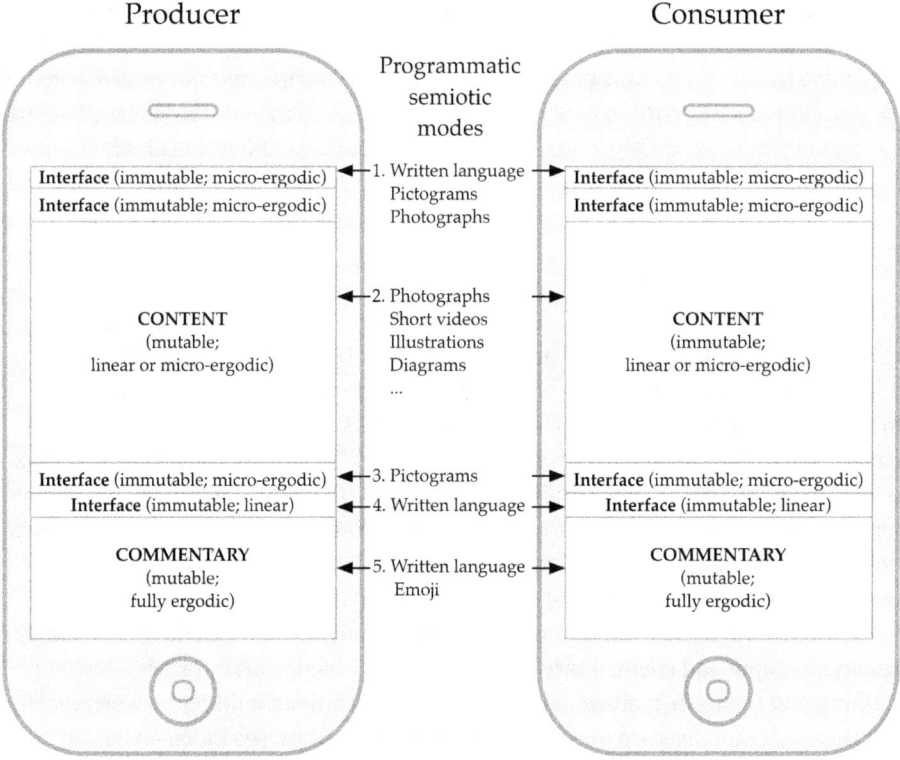

Fig. 16.2. Subcanvases on Instagram for both producers and consumers

with colour—as they appear to the users, change. This influences the dynamics of the textual artifact, but not its mutability.

However, when taking on the role of a producer (as illustrated on the left-hand side in Figure 16.2), the user has full control over the subcanvas dedicated to content (2), which can carry a wealth of semiotic modes ranging from photography, illustrations, audiovisual media to written language (as screenshot), diagrams and much more. This content may be manipulated by using the platform's own tools, which provide pre-defined filters that adjust the content's appearance.

The role of the producer

These filters change the properties of the image, for instance, by brightening or highlighting shadows, by desaturating colours or by imbuing tones whose effects are often described as vintage flavours or vignette effects (Alper 2014; Zappavigna 2016). Alternatively, the producers may draw on external applications to manipulate the content. One such application is *Layout*, which—as the name suggests—enables the producers to create more elaborate compositions with their content. This naturally brings along some affordances belonging to layout proper (→ §10 [p. 263]).

Filters

The semiotic modes available on the content canvas already provide the producer with considerable potential for meaning-making. However, this potential is greatly expanded by the fact that the producer is also able to describe anything realised on the *content canvas* using written language and emojis. This description is displayed on the *commentary canvas*, which opens up a wide range of possibilities for combinations of semiotic modes realised on the two canvases. Instagram users frequently exploit these combinations for a variety of purposes, such as to construe subjectivity (Zappavigna 2016).

The role of the consumer

For the consumer (depicted on the right-hand side of Figure 16.2), the communicative situation is slightly different: the consumer cannot manipulate the content canvas. In other words, this canvas appears immutable to the consumer, because the producer alone wields complete power over its content and the semiotic modes used to realise this content. The platform, however, allows the consumer to perform different actions with or related to the content canvas, which include conveying affect (liking), distributing the content (sharing) and engaging in further interaction (commenting). The last activity—commenting—takes place on the commentary canvas, which hosts mutable and fully ergodic communicative situation. This again typically takes the shape of multimodal dialogue using written language and emojis.

For the analyst, one fruitful target of further analysis would be the interrelations between content and commentary canvases. Several frameworks are available for describing text-image or more general intersemiotic relations holding between the semiotic modes mobilised on these canvases (for an overview, see Bateman 2014*b*). The combinations of written language, emojis and hashtags constitute another interesting avenue for further work. Previous studies have, for example, explored the use of specific hashtags such as #funeral (Gibbs et al. 2015), #death (Wildfeuer et al. 2015; Wildfeuer 2017) and #motherhood (Zappavigna 2016) and their potential to communicate rather intimate and private topics that have not been discussed in public before. These issues thus go back to the question of how the communicative patterns of social media change our way of communicating about certain topics.

Quantitative analyses

Besides the qualitative analyses of social media content, more extensive quantitative analyses are possible, especially if computational techniques and the aforementioned APIs are used for collecting and analysing the data—often in larger volumes: (→ §5.5 [p. 162]). To draw on an example, O'Halloran, Chua and Podlasov (2014) apply face recognition, an active area of computer vision research, to detect faces automatically in a collection of 301865 Instagram images. They estimate that 20–30% of their data features faces, which could be singled out for a further investigation of the practices and contexts of taking selfies. Other common motifs arising from banal everyday events include photos of home-made food or restaurant meals (often contextualised under the hashtag #foodporn or #instafood) or photos of impressive cloud formations (filed under the tag #cloudporn). How such common events begin to be grouped under newly coined hashtags represent another issue that could be investigated quantitatively.

Although computational approaches can be used to scale up the volume of data analysis, there are considerable challenges in automating analyses—principally because social media data is generally inherently 'noisy'. Instagram, for example, is cluttered with screenshots, memes and other non-photographic content, whereas the captions and comments can be multilingual and written in non-standard forms. All of these pose additional challenges when the linguistic contributions are to be analysed automatically.

For these reasons, automatic analyses of social media images have been performed using very low-level features of the image, such as the distribution of pixel-level properties to visualise their groupings (see, e.g., Manovich 2012). This level of detail naturally loses touch with much, if not all, meaningful content in the images. Not surprisingly, the engineering departments of various social media platforms are among those currently taking on the challenge of detecting and segmenting actual objects in photographs (Pinheiro et al. 2015). For multimodal analysts wishing to perform quantitative analyses, this should motivate collaboration with computer scientists, who may in turn benefit from new means of describing and annotating images, as human-annotated data provides the foundation for developing computer programs to perform similar tasks (Lin et al. 2014).

16.4 Summary

With the short overview of analytical canvases available for the study of Instagram, we have shown several aspects of the rich potential to create meaning in social media. This may count as an example analysis for the study of various platforms and networks all producing multimodal content. By first defining the medium itself under consideration—in this case an image-text platform—and then following the analytical steps suggested in Chapter 7, it is possible to approach each platform or network respectively and to determine the specific communicative situation and the canvases to be examined. This will not only help distinguish the concrete communicative behaviours present, but also differentiate, for example, between production and reception processes of the users and consumers involved.

17 Computer and video games

In this use case chapter, we turn our attention to computer and video games. Games have of course been studied for a long time and in their own right within the field of game studies, but they are now beginning to be picked up for analysis in fields closer to multimodal research as well. For instance, in an introduction to a recent book that explores the discourse analysis of video games, Gee (2015: 1) notes that game studies has not shown much interest in discourse analysis, that is, analysing games as a form of communication. Likewise, the interest of discourse analysis in games has been equally limited. According to Gee, however, there seems to be a growing interest in multimodality among both fields, which provides a useful point of departure for our discussion of computer and video games in this chapter.

Choice of perspective
To get started, we propose that approaching games from a clear analytical perspective —as, for example, the discourse analytical perspective or any other perspective for that matter— will benefit from being able to handle their multimodality. This may be considered equally beneficial for game studies, in which multimodality may be considered from the viewpoint of human-computer interaction (→ §2.4.2 [p. 41]), that is, focusing on how different senses may be used for input and output on devices and interfaces (see, e.g., Deng et al. 2014). Alternatively, the issue of multimodality may be taken up for discussion from a perspective that is closer to our approach in this book, examining the media, modes and genres deployed in games (Dunne 2014). Either way, pulling apart the communicative situations involved in *playing* games and those situations *represented* in them may be suggested to contribute to both fields and their corresponding research interests.

Previous work
That being said, games have not previously attracted much attention within the multimodal research community, at least when compared to some of the more popular artefacts taken up for analysis, such as documents, film and websites. Consequently, the previous work is rather scarce, and consolidating the limited body of available multimodal research with the already established field of game studies remains a long-term enterprise, which nevertheless appears a promising avenue of research. Before taking a small step in this direction, a caveat: the book medium restricts our capability to represent games and gameplay. To acquire a better sense of their multimodality, you can look up recordings of gameplay on websites that host videos. Even better, play some games yourself, as the activity of playing is always subjective and situated—to a certain extent—and therefore best experienced first-hand (Stenros 2015).

In the following case studies we will bring various pieces from our analytical toolkit to bear on two different kinds of computer and video games: turn-based strategy games and real-time, first-person action games. We will not be able to engage here with the extensive debates within game studies about the nature of games, play and questions of narrativity; much is available on these issues elsewhere (see, e.g., Aarseth 1997; Wolf 2001; Wardrip-Fruin and Harrigan 2004; Carr et al. 2006). Our goal is rather to shine some light on the multimodal foundations of computer and video games, and to show how to impose analytical control on spatiotemporal and interactive media. This is a crucial step for multimodality research, as "adapting theory for games analysis involves considering how games differ from other media forms" (Carr 2014: 505). However, as we will see, analysing the multimodality of games cannot be based solely on locating differences, but must actively build bridges to other areas of multimodal analysis as well.

17.1 Example analysis: turn-based strategy games

Civilization, created by Sid Meier and first published in 1991, is a series of *turn-based strategy games*, currently at the sixth edition, published in October 2016. The game concept involves leading a civilisation from prehistory to near future, taking control of scientific, economic, cultural, religious and militaristic development against a number of opponents controlled by the computer. Because the game involves both historical and contemporary civilisations, much has been written about the game's relation to history, its depiction of nationalities and the game's underlying cultural assumptions (see, e.g., Chapman 2013). In this case study, however, we focus exclusively on the multimodal characteristics of the game, and more specifically, on *Civilization V* and its subsequent add-ons (2010–2013), which we proceed to take apart using the analytical framework introduced in this book.

Positioned within our proposed typology of communicative situations, *Civilization V* naturally shares several basic characteristics with all computer games: they are inherently *mutable* and *ergodic* (see Figure 3.9 in chapter 3). This means that the sign consumer must take on a specific role, in this case that of a *player*, in order to 'extract' any meaning from the game, which is realised on a spatiotemporal, interactive canvas. In turn-based strategy games, the player is thus usually *involved* as a participant, whereas first-person shooters/action games, to which we will turn below, *immerse* the player into a medium such as a 3D environment. For this reason, games are likely to invest considerable effort in conveying this sense of immersion. As Lemke (2009: 147) states: "when the medium is not just interactive, but cumulatively responsive to user actions, not just in one moment, but over longer timescales as well, the sense of immersion is greatly enhanced". In order to become immersed into the game, the player must put in particular kinds of ergodic work to elicit a response from the game. Note, however, that immersion itself can be approached from several different angles (Ermi and Mäyrä 2007).For current purposes, we restrict our discussion to game design from the multimodal perspective proposed in this book.

The interaction of the player in our current example generates a coherent form of unfolding discourse—the experience of participating in the game (Voorhees 2009). In other words, the player takes on the role of an active participant *involved* in the game, manipulating game elements such as cities, units and terrain, instead of merely *ob-*

Interactivity and immersion

serving the unfolding events from a distance. For our analysis, identifying the player's role as an active participant serves as a useful starting point for sketching a picture of the communicative situations involved in playing the game.

Fig. 17.1. Normal game view in *Civilization V* ©Firaxis Games 2010–2013

Ergodic work of the player

As one communicative situation can be nested within another, this means that our analysis must also account for the possibility that as an active and involved participant, the player needs to perform several different kinds of ergodic work across several subcanvases. Note that the *configurational* work of playing the game, that is, manipulating the unfolding situations during gameplay, is performed with the help of a user interface. Because the user interface provides essential information about the state of the game, outlining the available actions and/or the consequences of previous actions, the player must also actively *explore* the interface laid out on top of the other canvas as well. This may be considered a different form of ergodic work, which is best explained by a screenshot of the so-called normal game view, shown in Figure 17.1. The configurational work, moving and manipulating the game elements, takes place on the subcanvas in the middle, while the immutable (and dynamic) interface laid out on top of this canvas must be explored to retrieve information on the state of the game.

Voorhees (2009: 262) has argued that *Civilization* "demands thought in the form of long-term planning and attention to detail", which characterises strategy games more generally, making them "a cognitive activity marked by reflection, contemplation and deliberate action." From the perspective of multimodality, we may therefore ask: how does the game support this kind of cognitive activity multimodally? To answer this question, we need to pull apart the various canvases on which the different forms of ergodic work take place.

Like many other computer games, a turn-based strategy game such as *Civilization V* is built on a dynamic, spatiotemporal 2D canvas, which is provided by the computer game medium. Having this kind of dynamic canvas at hand means that the game itself and all the ergodic work involved can unfold over time and space. However, because *Civilization V* is turn-based, the player maintains partial control of the temporal dimension, having the ability to decide when to turn the control over to the computer. This is one of the defining characteristics of turn-based games: controlling the temporal progress of events on the 2D canvas gives the player as much time as necessary to perform the ergodic work required. In their discussion of time in games, Zagal and Mateas (2010) refer to this as coordinated time, in which a single round constitutes the basic building block for events taking place in the game. Coordinated time stands in stark contrast to real-time, for instance in first-person shooters or real-time strategy games, in which the events unfold constantly while the game is in progress.

So far, we have already touched upon the issue that the ergodic work required for playing *Civilization V* is distributed over several subcanvases presented simultaneously on the gameplay screen. The player performs most of the configurational work on the animated, mutable representation of a 3D environment, which is shown from a top-down angle. This is the subcanvas on which the majority of the game events unfolds; for convenience, we will refer to this canvas as the 3D environment. The results of these events, in turn, are reflected on an immutable 2D canvas that hosts the user interface, which is superimposed on the 3D environment. We will refer to this subcanvas simply as the interface. Additional canvases are also brought into play during certain phases of the game in the form of cutscenes, which we will not take up for discussion here. We note, however, that they could be described using the methods presented elsewhere in this book; the same applies to the scripted, dialogic interaction with the computer-controlled leaders of other civilisations, which take place on another canvas (see, e.g., Chapters 9 and 13).

Now, having identified the canvases central to the gameplay, we may consider the kinds of activities that are performed on them and how they are supported multimodally. As pointed out above, much of the gameplay involves manipulating game elements in the 3D environment. Such tasks involve building cities, manipulating the surrounding terrain and moving units in order to achieve both short- and long-term goals. These goals may include, for instance, maintaining peaceful relations with rival civilisations through trading, waging war against them and developing the infrastructure of one's own civilisation. Generally, these activities may be described as *managing* "the careful use of limited resources to deal with internal and external pressures" (Voorhees 2009: 262). The task of helping the player to manage the gameplay situation is naturally the main responsibility of the user interface. This relies on combinations of icons, pictograms, written language, 3D drawings and a map, which together provide

Time

Interfaces and canvases

the player with a concise overview of the current state of the game. Other activities include diplomatic *negotiation* with rivals and *planning* scientific research, which we will address shortly below.

We can identify various combinations of semiotic modes that are being mobilised to help the player to perform all of the aforementioned activities across the identified canvases. To begin, the 3D environment draws on a combination of cartography, 3D illustrations and animations. In this case, the cartographic contribution consists of a grid, which structures the game space by constraining movements and actions. This environment, however, is not purely cartographic, but is also occupied by both 3D illustrations and animations: the terrain is typically represented using static illustrations, while the units moving across it are animated. Yet another layer with diagrammatic elements and labels is superimposed on top of the illustrations and animations to designate their properties. Performing the configurational work of playing the game requires the player to constantly consolidate the contributions of these semiotic modes.

Fig. 17.2. Strategic view in *Civilization V*
©Firaxis Games 2010–2013

To highlight the multimodality of the 3D environment, we can also consider the so-called strategic view, to which the user can switch in the main gameplay screen. This view, which is shown in Figure 17.2, renders the 3D environment using a map based on six-sided (hexagonal) grids, while simultaneously replacing the illustrations and animations with pictograms. Removing these two semiotic modes from the canvas simplifies the view by providing an abstraction of the environment, which allows the player to focus on the essential gameplay elements, such as resources (wheat, fish, etc.) and improvements (farms, mines, fisheries, etc.), which are occasionally obstructed in the normal game view. Generally, this kind of abstraction provides additional support for management, which we identified as an activity central to the gameplay, by drawing on different means of expression than the 3D environment.

In addition to the 2D canvas which hosts the 3D environment and the interface, another contribution to the game may be identified in the form of a linear canvas: the sound stream. As Cook (2014) notes in her analysis of *Civilization IV*, the preceding instalment in the series, various different 'sound objects' may be found in the game. These range from the score, or musical soundtrack, to sounds that reflect the properties of some particular location in the 3D environment (Cook 2014: 168). To draw on an example, zooming in on an area of the map with ocean will invoke the sound of waves, cutting down a forest will be accompanied by the sound of falling trees, and so on. The music score, in turn, reflects progress in the game: the music changes as the game proceeds from prehistory to future. In this way, how the player interacts with the 2D canvas largely determines what kinds of objects are rendered on the linear canvas of the sound stream (cf. Hart 2014). Together, these may be described as *synchronisation* across canvases.

Moreover, if we venture away from the normal and strategic game views, we can find additional semiotic modes being put to work to perform other activities as well. One such activity is planning a research agenda, which plays a major role in *Civilization*, as scientific advances introduce new units, resources, buildings, etc. into the game (Owens 2011). The other is diplomacy, which involves negotiating with rivals over trade, alliances and so forth. To accomplish these activities, the entire screen is given over to canvases that host combinations of semiotic modes that differ considerably from those deployed in the normal view. These canvases are shown in Figure 17.3.

Fig. 17.3. Technology tree and diplomacy views in *Civilization V* ©Firaxis Games 2010–2013

In the technology view on the left, we can find the diagrammatic mode in action: an undirected graph, whose nodes combine written language, 3D illustrations and pictograms to describe each technology, presents the so-called 'technology tree', which determines how the research agenda unfolds. Because acquiring new technologies in order to reap their benefits is a major aspect of the gameplay, the consequences of researching a particular technology are represented using pictograms, which specify the units and improvements enabled by the technology.

In the diplomacy view on the right, the entire screen is given over to a dynamic 3D animation of the rival leader, complemented by an interface that combines written language, tables (not shown) and pictograms. This interface is used for participating in a scripted dialogue with the rival, negotiating over trade, making demands or even declaring war. For this purpose, the 3D animation also draws on the resources provided by embodied communication, as facial expressions and gestures are used to convey the rival leader's prevailing attitude towards the player, which may range from negative to positive. The linear canvas of the sound stream is simultaneously used to play variations of songs associated with the culture, whose orchestration depends on the technological level achieved by the leader in question (Cook 2014: 174–175).

All of the activities described above may be performed during just one turn of the game, which requires the player to rapidly switch between different multimodal genres, that is, more or less staged activities performed on different canvases. Lemke (2005: 55) has described these fast-paced moves as 'traversals', which in the context of *Civilization V* contribute towards weaving together the various activities, providing a coherent gameplay experience. For the analyst, identifying these activities and the semiotic modes used for realising them opens up a space for more detailed analyses of multimodality from various perspectives. We could, for instance, analyse how rival leaders and their characteristics are construed multimodally and presented to the player in dialogic interaction (see, e.g., Maiorani 2009; Feng and O'Halloran 2013), in order to examine the claim that *Civilization* advocates a Western, imperialistic perspective to representing non-Western civilisations (see, e.g., Pobłoski 2002; Salter 2011).

Genre   An alternate avenue of research could involve taking on the notion of genre in computer game research, which Voorhees (2009: 263) characterises as "incredibly undertheorised". Mapping the canvases at play and the semiotic modes they carry, while also re-examining the interaction between these canvases over time, could lay a foundation for more empirically-grounded categories of game genres (cf. Elverdam and Aarseth 2007). Note, however, as a final word of warning, the multimodal complexity of both turn-based and real-time games can rapidly overwhelm the analyst: focusing on specific canvases may help both to limit the task and to sharpen analysis. To contextualise the analysis properly, however, one should become familiar with the entire game: continuing for just one more turn may be essential for mapping the canvases at work.

17.2 Example analysis: first-person, real-time games

So far we have focused on games with coordinated time, which proceed turn-by-turn, but how to work with multimodality in games that unfold in real time? In Chapter 13 we learned about methods for dissecting audiovisual media that also unfold in time, but the absence of shots or equivalent structural units in games can make it challenging to get a firm grasp on multimodality. From this perspective, the challenge lies in imposing analytical control on the dimension of time. Without clear boundaries, the best course of action is—once again—to pull the various canvases apart, in order to identify which semiotic modes are being mobilised and for what purpose. To do so, we shall now consider several issues that the analyst facing a real-time game is bound to encounter.

Increasing   To draw on an example of how media can work to increase immersion, a 3D repres-
immersion   entation on a 2D canvas will always have certain limitations, for instance, in relation

to the field of view. If you happen to read this book on a screen, consider the area that falls into your peripheral view and therefore outside the 'viewport' provided by the screen. Because the field of view into the 3D environment is equally limited, games often help the player to navigate and interact with this environment. This kind of assistance may include, for instance, the possibility to switch between internal first person and external third person views—commonly referred to as *cycling* between different views into the 3D environment (Leander and Vasudevan 2014: 159). Alternatively, the game interface may provide overlays that leverage diagrammatic or graphic elements to help the player to move around the 3D environment and to locate objects or targets relevant to the gameplay.

Fig. 17.4. Third-person view in *Armed Assault 3*. The points of interest are marked in red boxes. ©Bohemia Interactive 2013

Figure 17.4 shows the third-person view in *Armed Assault 3*—a military first-person shooter—which illustrates various semiotic modes contributing to the gameplay on this canvas. In addition to the 3D environment, the canvas is occupied by several instances of written language, which serve different purposes. The text on the right, whose purpose is to contextualise the situation, disappears shortly after the game begins (#1). The text on the left, however, appears and disappears as required, as it provides a visual rendition of the radio chatter realised on a different canvas, that is, on that of the sound stream (#2). This helps the player cope with the fast-paced game: synchronising these two semiotic modes, which exist on different canvases, helps to counter the inherent transience of the sound stream: what is uttered does not survive past the moment the utterance is finished. Rendering the spoken dialogue in written language counters the event's transience, giving the player more time to react to the communicative situation.

In addition to written text, the game uses pictograms (#3) to indicate the position of non-player characters. If the non-player characters do not fall within the player's line of sight, the pictograms will float around the screen to indicate the characters' position relative to the player. This means that they are not specific to the 3D environment, but rather belong to the game interface, which occupies its own subcanvas. The same applies to the combination of written language, graphic elements (lines) and two-dimensional drawings (#4), which delivers information about the state of the character to the player. These subcanvases, which we treat as a part of the overall interface, may be characterised as information displays.

Analysing sound
More generally, the game interface acts as a central access structure to the game by allowing the player to manipulate a number of subcanvases. We say a number, because the player is rarely able to manipulate subcanvases in the sound stream. From a multimodal perspective, the canvas of a sound stream makes a different contribution to the game than, for instance, allocating yet another part of the screen to a 2D subcanvas. Because the sound stream is not founded on a 2D canvas, but draws on an entirely different material substrate, the linear canvas of sound may be invoked without sacrificing any of the precious real estate on the screen. What is more, the sound stream provides access to spoken language, which is able to transmit a wealth of linguistic, emotional and physical information *simultaneously* using different frequencies of the sound stream. Drawing on this canvas adds considerably to immersion, whose potential the game producers frequently exploit (Collins 2008).

Following this, it may be argued that games such as *Armed Assault 3* actively build on a combination of semiotic modes deployed across various canvases. To illustrate how real-time games are construed multimodally, let us now look at the example in Figure 17.5, which features a sequence of screenshots showing gameplay events during the first two minutes of a mission in *Armed Assault 3*.

The entire sequence begins with a gradual transition from a black screen to the 3D environment, marked by the horizontal line on top of Figure 17.5. This transition, of course, bears close resemblance to those found in film, which we will take up for discussion shortly below. Numerous sound objects, such as footsteps, distant gunfire and explosions, and other ambient sounds, are simultaneously rendered on the sound stream, as indicated by the green arrow that points downwards.

Depicted media in video games
The actual gameplay begins with the player entering a specific location, Camp Maxwell, whose name is featured in the first instance of a depicted medium   encountered in the 3D environment—the sign at the entrance on the right-hand side of screen #1. The same information is also rendered in written language on the game interface, letter by letter, accompanied by a ticker-like sound until the entire text is rendered on the screen. In this way, the game draws on semiotic modes made available by three different subcanvases—the 3D environment, the game interface and the sound stream—to indicate the player's location.

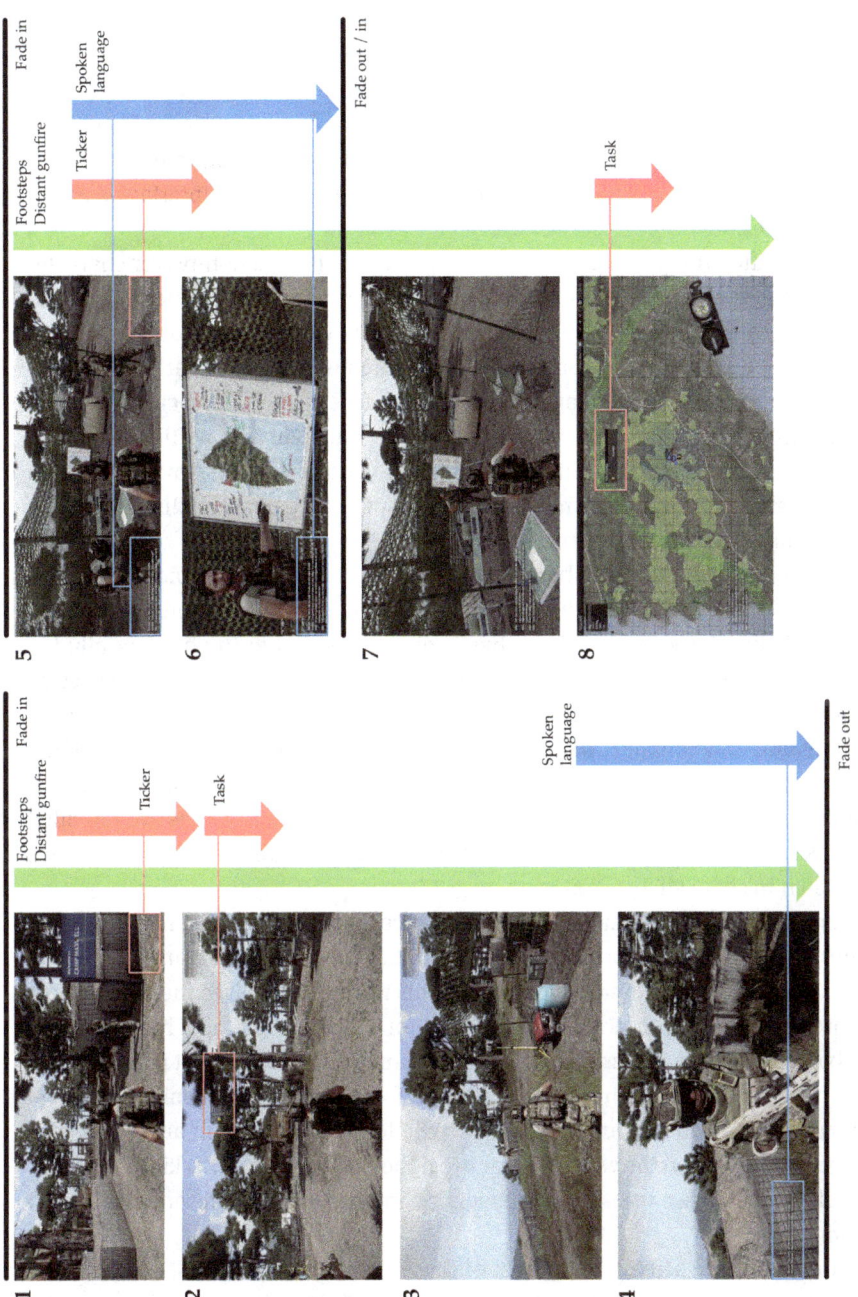

Fig. 17.5. A visual sequence protocol showing the orchestration of semiotic modes during two minutes of gameplay in *Armed Assault 3* ©Bohemia Interactive 2013

Upon entering the camp (#2), the player is assigned a task, which requires reporting for duty at a guard post at the camp perimeter. A simple non-naturalistic sound object ('bleep') is deployed to call the player's attention to the task, which is simultaneously rendered in written language on an information display superimposed on the 3D environment. Essentially, what we have here are two distinct kinds of sound objects synchronised with the 2D canvases on the screen: those pertaining to events occurring in the world depicted in the virtual 3D environment (footsteps and gunfire) and those originating in the game interface (tickers and task notifications). By pulling the sources of different sound objects apart we can also begin to explore how the semiotic work is divided between the 3D environment and the game interface realised on the 2D canvas.

Having located the guard post (#3) and reported for duty, a non-player character informs the player about the upcoming mission briefing, which constitutes the second task assigned to the player (#4). For this purpose, the game invokes another depicted medium, that is, embodied communication, which we described earlier as a combination of gesture and spoken language in face-to-face spoken interaction (→ §8 [p. 239]). This utterance is also rendered on an information display in written language, as marked by the blue box. As suggested above, this information display may be seen as working against the inherent transience of the sound stream, providing the player with more time to react to the event.

The player is then informed about completing the first task (not depicted in Figure 17.5) using the standard task notification shown in screen #2, before the screen fades into black. Fading back in, the player has been transported to the briefing alluded to by the non-player character (#5). The fade in/out transition may be suggested to be a borrowing from the medium of film, where it often signals a change in time or place (Bateman and Schmidt 2012: 80). As we have emphasised, the meaning of the transition is not static, but inferred dynamically from the immediate context with the help of discourse semantics (→ §4.1.1 [p. 116]). Here the necessary cues are provided by the task notification and the non-player character's utterance ("Sergeant Lacey's pinned down in some village. You ready to go?").

In screen #6 we encounter yet another instance of depicted media, namely an audiovisual presentation supported by a static artefact—in this case, an annotated map (→ §14 [p. 340]). Depicting a medium such as this naturally involves the entire range of synchronisation acts present in real-life situations as well, including hand gestures and verbal references to the visual aid. It is important to understand that making sense of the depicted medium requires the player to perform ergodic work similar to that required by their real-world counterparts; *listening* to the presentation and *composing* the content presented on the visual aid back together. Undertaking this kind of ergodic work is likely to increase the sense of immersion into the world represented in the 3D environment.

Semiotic
potential of
3D environ-
ment

At this point, the semiotic potential of the 3D environment needs to be underlined. This medium, which is often, but not always mobilised in games, draws on the available 2D and linear canvases for its capability to express a wide range of meanings. By exploiting these two canvases, the 3D environment can depict media that draw on similar canvases, ranging from signs in built environments ('Camp Maxwell') to different forms of spoken interaction (embodied face-to-face and mediated chatter over radio), and audiovisual presentations combining both 2D and linear canvases. However, the player is immersed into the 3D environment in the form of an avatar, and the resulting interactions are therefore limited by the avatar's capabilities to engage with the environment. This sets certain challenges to interacting with depicted media, which become evident in the final part of this brief analysis.

Following the conclusion of the audiovisual presentation in screen #7, the actual mission of assisting another group of soldiers is assigned to the player. Just as above, the fade in/out transition preceding screen #7 indicates the passing of time, as the non-player characters are no longer present following the briefing officer's order ("Okay. Regroup and report in."). Next, in screen #8 we find the player checking the map, which is provided by the game interface on a separate subcanvas to facilitate gameplay: although 3D environments may feature depicted media, they do not always provide the necessary interfaces for performing the required ergodic work in an efficient manner. This is where the game interface steps in as a replacement, providing the player with the means for performing the necessary ergodic work: an unobstructed top-down view to a map, complete with an interface to manipulate the cartographic representation.

17.3 Summary

In this chapter we have attempted to outline certain differences between real-time and turn-based games from a multimodal perspective. Whereas real-time games must *constantly* coordinate events across both 2D (screen) and linear (sound stream) canvases—an issue we are already familiar with from other audiovisual media such as film—turn-based games do not. Since there are no constantly unfolding events, interactions between 2D and linear canvases are often limited to specific events in turn-based games. Real-time games, however, must synchronise each occurring event across the 2D and linear canvases. To sum up, this results in a significant difference in the pace and scale of synchronisation.

Figure 17.6 attempts to illustrate this by showing how appropriate combinations of semiotic modes on each canvas are coordinated as they unfold in time. Additionally, the capability to cycle through different subcanvases—either in real time or within a single turn of the game—facilitates the ergodic work of undertaking different gaming-related activities. Moreover, for the analyst, the switches between subcanvases can act as a boundary for tracing the activities undertaken within the game.

As in other use case examples before, our analysis made it again possible to highlight the complexity of multimodality, which is similarly high in computer and video

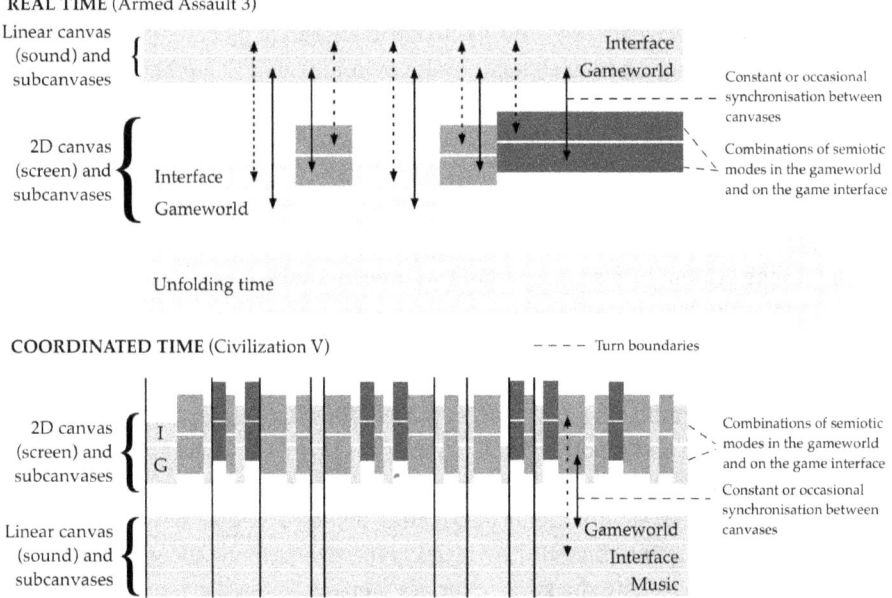

REAL TIME (Armed Assault 3)

Unfolding time

COORDINATED TIME (Civilization V)

Fig. 17.6. Game time and the synchronisation of semiotic modes across multiple canvases

games. Not only may they embed *depictions* of various media found in real life, ranging from posters, advertisements or the use of *radio*, for example, as in the *Grand Theft Auto* series (→ §4.2 [p. 126]), but these depictions also require the player to put in a significantly high effort of ergodic work, which may range from *composition*—making sense of semiotic modes on a 2D canvas depicted in the game—to *listening* to the gameworld sounds and radio. To do so, the players naturally draw on their experience of the depicted media in the real world (Rambusch and Susi 2008).

To conclude, computer and video games present an elusive but by no means unattainable target for multimodal analysis. As we have attempted to show, any activities involving ergodic work are likely to become far more manageable if the canvases on which they take place are mapped properly. Moreover, as we saw above, some canvases may be depicted in the gameworld: recognising these canvases is especially important for determining how the canvases relate to each other. This must be followed by identifying the semiotic modes, before analysing their contribution to the communicative situation at hand.

18 Final words: ready, steady, analyse!

In this final chapter, we will bring the discussion to a (temporary!) close and provide some pointers for taking multimodality research and practice further. We saw in our opening chapters how multimodality is an issue that is forcing itself on a broad range of disciplines concerned with communication. We also saw there, however, that the diversity of approaches had so far not coalesced into a broader working practice with established methods and paradigms. The central chapters of our book were then intended to provide a more robust basis for taking multimodal research further—with the slightly longer term goal then to help a genuine 'field' of multimodality to emerge. The use case chapters of the last part of the book set about giving you examples of how a broader foundational framework for multimodal research can assist you in organising intrinsically complex communication situations and encouraging greater reuse and transfer of results, while still respecting the particularities of individual semiotic modes, media and communicative situations.

18.1 Lessons learned: the take-home messages

It will be useful now to summarise what the overall 'take-home' messages of the book are: there are certain important lessons to be learned, not only for individual pieces of practical research or analysis, but also for more general approaches to the tasks and challenges of multimodality as such.

First, then, it is good to emphasise once again that multimodality, in our view, is *essentially a collaborative enterprise*. This means that we always have to make the most of work that has already been done in any of the many areas that multimodal research will, of necessity, come in contact with. This has several very practical, methodological and scientific consequences. Perhaps most important of all, it is necessary to **avoid attempts to reinvent the wheel**. Enthusiasm over a new field of study and the subsequent haste to engage with it may result in rushing past bodies of knowledge which possess a far more developed understanding of the terrain ahead, often based on long histories of engagement with that terrain and sometimes also already drawing on empirical research. For this reason, we have attempted to underline the value of such empirically-founded knowledge over preliminary conceptions derived solely through theory.

Considered more generally and methodologically, the situation with multimodality research occasionally has resembled what David Machin has observed in relation to growing interest in visual communication:

> "As it happens in other fields where a 'new' realm of investigation is 'discovered' it can then herald a new flurry of activity that can, to those outside looking in, appear rather arbitrary. New network leaders will emerge in this pioneering area of research. New terminologies will appear to account

Collaboration first

for the very same things already documented decades before in a different field." (Machin 2014: 5)

This is the danger we pointed to at the end of our navigator chapter as the 're-description syndrome' (→ §7.4 [p. 232]). When undertaken from 'outside' of a discipline that may already have very substantial bodies of work to draw on, relabelling of this kind always runs the risk of seeming superficial, picking out phenomena that the older discipline will see as 'obvious'. When relabellings become valued in their own right from within a specific perspective, they readily serve to reinforce borders that may have been originally established more or less by accident or without the benefit of deeper studies. Caution and a willingness to listen and engage with diverse contributions is therefore always going to be preferable.

Such an engagement also requires a willingness to immerse oneself in the research traditions of whatever media and modes contribute to one's areas of interest. This can be a substantial effort of course, but if one is genuinely interested in the area of study in focus, such broadening of foundations will also be very enjoyable— especially when combinations or parallels across fields begin to emerge. When we are able to see these combinations as well as the broader background, and are ready to work from there, the terrain may be navigated with far greater insight.

Recombining methods

And second, we have also suggested that sometimes it will be necessary to reconsider established practices. For example, when addressing methods, we have encouraged you to consider combinations and to pick methods as may be beneficial for particular tasks and areas of study. We do **not**, however, intend by this to incite a revolution! It is an in many ways unfortunate feature of the recent history of scientific and academic research, particularly in areas that do not have empirical methods as a natural 'brake' on speculation, that progress is conflated with rejecting what has been done before. As a consequence, the period has been marked by numerous so-called *turns*.

Turns?

These include the *linguistic turn* in the mid-20th century (Rorty 1979), followed by a counterbalancing move lending weight to the visual as a part of the *pictorial turn* (Mitchell 1992, 1994) or *iconic turn* (Boehm 1995; Moxey 2008). These have been followed in their own right by a *cultural turn* (Jameson 1998; Jacobs and Spillman 2005) and a *performative turn* (Kapchan 1995). Such turns lend themselves too easily to marking out revolutionary 'moments' that have arisen as disciplines attempt to assert, or have asserted for them, relative 'dominance' concerning their rights to offer more inclusive accounts of, often multimodal, cultural phenomena.

It is important not to let the discourse of any of these 'turns' distract attention from some of the unchanging characteristics of the kinds of phenomena addressed when focusing on multimodality. In fact, core issues pertaining to multimodality (such as the complexity of the media involved or their combination of resources as well as the definition of modes), have remained in place regardless of any particular 'turn's' promotion as a comprehensive perspective replacing those available previously. Consequently, we have not in this book adopted this kind of historical view on the development of multimodality, whereby each 'turn' emphasises its own practices often to the expense of others. We focused rather on showing that, on the side of academic research, quite different disciplines have all found themselves in the position of attempting moves beyond their traditional concerns, addressing facets of multimodality as they go. This has commonly involved making observations and drawing conclusions from a single perspective as if these were properties that multimodality in general must exhibit. This is often not the case, however.

18.2 Our goals in the book

We can also usefully reiterate what we have wanted to achieve in the book, rather than what we wanted to avoid. And, standing again here in first place, is the emphasis on demonstrating **collaboration rather than competition**. This entails a more useful model than the 'replacement' approach of 'turns' mentioned above. Indeed, this is also sometimes proclaimed by the more reasonable of those proposing such turns as well—a 'turn' should not be a turn 'away' from what has gone before, but rather should function as an extension or expansion. When seen more in this light, each turn mentioned above brings important new insights, theories and methods. It is precisely these extensions and expansions that we have been concerned with throughout this book, making sure that no aspect of multimodality is left out along the way. Similarly, as we have emphasised at several points, the ideas set out in this book are intended to complement, not replace, existing disciplinary approaches. Sometimes approaches to multimodality have appeared to give the impression that they seek to replace what has gone before; they are then, quite rightly, criticised.

Having characterised a broad range of characteristics of the various kinds of expressive resources that appear and re-appear across multimodal communication of all kinds, we sought then to **provide a solid foundation for research**. We argued that any investigation of multimodality—wherever it appears—must be placed on such a solid foundation in order not to be overwhelmed by detail and to make the most of existing work in other areas. Therefore, we have placed communicative situations first; these lead to the possibility of different forms of expression, and various disciplines may need to be drawn on depending on just what those forms of expression are. In short, we have attempted to introduce the theoretical foundations and practical methodologies necessary for investigating communicative artefacts, materials, events and performances on their own merits, rather than by pushing them into categories designed for other objects of investigation.

The proposed framework may now be put to use in exploring the entire territory of multimodality. Nevertheless, this framework is not a map; a map would entail that we already have a good sense of the landscape and that landscape's shape and features. The framework we have provided is instead a tool for creating more detailed sketches based on the foundation we provide. The rest will rely on the expertise and enthusiasm of the individuals exploring the territory, an exploration that can only be carried out by actually doing research and practical work of the kinds we have pointed to.

Figure 18.1 suggests graphically this 'interventional', or 'mediating', role that we see for the detailed account of multimodality we have presented throughout this book. Rather than having individual disciplines or research questions repeatedly come up against the problems and challenges of multimodality from within their own perspectives, we have argued that a methodological approach focusing on the role of modalities and their interaction in their own right provides a more systematic path from research questions to the multimodal phenomena lying at the heart of the enterprise.

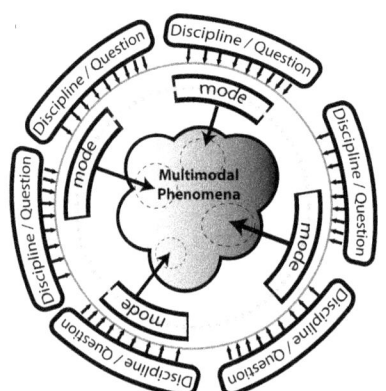

Fig. 18.1. Final revision of our ways into multimodality diagrams. Now we see that defining modalities acts as a buffering mechanism that mediates between individual disciplines and research questions (the outer circle), which will usually be concerned with multiple modes and their combinations, and the detailed internal accounts of those modalities. Multimodality (the inner circle) concerns itself centrally with how those modalities are to be defined so they can interact with each other and be used.

18.3 Be a multimodalist: have fun and explore!

And this then brings us to the last point of all: **have fun and explore!** Students introduced to multimodality may occasionally wonder about the value of pursuing painstaking analyses of everyday communicative situations. After all, is there a point in deconstructing the messages you send to friends on WhatsApp or your favourite film or comics album? We argue that bringing these situations under the microscope can indeed reveal important aspects of our quite extraordinary capability to manage complex multimodality in everyday life, both as consumers and producers of signs.

Whether you are a student, researcher or a practitioner, understanding multimodality better opens up new possibilities. By following the methodological steps we have suggested, the foundation is prepared for driving systematic and reproducible research into multimodal phenomena from diverse starting points, while maintaining the culminative progress and results necessary for understanding to grow.

Bibliography

Aarseth, E. (2004), Genre trouble: narrativism and the art of simulation, *in* N. Wardrip-Fruin and P. Harrigan, eds, 'FirstPerson: New Media as Story, Performance, and Game', MIT Press, Cambridge, MA, pp. 45–55.

Aarseth, E. J. (1997), *Cybertext: perspectives on ergodic literature*, John Hopkins University Press, Baltimore, MD.

Abbott, L. L. (1986), 'Comic Art: Characteristics and Potentialities of a Narrative Medium', *Journal of Popular Culture* **19**(4), 155–176.

Adami, E. (2015), 'What's in a click? A social semiotic framework for the multimodal analysis of website interactivity', *Visual Communication* **14**(2), 133–153.

Adams, A., Boersema, T. and Mijksenaar, M. (2010), 'Warning symbology: difficult concepts may be successfully depicted with two-part signs', *Information Design Journal* **18**(2), 94–106.

Allwood, J. (2008), Multimodal corpora, *in* A. Lüdeling and M. Kytö, eds, 'Corpus Linguistics. An International Handbook', Vol. 1, Mouton de Gruyter, Berlin, pp. 207–225.

Allwood, J., Cerrato, L., Jokinen, K., Navarretta, C. and Paggio, P. (2007), 'The MUMIN coding scheme for the annotation of feedback, turn management and sequencing phenomena', *Language Resources and Evaluation* **41**(3), 273–287. http://dx.doi.org/10.1007/s10579-007-9061-5

Allwood, J., Kopp, S., Grammer, K., Ahlsén, E., Oberzaucher, E. and Koppensteiner, M. (2007), 'The analysis of embodied communicative feedback in multimodal corpora: a prerequisite for behavior simulation', *Journal on Language Resources and Evaluation* **41**(2–3), 325–339. Special Issue on Multimodal Corpora for Modeling Human Multimodal Behaviour.

Alper, M. (2014), 'War on Instagram: Framing conflict photojournalism with mobile photography apps', *New Media & Society* **16**(8), 1233–1248.

Altman, R. (1999), *Film/Genre*, British Film Institute, London.

Altman, R. (2008), *A theory of narrative*, Columbia University Press, New York.

American Psychological Association (2016), 'APA Style Blog'. Last accessed: 12 November 2016. http://blog.apastyle.org/apastyle/social-media/

Andersen, P. B. (1997), *A Theory of Computer Semiotics*, Cambridge University Press, Cambridge.

Andersson, H. and Machin, D. (2016), A Multimodal Approach to Song, *in* N.-M. Klug and H. Stöckl, eds, 'Sprache im multimodalen Kontext / Language and Multimodality', number 7 *in* 'Handbooks of Linguistics and Communication Science (HSK)', de Gruyter Mouton, Berlin, pp. 372–391.

André, E. and Rist, T. (1993), The design of illustrated documents as a planning task, *in* M. T. Maybury, ed., 'Intelligent Multimedia Interfaces', AAAI Press/The MIT Press, Menlo Park (CA), Cambridge (MA), London (England), pp. 94–116.

Archer, A. and Breuer, E., eds (2015), *Multimodality in Writing: The state of the art in theory, methodology and pedagogy*, Brill N.V., Leiden, The Netherlands.

Arens, Y. and Hovy, E. H. (1990), How to describe what? Towards a theory of modality utilization, *in* 'The Twelfth Annual Conference of the Cognitive Science Society', Lawrence Erlbaum Associates, Hillsdale, New Jersey, pp. 487–494.

Asher, N. and Lascarides, A. (2003), *Logics of conversation*, Cambridge University Press, Cambridge.

Atkinson, P. (2010), 'The Graphic Novel as Metafiction', *Studies in Comics* **1**(1), 107–125.

Azcárate, A. L.-V. and Sukla, A. C., eds (2015), *The Ekphrastic Turn: Inter-art Dialogues*, Common Ground Publishing, Champaign, IL.

Bachmann, C. A. (2016), Materiality and Mediality of Comics: The Case of éditions polystrène. Symposium "The Empirical Study of Comics", Delmenhorst.

Baetens, J. (2004), 'Autobiographies et Bandes Dessinées: Problèmes, Enjeux, Exemples', *Belphégor. Littérature Populaire et Culture Mediatique* 4(1).

Baetens, J. (2015), *The graphic novel: an introduction*, Cambridge University Press, Cambridge and New York.

Baldry, A. and Thibault, P. J. (2006), *Multimodal Transcription and Text Analysis: a multimedia toolkit and coursebook with associated on-line course*, Textbooks and Surveys in Linguistics, Equinox, London and New York.

Barnes, D. (2009), 'Time in the Gutter: Temporal Structures in Watchmen', *KronoScope* 9(1-2), 51–60.

Barthes, R. (1977), *Image – Music – Text*, Fontana Press, London. edited and Translated by Stephen Heath.

Bateman, J. A. (2007), 'Towards a *grande paradigmatique* of film: Christian Metz reloaded', *Semiotica* 167(1/4), 13–64.

Bateman, J. A. (2008), *Multimodality and Genre: a foundation for the systematic analysis of multimodal documents*, Palgrave Macmillan, Basingstoke.

Bateman, J. A. (2009), Discourse across semiotic modes, *in* J. Renkema, ed., 'Discourse, of course: an overview of research in discourse studies', John Benjamins, Amsterdam / Philadelphia, pp. 55–66.

Bateman, J. A. (2011), The Decomposability of Semiotic Modes, *in* K. L. O'Halloran and B. A. Smith, eds, 'Multimodal Studies: Multiple Approaches and Domains', Routledge Studies in Multimodality, Routledge, London, pp. 17–38.

Bateman, J. A. (2014*a*), Genre in the age of multimodality: some conceptual refinements for practical analysis, *in* P. Evangelisti Allori, J. A. Bateman and V. K. Bhatia, eds, 'Evolution in Genres: Emergence, Variation, Multimodality', Linguistic insights, Peter Lang, Frankfurt am Main, pp. 237–269.

Bateman, J. A. (2014*b*), *Text and Image: a critical introduction to the visual/verbal divide*, Routledge, London and New York.

Bateman, J. A. (2014*c*), Using Multimodal Corpora for Empirical Research, *in* C. Jewitt, ed., 'The Routledge Handbook of multimodal analysis', 2 edn, Routledge, London, pp. 238–252.

Bateman, J. A. (2016), Methodological and theoretical issues for the empirical investigation of multimodality, *in* N.-M. Klug and H. Stöckl, eds, 'Sprache im multimodalen Kontext / Language and Multimodality', number 7 *in* 'Handbooks of Linguistics and Communication Science (HSK)', de Gruyter Mouton, Berlin, pp. 36–74.

Bateman, J. A. (2017*a*), Multimodality and genre: issues for information design, *in* A. Black, P. Luna, O. Lund and S. Walker, eds, 'Information Design: Research and Practice', Routledge, London, pp. 221–241.

Bateman, J. A. (2017*b*), The place of Systemic-Functional Linguistics as a linguistic theory in the 21st Century, *in* T. Bartlett and G. O'Grady, eds, 'The Routledge Handbook of Systemic Functional Linguistics', Routledge, London.

Bateman, J. A., Delin, J. L. and Henschel, R. (2004), Multimodality and empiricism: preparing for a corpus-based approach to the study of multimodal meaning-making, *in* E. Ventola, C. Charles and M. Kaltenbacher, eds, 'Perspectives on Multimodality', John Benjamins, Amsterdam, pp. 65–87.

Bateman, J. A., Delin, J. L. and Henschel, R. (2007), Mapping the multimodal genres of traditional and electronic newspapers, *in* T. D. Royce and W. L. Bowcher, eds, 'New Directions in the Analysis of Multimodal Discourse', Lawrence Erlbaum Associates, Mahwah, NJ, pp. 147–172.

Bateman, J. A., Kamps, T., Kleinz, J. and Reichenberger, K. (2001), 'Constructive text, diagram and layout generation for information presentation: the DArt$_{bio}$ system', *Computational Linguistics* 27(3), 409–449.

Bateman, J. A. and Schmidt, K.-H. (2012), *Multimodal Film Analysis: How Films Mean*, Routledge Studies in Multimodality, Routledge, London.

Bateman, J. A., Tseng, C., Seizov, O., Jacobs, A., Lüdtke, A., Müller, M. G. and Herzog, O. (2016), 'Towards next-generation visual archives: image, film and discourse', *Visual Studies* 31(2), 131–154.

Bateman, J. A. and Veloso, F. O. D. (2013), 'The Semiotic Resources of Comics in Movie Adaptation: Ang Lee's *Hulk* (2003) as a Case Study', *Studies in Comics* 4(1), 137–159.

Bateman, J. A., Veloso, F. O., Wildfeuer, J., Cheung, F. H. and Guo, N. S. (2016), 'An open multi-level classification scheme for the visual layout of comics and graphic novels: motivation and design', *Journal of Digital Scholarship in the Humanities* .

Bateman, J. A. and Wildfeuer, J. (2014a), 'A multimodal discourse theory of visual narrative', *Journal of Pragmatics* 74, 180–218.
http://dx.doi.org/10.1016/j.pragma.2014.10.001

Bateman, J. A. and Wildfeuer, J. (2014b), 'Defining units of analysis for the systematic analysis of comics: A discourse-based approach', *Studies in Comics* 5(2), 371–401.
http://dx.doi.org/10.1386/stic.5.2.371_1

Bell, P. (2001), Content analysis of visual images, *in* T. van Leeuwen and C. Jewitt, eds, 'Handbook of visual analysis', Sage, London, chapter 2, pp. 10–24.

Bellour, R. (2000 [1976]), To segment / to analyze (on *Gigi*), *in* C. Penley, ed., 'The analysis of film', Indiana University Press, Bloomington and Indianapolis, pp. 193–216. Originally published in *Quarterly Review of Film Studies* 1(3), August 1976.

Benson, P. (2017), *The Discourse of YouTube: Multimodal Text in a Global Context*, Vol. 15 of *Routledge Studies in Multimodality*, Routledge, New York and London.

Benyon, D. (2014), *Designing Interactive Systems: A comprehensive guide to HCI, UX and interaction design*, 3 edn, Pearson, Harlow, England.

Berger, P. L. and Luckmann, T. (1966), *The social construction of reality: a treatise in the sociology of knowledge*, Anchor Books, Garden City, New York.

Bergs, A. and Zima, E., eds (2017), *Towards a multimodal construction grammar*, de Gruyter Mouton, Berlin. Linguistic Vanguard. A Multimodal Journal for the Language Sciences: Special Issue.

Bernsen, N. O. (2002), Multimodality in language and speech systems – from theory to design support tool, *in* B. Granström, D. House and I. Karlsson, eds, 'Mutlimodality in language and speech systems', Kluwer Academic Publishers, Dordrecht, pp. 93–148.

Bertin, J. (1981), *Graphics and graphic information processing*, Walter de Gruyter, Berlin.

Bertin, J. (1983), *Semiology of graphics*, University of Wisconsin Press, Madison, Wisconsin. Translated *Sémiologie graphique* (1967) by William J. Berg.

Bertin, J. (2001), 'Matrix theory of graphics', *Information Design Journal* 10(1), 5–19.

Bezemer, J. (2014), Multimodal transcription: a case study, *in* S. Norris and C. D. Maier, eds, 'Interactions, Images and Texts: A Reader in Multimodality', Mouton de Gruyter, Berlin, pp. 155–170.

Bezemer, J., Cope, A., Faiz, O. and Kneebone, R. (2012), 'Participation of Surgical Residents in Operations: Challenging a Common Classification', *World Journal of Surgery* 36(9), 2011–2014.

Bezemer, J., Cope, A., Kress, G. and Kneebone, R. (2014), 'Holding the Scalpel: Achieving Surgical Care in a Learning Environment', *Journal of Contemporary Ethnography* 43(1), 38–63.

Bezemer, J. and Kress, G. (2008), 'Writing in multimodal texts: a social semiotic account of designs for learning', *Written Communication* 25(2), 165–195.

Bezemer, J. and Kress, G. (2016), *Multimodality, Learning and Communication: A Social Semiotic Frame*, Routledge, New York and London.

Bezemer, J. and Mavers, D. (2011), 'Multimodal transcription as academic practice: a social semiotic perspective', *International Journal of Social Research Methodology* 14(3), 191–206.

Birch, D. (1991), 'Drama, performance, praxis', *Social Semiotics* 1(2), 170–193.

Birdwhistell, R. (1970), *Kinesics and Context: Essays on Body Motion Communication*, Ballentine Books, New York.

Black, A., Luna, P., Lund, O. and Walker, S. (2017), *Information Design: Research and Practice*, Routledge, London.

Blank, J. (2010), 'Alles ist zeigbar? Der Comic als Medium der Wissensvermittlung nach dem iconic turn', *KulturPoetik* 10(2), 214–233.

Blommaert, J. (2005), *Discourse: a critical introduction*, Key topics in sociolinguistics, Cambridge University Press, Cambridge.

Boehm, G. (1995), Die Wiederkehr der Bilder?, *in* G. Boehm, ed., 'Was ist ein Bild?', 2 edn, Wilhelm Fink Verlag, München, pp. 11–38.

Boeriis, M. and Holsanova, J. (2012), 'Tracking visual segmentation: connecting semiotic and cognitive perspectives', *Visual Communication* 11(3), 259–281.

Bogost, I. (2006), *Unit Operations: An Approach to Videogame Criticism*, The MIT Press, Cambridge, MA.

Bolter, J. D. and Grusin, R. (2000), *Remediation: Understanding New Media*, MIT Press, Cambridge, MA.

Bordwell, D. (1985), *Narration in the fiction film*, University of Wisconsin Press, Madison, WI.

Bordwell, D. (2007), *Poetics of Cinema*, Routledge, London and New York.

Bordwell, D. and Thompson, K. (2013), *Film Art: An Introduction. Tenth Edition*, 10 edn, McGraw-Hill, New York.

Bordwell, D., Thompson, K. and Staiger, J. (1985), *The Classical Hollywood Cinema. Film Style and Mode of Production to 1960*, Columbia University Press, New York.

Bounegru, L., Venturini, T., Gray, J. and Jacomy, M. (2016), 'Narrating networks: Exploring the affordances of networks as storytelling devices in journalism', *Digital Journalism* . http://dx.doi.org/10.1080/21670811.2016.1186497

Bramlett, F., Cook, R. T. and Meskin, A., eds (2016), *The Routledge Companion to Comics*, Routledge, London and New York.

Branigan, E. (1984), *Point of view in the cinema: a theory of narration and subjectivity in classical film*, number 66 *in* 'Approaches to semiotics', Mouton, Berlin.

Bremond, C. (1964), 'Le Message narratif', *Communications* pp. 4–32.

Brenger, B. and Mittelberg, I. (2015), Shakes, nods and tilts. Motion-capture data profiles of speakers' and listeners' head gestures, *in* 'Proceedings of the 3rd Gesture and Speech in Interaction (GESPIN) Conference', pp. 43–48.

Bressem, J. (2013), A linguistic perspective on the notation of form features in gestures, *in* C. Müller, A. Cienki, E. Fricke, S. Ladewig, D. McNeill and S. Tessendorf, eds, 'Body–Language–Communication. An International Handbook on Multmodality in Human Interaction', number 38/1 *in* 'Handbücher zur Sprach- und Kommunikationswissenschaft / Handbooks of Linguistics and Communication Science (HSK)', Mouton de Gruyter, Berlin and New York, pp. 1079–1098.

Bressem, J., Ladewig, S. H. and Müller, C. (2013), Linguistic annotation system for gestures (LASG), *in* C. Müller, A. Cienki, E. Fricke, S. Ladewig, D. McNeill and S. Tessendorf, eds, 'Body–Language–Communication. An International Handbook on Multmodality in Human Interaction', number 38/1 *in* 'Handbücher zur Sprach- und Kommunikationswissenschaft / Handbooks of Linguistics and Communication Science (HSK)', Mouton de Gruyter, Berlin and New York, pp. 1098–1125.

Brill, J., Kim, D. and Branch, R. (2007), 'Visual literacy defined – The results of a Delphi study', *Journal of Visual Literacy* **27**, 47–60.

Bucher, H.-J. (2007), Textdesign und Multimodalität. Zur Semantik und Pragmatik medialer Gestaltungsformen, *in* K. S. Roth and J. Spitzmüller, eds, 'Textdesign und Textwirkung in der massenmedialen Kommunikation', UVK, Konstanz, pp. 49–76.

Bucher, H.-J. (2010), Multimodalität – eine Universalie des Medienwandels. Problemstellung und Theorien der Multimodalitätsforschung, *in* H.-J. Bucher, T. Gloning and K. Lehnen, eds, 'Neue Medien – neue Formate. Ausdifferenzierung und Konvergenz in der Medienkommunikation', number 10 *in* 'Interaktiva. Schriftenreihe des Zentrums für Medien und Interaktivität (ZMI), Gießen', Campus Verlag, Frankfurt / New York, pp. 41–79.

Bucher, H.-J. (2011), Multimodales Verstehen oder Rezeption als Interaktion. Theoretische und empirische Grundlagen einer systematischen Analyse der Multimodalität, *in* H.-J. Diekmannshenke, M. Klemm and H. Stöckl, eds, 'Bildlinguistik. Theorien – Methoden – Fallbeispiele', Erich Schmidt, Berlin, pp. 123–156.

Bucher, H.-J., Krieg, M. and Niemann, P. (2010), Die wissenschaftliche Präsentation als multimediale Kommunikationsform, *in* H.-J. Bucher, T. Gloning and K. Lehnen, eds, 'Neue Medien – neue Formate. Ausdifferenzierung und Konvergenz in der Medienkommunikation', number 10 *in* 'Interaktiva. Schriftenreihe des Zentrums für Medien und Interaktivität (ZMI), Gießen', Campus Verlag, Frankfurt / New York, pp. 381–412.

Bucher, H.-J. and Niemann, P. (2012), 'Visualizing science: the reception of PowerPoint presentations', *Visual Communication* **11**(3), 283–306.

Burgoon, J. K., Guerrero, L. K. and Manusov, V. (2011), Nonverbal signals, *in* M. L. Knapp and J. A. Daly, eds, 'Sage Handbook of Interpersonal Communication', 4 edn, Sage Publications, Inc., Los Angeles, chapter 8, pp. 239–281.

Burke, L. (2015), *The Comic Book Film Adaptation*, University Press of Mississippi, Jackson.

Burn, A. (2014), The kineikonic mode: towards a multimodal approach to moving-image media, *in* C. Jewitt, ed., 'The Routledge Handbook of multimodal analysis', 2 edn, Routledge, London, pp. 375–385.

Caldwell, D. and Zappavigna, M. (2011), Visualizing multimodal patterning, *in* S. Dreyfus, S. Hood and M. Stenglin, eds, 'Semiotic Margins: reclaiming meaning', Continuum, London, pp. 229–242.

Cao, Y. and O'Halloran, K. L. (2015), 'Learning human photo shooting patterns from large-scale community photo collections', *Multimedia Tools and Applications* **74**(24), 11499–11516.

Carr, D. (2014), Interpretation, representation and methodology: issues in computer game analysis, *in* D. Machin, ed., 'Visual Communication', Mouton de Gruyter, Berlin, pp. 501–516.

Carr, D., Buckingham, D., Burn, A. and Schott, G., eds (2006), *Computer Games: Text, Narrative and Play*, Polity Press, Cambridge.

Chapman, A. (2013), 'Is Sid Meier's Civilization history?', *Rethinking History* **17**(3), 312–332.

Chatman, S. (1978), *Story and discourse: narrative structure in fiction and film*, Cornell University Press, Ithaca and London.

Choi, Y. S., Gray, H. M. and Ambady, N. (2006), The glimpsed world: Unintended communication and unintended perception, *in* R. R. Hassin, J. S. Uleman and J. A. Bargh, eds, 'The New Unconscious', Oxford University Press, Oxford and New York, pp. 309–333.

Christiansen, H.-C. (2000), Comics and film: a narrative perspective, *in* A. Magnussen and H.-C. Christiansen, eds, 'Comics & culture: Analytical and theoretical approaches to comics', Museum Tusculanum Press University of Copenhagen, Copenhagen, Denmark, pp. 107–122.

Cobley, P. and Schulz, P. J., eds (2013), *Theories and Models of Communication*, De Gruyter Mouton, Berlin and Boston.

Cohen, D. L. and Snowden, J. L. (2008), 'The relations between document familiarity, frequency, and prevalence and document literacy performance among adult readers', *Reading Research Quarterly* 43(1), 9–26.

Cohen, J. (1988), *Statistical power analysis for the behavioral sciences*, 2 edn, Erlbaum, Hillsdale, NJ.

Cohn, N. (2013*a*), 'Beyond speech balloons and thought bubbles: The integration of text and image', *Semiotica* 197, 35–63.

Cohn, N. (2013*b*), 'Navigating comics: an empirical and theoretical approach to strategies of reading comic page layouts', *Frontiers in Psychology* 4(186), 1–15.

Cohn, N. (2013*c*), *The Visual Language of Comics: Introduction to the structure and cognition of sequential images*, Bloomsbury, London.

Cohn, N. (2016), 'A multimodal parallel architecture: A cognitive framework for multimodal interactions', *Cognition* 146, 304–323.

Cohn, N. and Campbell, H. (2015), 'Navigating Comics II: Constraints on the Reading Order of Comic Page Layouts', *Applied Cognitive Psychology* 29(2), 193–199.

Cohn, N. and Ehly, S. (2016), 'The vocabulary of manga: Visual morphology in dialects of Japanese Visual Language', *Journal of Pragmatics* 92, 17–29.

Cohn, N., Paczynski, M., Jackendoff, R., Holcomb, P. J. and Kuperberg, G. R. (2012), '(Pea)nuts and bolts of visual narrative: structure and meaning in sequential image comprehension', *Cognitive Psychology* 65(1), 1–38.

Collier, M. (2001), Approaches to analysis in visual anthropology, *in* T. van Leeuwen and C. Jewitt, eds, 'Handbook of visual analysis', Sage, London, chapter 3, pp. 35–60.

Collins, K. (2008), *Game Sound: An Introduction to the History, Theory, and Practice of Video Game Music and Sound Design*, MIT Press, Cambridge, MA.

Colomer, T., Kümmerling-Meibauer, B. and Silva-Díaz, C., eds (2010), *New directions in picturebook research*, Routledge, London.

Connors, S. P. (2012), 'Weaving Multimodal Meaning in a Graphic Novel Reading Group', *Visual Communication* 12(1), 27–53.

Cook, K. M. (2014), Music, history, and progress in Sid Meier's *Civilization IV*, *in* K. Donnelly, W. Gibbons and N. Lerner, eds, 'Music in Video Games: Studying Play', Routledge, New York and London, pp. 166–182.

Cook, R. T. (2011), 'Do comics require pictures? Or why *Batman* #663 is a comic', *The Journal of Aesthetics and Art Criticism* 69(3), 285–296.

Cook, R. T. (2012), Why Comics Are Not Films. Metacomics and Medium-Specific Conventions, *in* A. Meskin and R. T. Cook, eds, 'The Art of Comics: A Philosophical Approach', Blackwell, Maiden and Oxford, pp. 165–187.

Cook, R. T. (2016), Metacomics, *in* F. Bramlett, R. Cook and A. Meskin, eds, 'The Routledge Companion to Comics', Routledge, New York and London, pp. 257–266.

Cope, B. and Kalantzis, M. (2000), Designs for social futures, *in* M. Kalantzis and B. Cope, eds, 'Multiliteracies: Literacy Learning and the Design of Social Futures', Routledge, London, chapter 10, pp. 203–234.

Cotter, C. (2015), Discourse and Media, *in* D. Schiffrin, D. Tannen and H. Hamilton, eds, 'The Handbook of Discourse Analysis', Blackwell, pp. 795–821.

Couldry, N. and Hepp, A. (2013), 'Conceptualizing mediatization: Contexts, traditions, arguments', *Communication Theory* 23(3), 191–202.

Couldry, N. and Hepp, A. (2017), *The Mediated Construction of Reality*, Polity, Cambridge, UK and Malden, MA.

Coventry, K. R., Lynott, D., O'Ceallaigh, R. and Miller, J. (2008), 'Happy is Up and Sad is Down; the facial emotion-spatial congruity effect', *Psychonomic Bulletin and Review* .

Coy, W. and Pias, C., eds (2009), *PowerPoint: Macht und Einfluss eines Präsentationsprogramms*, Fischer Taschenbuch Verlag, Frankfurt am Main.

Crawford, L. (1985), 'Monstrous criticism: finding, citing – analyzing film', *Diacritics* **15**(1), 58–70. http://www.jstor.org/stable/464631

David, G. and Cambre, C. (2016), 'Screened intimacies: Tinder and the swipe logic', *Social Media and Society* **2**(2).

Davidov, D., Tsur, O. and Rappoport, A. (2010), Enhanced sentiment learning using Twitter hashtags and smileys, *in* 'Proceedings of the 23rd International Conference on Computational Linguistics (COLING'10)', Association for Computational Linguistics, pp. 241–249.

Davidson, J. and di Gregorio, S. (2011), Qualitative Research and Technology. In the Midst of a Revolution, *in* N. K. Denzin and Y. S. Lincoln, eds, 'The Sage handbook of qualitative research', 4 edn, Sage, Los Angeles, pp. 627–643.

de Marinis, M. (1993), *The semiotics of performance*, Indiana University Press, Bloomington, IN.

de Saint-Georges, I. (2004), Materiality in discourse: the influence of space and layout in making meaning, *in* P. LeVine and R. Scollon, eds, 'Discourse and technology: multimodal discourse analysis', Georgetown University Press, Washington, D.C., pp. 71–87.

de Toro, F. (1995), *Theatre Semiotics: Text and staging in modern theatre*, Toronto University Press, Toronto.

Delin, J. L. and Bateman, J. A. (2002), 'Describing and critiquing multimodal documents', *Document Design* **3**(2), 140–155.

Deng, S., Kirkby, J. A., Chang, J. and Zhang, J. J. (2014), 'Multimodality with eye tracking and haptics: A new horizon for serious games?', *International Journal of Serious Games* **4**(1), 17–34.

Department of Education (GOV.UK) (2013), 'National curriculum in England: art and design programmes of study', online.
https://www.gov.uk/government/publications/national-curriculum-in-england-art-and-design-programmes-of-study/national-curriculum-in-england-art-and-design-programmes-of-study

Deppermann, A. (2013), 'Multimodal interaction from a conversation analytic perspective', *Journal of Pragmatics* **46**(1), 1–7.

Derrida, J. (1980), 'The Law of Genre', *Critical Inquiry* **7**(1), 55–81.

Dick, M. (2015), 'Just fancy that: An analysis of infographic propaganda in The Daily Express, 1956–1959', *Journalism Studies* **16**(2), 152–174.

Djonov, E. and Knox, J. (2014), How to analyze webpages, *in* S. Norris and C. D. Maier, eds, 'Interactions, Images and Texts: A Reader in Multimodality', Mouton de Gruyter, Berlin, pp. 171–194.

Djonov, E. and van Leeuwen, T. (2011), 'The semiotics of texture: from tactile to visual', *Visual Communication* **10**(4), 541–564.

Djonov, E. and van Leeuwen, T. (2013), 'Between the grid and composition: Layout in in Powerpoint's design and use', *Semiotica* **197**, 1–34.

Doelker, C. (1988), Das Bild in der Kommunikation, *in* L. Bosshart and J.-P. Chuard, eds, 'Communication visuelle: l'image dans la presse et la publicité', number 19 *in* 'Communication sociale, Cahiers de travaux pratiques', Editions Universitaires Fribourg Suisse, Fribourg, Suisse, pp. 119–142.

Drucker, J. (2013), 'Performative materiality and theoretical approaches to interface', *Digital Humanities Quarterly* **7**(1).
http://www.digitalhumanities.org/dhq/vol/7/1/000143/000143.html

Duncan, R. and Smith, M. J. (2009), *The power of comics: history, form and culture*, Continuum, New York and London.

Duncan, S., Cassell, J. and Levy, E. (2007), *Gesture and the Dynamic Dimension of Language*, John Benjamins, Amsterdam.

Dunmire, P. L. (2012), 'Political Discourse Analysis: Exploring the Language of Politics and the Politics of Language', *Linguistics and Language Compass* **6**(qq), 735–751–664.

Dunne, D. (2014), Multimodality or ludo-narrative dissonance: Duality of presentation in fringe media, *in* 'Proceedings of the Conference on Interactive Entertainment (IE2014)', Association for Computing Machinery, New York, NY, USA, pp. 29: 1–4.

Dürscheid, Christa und Frick, K. (2014), Keyboard-to-Screen-Kommunikation gestern und heute: SMS und WhatsApp im Vergleich, *in* A. Mathias, J. Runkehl and T. Siever, eds, 'Sprachen? Vielfalt! Sprache und Kommunikation in der Gesellschaft und den Medien. Eine Online-Festschrift zum Jubiläum von Peter Schlobinski', Vol. 64, Networx, Hanover, pp. 149–182.

Eisenlauer, V. (2013), *A Critical Hypertext Analysis of Social Media. The True Colours of Facebook*, Bloomsbury, London and New York.

Eisenlauer, V. (2014), 'Facebook as a third author—(semi-)automated participation framework in social network sites', *Journal of Pragmatics* **72**, 73–85.

Eisenstein, S. (1943), *The film sense*, Faber, London. Translated by J. Leyda.

Eisner, W. (1992), *Comics and sequential art*, Kitchen Sink Press Inc., Princeton, WI.

Eisner, W. (1996), *Graphic storytelling and visual narrative*, Poorhouse Press, Tamarac, FL.

Elkins, J. (1999), *The domain of images*, Cornell University Press, Ithaca, NY.

Elkins, J. (2003), *Visual studies: a skeptical introduction*, Routledge, London.

Elleström, L. (2010), The modalities of media: a model for understanding intermedial relations, *in* L. Elleström, ed., 'Media Borders, Multimodality and Intermediality', Palgrave Macmillan, Basingstoke, pp. 11–50.

Elleström, L. (2014), *Media transformation: the transfer of media characteristics among media*, Palgrave Macmillan, Basingstoke.

Elverdam, C. and Aarseth, E. (2007), 'Game classification and game design: construction through critical analysis', *Games and Culture* **2**(1), 3–22.

Engelhardt, Y. (2002), The language of graphics, ILLC dissertation series 2002–03, Institute for Logic, Language and Computation, Amsterdam, Amsterdam.

Ermi, L. and Mäyrä, F. (2007), Fundamental components of the gameplay experience: Analysing immersion, *in* S. de Castell and J. Jenson, eds, 'Worlds in Play: International Perspectives on Digital Games Research', Peter Lang, New York, pp. 37–53.

Escalera, S., Athitsos, V. and Guyon, I. (2016), 'Challenges in multimodal gesture recognition', *Journal of Machine Learning Research* **17**, 1–54.

Fann, K. T. (1970), *Peirce's theory of abduction*, Nijhoff, The Hague.

Feng, D. and O'Halloran, K. L. (2013), 'The multimodal representation of emotion in film: Integrating cognitive and semiotic approaches', *Semiotica* **197**, 79–100.

Fernandes, C., ed. (2016), *Multimodality and Performance*, Cambridge Scholars Publishing, Newcastle upon Tyne.

Field, A. (2016), *An Adventure in Statistics: the Reality Enigma*, Sage, London. Illustrated by James Iles.

Fischer-Lichte, E. (1992), *The semiotics of theatre*, Indiana University Press, Bloomington, IN.

Flewitt, R., Hampel, R., Hauck, M. and Lancaster, L. (2014), What are multimodal data collection and transcription?, *in* C. Jewitt, ed., 'The Routledge Handbook of multimodal analysis', 2 edn, Routledge, London, pp. 44–59.

Flowerdew, J. and Richardson, J., eds (2017), *Routledge Handbook of Critical Discourse Studies*, Routledge, London.

Forceville, C. J. (2006), Non-verbal and multimodal metaphor as a cognitivist framework: agendas for research, *in* G. Kristiansen, M. Achard, R. Dirven and F. R. de Mendoza Ibanez, eds, 'Cognitive Linguistics: Current Applications and Future Perspectives', Mouton de Gruyter, Berlin/New York, pp. 379–402. Republished in slightly amended form as ?.

Forceville, C. J. (2010), Balloonics: the visuals of balloons in comics, *in* J. Goggin and D. Hassler-Forest, eds, 'The rise and reason of comics and graphic literature: critical essays on the form', McFarland & Co., Jefferson, NC and London, pp. 56–73.

Forceville, C. J. (2011), 'Pictorial runes in *Tintin and the Picaros*', *Journal of Pragmatics* **43**(3), 875–890.

Forceville, C. J., El Rafaie, E. and Meesters, G. (2014), Stylistics and comics, *in* M. Burke, ed., 'The Routledge Handbook of Stylistics', Routledge, London and New York, pp. 485–499.

Forsey, J. (2013), *The aesthetics of design*, Oxford University Press, Oxford and New York.

Foucault, M. (1969), *The Archaeology of Knowledge*, Routledge, London.

Fricke, E. (2007), *Origo, Geste und Raum: Lokaldeixis im Deutschen*, de Gruyter, Berlin.

Fricke, E. (2012), *Grammatik multimodal: Wie Wörter und Gesten zusammenwirken*, Mouton de Gruyter, Berlin and New York.

Fricke, E. (2013), Towards a unified grammar of gesture and speech: A multimodal approach, *in* C. Müller, A. Cienki, E. Fricke, S. Ladewig, D. McNeill and S. Tessendorf, eds, 'Body – Language – Communication / Körper – Sprache – Kommunikation', number 38/1 *in* 'Handbücher zur Sprach- und Kommunikationswissenschaft / Handbooks of Linguistics and Communication Science (HSK)', Mouton de Gruyter, Berlin and New York, pp. 733–754.

Fricke, E., Bressem, J. and Müller, C. (2014), Gesture families, *in* C. Müller, A. Cienki, E. Fricke, S. H. Ladewig, D. McNeill and J. Bressem, eds, 'Body–Language–Communication. An International Handbook on Multmodality in Human Interaction', number 38/2 *in* 'Handbücher zur Sprach- und Kommunikationswissenschaft / Handbooks of Linguistics and Communication Science (HSK)', Mouton de Gruyter, Berlin and New York, pp. 1630–1640.

Frow, J. (2006), *Genre*, Routledge, London.

Garfinkel, H. (1972 [1967]), Studies in the routine grounds of everyday activities, *in* D. Sudnow, ed., 'Studies in Social Interaction', The Free Press, New York, pp. 1–30.

Gee, J. P. (2015), *Unified Discourse Analysis: Language, Reality, Virtual Worlds, and Video Games*, Routledge, London and New York.

Gendolla, P. and Schäfer, J., eds (2007), *The Aesthetic of Net Literature: Writing, Reading and Playing in Programmable Media*, transcript, Bielefeld.

Gibbons, A. (2011), *Multimodality, Cognition, and Experimental Literature*, Routledge Studies in Multimodality, Routledge, London.

Gibbs, M., Meese, J., Arnold, M., Nansen, B. and Carter, M. (2015), '#Funeral and Instagram: Death, Social Media, and Platform Vernacular', *Information, Communication & Society* **18**(3), 255–268.

Gibson, J. J. (1977), The theory of affordances, *in* R. Shaw and J. Bransford, eds, 'Perceiving, Acting, and Knowing: Toward and Ecological Psychology', Erlbaum, Hillsdale, NJ, pp. 62–82.

Giddens, A. (1984), *The constitution of society: outline of the theory of structuration*, Polity Press, Cambridge.

Gilje, Oystein. (2011), 'Working in tandem with editing tools – iterative meaning-making in filmmaking practices', *Visual Communication* **10**(1), 45–62.

Gilje, Oystein. (2015), Writing within modes and across Modes in Filmmaking, *in* A. Archer and E. Breuer, eds, 'Multimodality in Writing: The state of the art in theory, methodology and pedagogy', Brill N.V., Leiden, The Netherlands, pp. 154–172.

Goggin, J. and Hassler-Forest, D. (2010*a*), Out of the Gutter: Reading Comics and Graphic Novels, *in* J. Goggin and D. Hassler-Forest, eds, 'The Rise and Reason of Comics and Graphic Literature. Critical Essays on the Form', McFarland & Co., Jefferson and London, pp. 5–24.

Goggin, J. and Hassler-Forest, D., eds (2010*b*), *The rise and reason of comics and graphic literature: critical essays on the form*, McFarland & Co., Jefferson, NC and London.

Gombrich, E. (1960), *Art and Illusion: A Study in the Psychology of Pictorial Representation*, Pantheon Books, New York.

Gombrich, E. (1982*a*), *The image and the eye: further studies in the psychology of pictorial representation*, Phaidon, Oxford, chapter Image and code: scope and limits of conventionalism in pictoral representation, pp. 278–297.

Gombrich, E. H. (1982*b*), *The image and the eye: further studies in the psychology of pictorial representation*, Phaidon, Oxford.

Gonzalez, J. (2014), 'Living in the Funnies: Metafiction in American Comic Strips', *The Journal of Popular Culture* **47**(4), 838–856.

Goodling, L. B. (2014), 'The Multimodal Turn in Higher Education: On Teaching, Assessing, Valuing Multiliteracies', *Pedagogy* **14**(3), 561–568.

Goodman, N. (1969), *Languages of Art. An approach to a theory of symbols*, Oxford University Press, London.

Goodwin, C. (2000), 'Action and embodiment within situated human interaction', *Journal of Pragmatics* **32**, 1489–1522.

Goodwin, C. (2001), Practices of seeing visual analysis: an ethnomethodological approach, *in* T. van Leeuwen and C. Jewitt, eds, 'Handbook of visual analysis', Sage, London, chapter 8, pp. 157–182.

Gordon, I., Jancovich, M. and McAllister, M., eds (2010), *Film and Comic Books*, University Press of Mississippi, Jackson.

Graaf, A., Sanders, J. and Hoeken, J. (2016), 'Characteristics of narrative interventions and health effects: A review of the content, form, and context of narratives in health-related narrative persuasion research', *Review of Communication Research* **4**, 88–131.

Granström, B., House, D. and Karlsson, I., eds (2002), *Mutlimodality in language and speech systems*, Kluwer Academic Publishers, Dordrecht.

Grant, A. E. and Wilkinson, J. S., eds (2009), *Understanding media convergence: the state of the field*, Oxford University Press, New York and Oxford.

Grant, B. K., ed. (2012), *Film Genre Reader IV*, University of Texas Press, Austin, TX.

Grau, O. (2003), *Virtual Art: From Illusion to Immersion*, The MIT Press, Cambridge, MA.

Grodal, T. (2009), *Embodied Visions: Evolution, Emotions, Culture, and Film*, Oxford University Press, Oxford.

Groensteen, T. (2007 [1999]), *The system of comics*, University Press of Mississippi, Jackson, Miss. translated by Bart Beaty and Nick Nguyen, from the original French *Système de la bande desinée* (1999).

Guo, L. (2004), Multimodality in a biology textbook, *in* K. L. O'Halloran, ed., 'Multimodal discourse analysis: systemic functional perspectives', Open Linguistics Series, Continuum, London, pp. 196–219.

Gwet, K. L. (2014), *Handbook of Inter-Rater Reliability: The Definitive Guide to Measuring the Extent of Agreement Among Raters*, 4 edn, Advanced Analytics, LLC, Gaithersburg, MD.

Haahr, M. (2015), Everting the Holodeck. Games and Storytelling in Physical Space, *in* H. Koenitz, G. Ferri, M. Haahr, D. Sezen and T. İbrahim Sezen, eds, 'Interactive Digital Narrative. History, Theory and Practice', Routledge, New York and London, pp. 211–226.

Habermas, J. (1981), *Theorie der kommunikativen Handelns*, Suhrkamp, Frankfurt a.M.

Hall, E. T. (1968), 'Proxemics', *Current Anthropology* **9**(2/3), 83–108.

Hallenberger, G. (2002), Das Konzept "Genre": Zur Orientierung von Medienhandeln, *in* P. Gendolla, P. Ludes and V. Roloff, eds, 'Bildschirm – Medien – Theorien', Wilhelm Fink Verlag, München, pp. 83–110.

Halliday, M. A. K. (1978), *Language as social semiotic*, Edward Arnold, London.

Halliday, M. A. K. (1985), *An Introduction to Functional Grammar*, Edward Arnold, London.

Hannus, M. and Hyönä, J. (1999), 'Utilization of illustrations during learning of science textbook passages among low- and high-ability children', *Contemporary Educational Psychology* **24**(2), 95–123.

Harms, W., ed. (1990), *Text und Bild, Bild und Text: DFG-Symposion 1988*, J.B. Metzlersche Verlagsbuchhandlung, Stuttgart.

Harrigan, J. A. (2008), Proxemics, kinesics and gaze, *in* J. A. Harrigan, R. Rosenthal and K. R. Scherer, eds, 'The New Handbook of Methods in Nonverbal Behavioral Research', Oxford University Press, Oxford, pp. 137–198.

Harrison, C., Hsieh, G., Willis, K. D., Forlizzi, J. and Hudson, S. E. (2011), Kineticons: Using iconographic motion in graphical user interface design, *in* 'Proceedings of the SIGCHI Conference on Human Factors in Computing Systems', CHI '11, ACM, New York, NY, USA, pp. 1999–2008. http://doi.acm.org/10.1145/1978942.1979232

Hart, I. (2014), 'Meaningful play: Performativity, interactivity and semiotics in video game music', *Musicology Australia* **36**(2), 273–290.

Hassler-Forest, D. and Nicklas, P. (2015), *The Politics of Adaptation: Media Convergence and Ideology*, Palgrave Macmillan, Basingstoke.

Hasson, U., Landesman, O., Knappmeyer, B., Vallines, I., Rubin, N. and Heeger, D. J. (2008), 'Neurocinematics: The Neuroscience of Film', *Projections: the Journal for Movies and Mind* **2**(1), 1–26.

Heath, C., Hindmarsh, J. and Luff, P. (2010), *Video in qualitative research: analysing social interaction in everyday life*, Sage, Los Angeles and London.

Heath, C., Knoblauch, H. and Luff, P. (2000), 'Technology and social interaction: the emergence of 'workplace studies", *The British Journal of Sociology* **51**(2), 299–320.

Heer, J. and Worcester, K., eds (2009), *A comics studies reader*, Mississippi University Press, Jackson, Miss.

Hegarty, M. and Just, M. (1993), 'Constructing mental models of machines from text and diagrams', *Journal of Memory and Language* **32**(6), 717–742.

Helbo, A. (1987), *Theory of Performing Arts*, John Benjamins, Amsterdam and Philadelphia.

Helbo, A. (2016), 'Semiotics and Performing Arts: Contemporary Issues', *Social Semiotics* **26**(4), 341–350.

Helms, J. M. (2015), 'Is This Article a Comic?', *Digital Humanities Quarterly* **9**(3).

Hepburn, A. and Bolden, G. B. (2013), The Conversation Analytic Approach to Transcription, *in* J. Sidnell and T. Stivers, eds, 'THe Handbook of Conversation Analysis', Blackwell, Malden, MA, pp. 57–76.

Hepp, A. (2012), 'Mediatization and the 'molding force' of the media', *Communications* **37**, 1–28.

Heritage, J. and Atkinson, M., eds (1984), *Structures of Social Action: Studies in Conversation Analysis*, Cambridge University Press, Cambridge.

Herman, D. (2010), Word-Image/Utterance-Gesture: Case Studies in Multimodal Storytelling, *in* R. Page, ed., 'New Perspectives on Narrative and Multimodality', Routledge, New York and London, pp. 78–98.

Herman, D., ed. (2003), *Narrative Theory and the Cognitive Sciences*, CSLI, Stanford, CA.

Herring, S. C., Stein, D. and Virtanen, T., eds (2013), *Pragmatics of Computer-Mediated Communication*, de Gruyter Mouton, Berlin.

Hiippala, T. (2012), 'Reading paths and visual perception in multimodal research, psychology and brain sciences', *Journal of Pragmatics* **44**(3), 315–327.

Hiippala, T. (2015*a*), Combining computer vision and multimodal analysis: a case study of layout symmetry in bilingual in-flight magazines, *in* J. Wildfeuer, ed., 'Building Bridges for Multimodal Research: International Perspectives on Theories and Practices of Multimodal Analysis', Peter Lang, Bern and New York, pp. 289–307.

Hiippala, T. (2015*b*), *The Structure of Multimodal Documents: An Empirical Approach*, Vol. 13 of *Routledge Studies in Multimodality*, Routledge, London.

Hiippala, T. (2016*a*), Aspects of multimodality in higher education monographs, *in* A. Archer and E. Breuer, eds, 'Multimodality in Higher Education', Vol. 33 of *Studies in Writing*, Brill, Leiden, pp. 53–78.

Hiippala, T. (2016*b*), 'The multimodality of digital longform journalism', *Digital Journalism* . http://dx.doi.org/10.1080/21670811.2016.1169197

Hiippala, T. (2016*c*), Semi-automated annotation of page-based documents within the Genre and Multimodality framework, *in* 'Proceedings of the 10th SIGHUM Workshop on Language Technology for Cultural Heritage, Social Sciences, and Humanities', Association for Computational Linguistics, Berlin, Germany, pp. 84–89.

Hill, S. (1983), *The Woman in Black*, Vintage Books, London.

Hjelmslev, L. (1961 [1943]), *Prolegomena to a theory of language*, University of Wisconsin Press, Madison, Wisconsin. Originally published 1943; Translated by F.J.Whitfield.

Hobbs, J. R. (1990), *Literature and cognition*, CSLI, Stanford, California. CSLI Lecture Notes.

Hochman, N. and Manovich, L. (2013), 'Zooming into an Instagram city: Reading the local through social media', *First Monday* **18**(7). http://firstmonday.org/article/view/4711/3698

Hockett, C. F. (1958), *A course in modern linguistics*, Macmillan, New York.

Holler, J. and Kendrick, K. H. (2015), 'Unaddressed participants' gaze in multi-person interaction: optimizing recipiency', *Frontiers in Psychology* **6**(98).

Holly, W. (2011), Bildüberschreibungen. Wie Sprechtexte Nachrichtenfilme lesbar machen, *in* H.-J. Diekmannshenke, M. Klemm and H. Stöckl, eds, 'Bildlinguistik. Theorien – Methoden – Fallbeispiele', Erich Schmidt, Berlin, pp. 235–256.

Holmqvist, K., Nyström, M., Andersson, R., Dewhurst, R., Jarodzka, H. and van de Weijer, J. (2011), *Eye Tracking: A Comprehensive Guide to Methods and Measures*, Oxford University Press, Oxford.

Holsanova, J. (2008), *Discourse, Vision, and Cognition*, number 23 *in* 'Human Cognitive Processes', Benjamins, Amsterdam.

Holsanova, J. (2012), 'New methods for studying visual communication and multimodal integration', *Visual Communication* **11**(3), 251–257.

Holsanova, J. (2014*a*), In the eye of the beholder: Visual communication from a recipient perspective, *in* D. Machin, ed., 'Visual Communication', Mouton de Gruyter, Berlin, pp. 331–355.

Holsanova, J. (2014*b*), Reception of multimodality: applying eyetracking methodology in multimodality research, *in* C. Jewitt, ed., 'The Routledge Handbook of multimodal analysis', 2 edn, Routledge, London, pp. 287–298.

Holsanova, J., Holmberg, N. and Holmqvist, K. (2008), 'Reading information graphics: the role of spatial contiguity and dual attentional guidance', *Applied Cognitive Psychology* **23**(9), 1215–1226.

Holsanova, J. and Nord, A. (2010), Multimedia design: media structures, media principles and users' meaning-making in newspapers and net papers, *in* H.-J. Bucher, T. Gloning and K. Lehnen, eds, 'Neue Medien – neue Formate. Ausdifferenzierung und Konvergenz in der Medienkommunikation', number 10 *in* 'Interaktiva. Schriftenreihe des Zentrums für Medien und Interaktiv-ität (ZMI), Gießen', Campus Verlag, Frankfurt / New York, pp. 81–103.

Holsanova, J., Rahm, H. and Holmqvist, K. (2006), 'Entry points and reading paths on newspaper spreads: comparing a semiotic analysis with eye-tracking measurements', *Visual Communication* 5(1), 65–93.

Humphrey, A. (2014), 'Beyond Graphic Novels: Illustrated Scholarly Discourse and the History of Educational Comics', *Media International Australia* 151(1), 73–80.

Humphrey, A. (2015*a*), 'Visual and Spatial Language: The Silent Voice of Woodstock', *Composition Studies* 43(1), 19–30.

Humphrey, A. S. (2015*b*), 'Multimodal Authoring and Authority in Educational Comics: Introducing Derrida and Foucault for Beginners', *Digital Humanities Quarterly* 9(3).

Hutcheon, M. and Hutcheon, L. (2010), Opera. Forever and Always Multimodal, *in* R. Page, ed., 'New Perspectives on Narrative and Multimodality', Routledge, New York and London, pp. 65–77.

Iedema, R. (2001), 'Resemiotization', *Semiotica* 37(1/4), 23–40.

Iedema, R. (2003), 'Multimodality, resemiotization: extending the analysis of discourse as multi-semiotic practice', *Visual Communication* 2(1), 29–58.

Iedema, R., Feez, S. and White, P. (1995), *Media literacy*, Vol. II of *Write It Right publications: Industry Research Monograph*, Metropolitan East Disadvantaged Schools Program, Sydney. NSW Department of School Education.

Iversen, M. (1986), Saussure v. Peirce: Models for a semiotics of visual art, *in* A. Rees and F. Borzello, eds, 'The new art history', Camden Press, London, pp. 82–94.

Jackendoff, R. (1987), 'On beyond Zebra: the relation of linguistic and visual information', *Cognition* 26(2), 89–114.

Jackendoff, R. (1991), 'Musical parsing and musical affect', *Music Perception* 9(2), 199–230.

Jackendoff, R. and Lerdahl, F. (2006), 'The capacity for music: what is it, and what's special about it?', *Cognition* **100**, 33–72.

Jacobs, D. (2013), *Graphic encounters: Comics and the sponsorship of multimodal literacy*, Blooms-bury Academic, London.

Jacobs, M. and Spillman, L. (2005), 'Cultural sociology at the crossroads of the discipline', *Poetics* 33(1), 1–14.

Jaeger, T. F. (2008), 'Categorical Data Analysis: Away from ANOVAs (transformation or not) and to-wards Logit Mixed Models', *Journal of Memory and Language* 59(4), 434–446.

Jameson, F. (1998), *The Cultural Turn: Selected Writings on the Postmodern, 1983–1998*, Verso, Brooklyn.

Jappy, T. (2013), *Introduction to Peircean visual semiotics*, Bloomsbury Advances in Semiotics, Bloomsbury, London and New York.

Jenkins, H. (2008), *Convergence Culture: Where Old and New Media Collide*, NYU Press, New York.

Jewitt, C. (2014*a*), An introduction to multimodality, *in* C. Jewitt, ed., 'The Routledge Handbook of multimodal analysis', Routledge, London, pp. 15–30.

Jewitt, C., Bezemer, J. and O'Halloran, K. (2016), *Introducing multimodality*, Routledge, London.

Jewitt, C., ed. (2014*b*), *The Routledge Handbook of multimodal analysis*, 2 edn, Routledge, London.

Jewitt, C. and Kress, G. (2003), *Multimodal literacy*, number 4 *in* 'New literacies and digital epistem-ologies', Peter Lang, Frankfurt a.M. / New York.

Johnson, J., Karpathy, A. and Li, F.-F. (2016), DenseCap: Fully convolutional localization networks for dense captioning, *in* 'Proceedings of the IEEE Conference on Computer Vision and Pattern Recognition (CVPR)'.

Jones, R. and Hafner, C. (2012), *Understanding Digital Literacies. A Practical Introduction*, Routledge, London and New York.

Kapchan, D. A. (1995), 'Performance', *The Journal of American Folklore* **108**(430), 479–508.

Kaplan, A. M. and Henlein, M. (2009), 'Users of the world, unite! The challenges and opportunities of Social Media', *Business Horizons* **53**(1), 59–68.

Karpathy, A. and Fei-Fei, L. (2015), Deep visual-semantic alignments for generating image descriptions, *in* 'Proceedings of the IEEE Conference on Computer Vision and Pattern Recognition (CVPR)'.

Kashtan, A. J. (2015), 'Materiality Comics', *Digital Humanities Quarterly* **9**(4). Online. http://www.digitalhumanities.org/dhq/vol/9/4/000212/000212.html

Kendon, A. (1972), Some Relationships between Body Motion and Speech: An Analysis of an Example, *in* A. W. Siegman and B. Pope, eds, 'Studies in Dyadic Communication', Elsevier, New York, pp. 177–210.

Kendon, A. (1980), Gesture and speech: two aspects of the process of utterance, *in* M. R. Key, ed., 'Nonverbal Communication and Language', Mouton, The Hague, pp. 207–227.

Kendon, A. (2004), *Gesture: visible action as utterance*, Cambridge University Press, Cambridge.

Kennedy, H., Hill, R. L., Aiello, G. and Allen, W. (2016), 'The work that visualisation conventions do', *Information, Communication & Society* **19**(6), 715–735.

Kipp, M. (2012), Multimedia annotation, querying and analysis in ANVIL, *in* M. T. Maybury, ed., 'Multimedia information extraction', John Wiley & Sons, Hoboken, NJ, chapter 21, pp. 351–368. IEEE Computer Society.

Kipp, M., Martin, J.-C., Paggio, P. and Heylen, D. (2009), *Multimodal corpora. From models of natural interaction to systems and applications*, Springer-Verlag, Berlin, Heidelberg.

Kipp, M., Neff, M. and Albrecht, I. (2007), 'An Annotation Scheme for Conversational Gestures: How to economically capture timing and form', *Journal on Language Resources and Evaluation* **41**(2–3), 325–339. Special Issue on Multimodal Corpora for Modeling Human Multimodal Behaviour.

Kjeldsen, J. (2013), 'Strategies of visual argumentation in slideshow presentations: The role of the visuals in an Al Gore presentation on climate change', *Argumentation* **27**(4), 425–443.

Kjeldsen, J. (2015), 'The study of visual and multimodal argumentation', *Argumentation* **29**(2), 115–132.

Klemm, M. and Michel, S. (2014), Medienkulturlinguistik. Plädoyer für eine holistische Analyse von (multimodaler) Medienkommunikation, *in* N. Benitt, C. Koch, K. Müller, L. Schüler and S. Saage, eds, 'Korpus – Kommunikation – Kultur. Ansätze und Konzepte einer kulturwissenschaftlichen Linguistik', WVT, Trier, pp. 183–215.

Klug, N.-M. and Stöckl, H., eds (2016), *Handbuch Sprache im multimodalen Kontext*, de Gruyter, Berlin.

Kluss, T., Bateman, J., Preußer, H.-P. and Schill, K. (2016), Exploring the role of narrative contextualization in film interpretation: issues and challenges for eye-tracking methodology, *in* C. D. Reinhard and C. J. Olson, eds, 'Making Sense of Cinema: empirical studies into film spectators and spectatorship', Bloomsbury Academic, New York and London, pp. 257–284.

Knoblauch, H. (2008), 'The performance of knowledge: Pointing and knowledge in PowerPoint presentations', *Cultural Sociology* **2**(1), 75–97.

Knoblauch, H. (2013), 'Communicative constructivism and mediatization', *Communication Theory* **23**(3), 297–315.

Knox, J. S. (2007), 'Visual-verbal communication on online newspaper home pages', *Visual Communication* **6**(1), 19–53.

Koenitz, H., Ferri, G., Haahr, M., Sezen, D. and İbrahim Sezen, T., eds (2015), *Interactive Digital Narrative. History, Theory and Practice*, Routledge, New York and London.

Koffka, K. (1935), *Principles of Gestalt Psychology*, Harcourt-Brace, New York.

Köhler, W. (1947), *Gestalt psychology*, Liveright, Liverpool.

Kong, K. C. (2013), 'A corpus-based study in comparing the multimodality of Chinese- and English-language newspapers', *Visual Communication* **12**(2), 173–196.

Kopp, S. and Wachsmuth, I., eds (2009), *Gesture in Embodied Communication and Human-Computer Interaction*, Springer, Berin.

Kornhaber, D. (2015), 'Every Text is a Performance: A Pre-History of Performance Philosophy', *Performance Philosophy* **1**, 24–35.

Kostelnick, C. and Hassett, M. (2003), *Shaping Information: The Rhetoric of Visual Conventions*, Southern Illinois University Press, Carbondale, Illinois.

Kranstedt, A., Kopp, S. and Wachsmuth, I. (2002), MURML: A Multimodal Utterance Representation Markup Language for Conversational Agents, Technical Report 2002/05, SFB 360 Situated Artificial Communicators, Universität Bielefeld, Bielefeld, Germany.
http://www.sfb360.uni-bielefeld.de/reports/2002/2002-5.html

Krejcie, R. V. and Morgan, D. W. (1970), 'Determining sample size for research activities', *Educational and Psychological Measurement* **30**, 607–610.

Kress, G. (2003), *Literacy in the New Media Age*, Routledge, London.

Kress, G. (2010), *Multimodality: a social semiotic approach to contemporary communication*, Routledge, London.

Kress, G. (2014), What is mode?, *in* C. Jewitt, ed., 'The Routledge Handbook of multimodal analysis', 2 edn, Routledge, London, pp. 60–75.

Kress, G., Jewitt, C., Ogborn, J. and Tsatsarelis, C. (2000), *Multimodal teaching and learning*, Continuum, London.

Kress, G. and Trew, A. A. (1978), 'Ideological transformation of discourse: or how the *Sunday Times* got its message across', *Journal of Pragmatics* **2**, 311–329. Also available in *Sociological Review*, **26**(4), 755-776.

Kress, G. and van Leeuwen, T. (1998), Front pages: the (critical) analysis of newspaper layout, *in* A. Bell and P. Garrett, eds, 'Approaches to Media Discourse', Blackwell, Oxford, pp. 186–219.

Kress, G. and van Leeuwen, T. (2001), *Multimodal discourse: the modes and media of contemporary communication*, Arnold, London.

Kress, G. and van Leeuwen, T. (2002), 'Colour as a semiotic mode: notes for a grammar of colour', *Visual Communication* **1**(3), 343–368.

Kress, G. and van Leeuwen, T. (2006), *Reading Images: the grammar of visual design*, 2 edn, Routledge, London and New York.

Kress, G. and van Leeuwen, T. (2006 [1996]), *Reading Images: the grammar of visual design*, Routledge, London and New York.

Krippendorff, K. (2004), *Content analysis: an introduction to its methodology*, 2 edn, Sage, London and Thousand Oaks, CA.

Krishna, R., Zhu, Y., Groth, O., Johnson, J., Hata, K., Kravitz, J., Chen, S., Kalantidis, Y., Li, L., Shamma, D. A., Bernstein, M. S. and Li, F. (2016), 'Visual genome: Connecting language and vision using crowdsourced dense image annotations', *CoRR* **abs/1602.07332**.
http://arxiv.org/abs/1602.07332

Krois, J. M. (2011), Image Science and Embodiment or: Peirce as Image Scientist, *in* H. Bredekamp and M. Lauschke, eds, 'John M. Krois: Bildkörper und Körperschema. Schriften zur Verkörperungstheorie ikonischer Formen', Akademie Verlag, Berlin, pp. 195–209.

Krotz, F. (2009), Mediatization. A concept with which to grasp media and societal change., *in* K. Lundby, ed., 'Mediatization: Concept, changes, consequences', Peter Lang, New York, pp. 19–38.

Krotz, F. (2012), Von der Entdeckung der Zentralperspektive zur Augmented Reality, *in* F. Krotz and A. Hepp, eds, 'Mediatisierte Welten', VS, Wiesbaden, pp. 27–58.

Kuhn-Osius, E. (2016), Before They Were Art: (WestGerman Proto-Comics and Comics: A Brief and Somewhat Subjective Survey, *in* L. M. Kutch, ed., 'Novel Perspectives on German-Language Comics Studies: History, Pedagogy, Theory', Lexington Books, London, pp. 37–65.

Labov, W. (1994), *Principles of linguistic change, Volume 1: internal factors*, Blackwell, Oxford.

Labov, W. (2001), *Principles of linguistic change, Volume 2: social factors*, Blackwell, Oxford.

Lacassin, F. (1972), 'The comic strip and film language', *Film Quarterly* **26**(1), 11–23. original articles translated by David Kunzle.

Lacey, N. (2004), *Introduction to Film*, Palgrave, Basingstoke and New York.

Laihonen, P. and Szabó, T. P. (2017), Investigating visual practices in educational settings: schoolscapes, language ideologies and organizational cultures, *in* M. Martin-Jones and D. Martin, eds, 'Researching Multilingualism: Critical and Ethnographic Approaches', Routledge, New York and London, pp. 121–138.

Lakens, D. (2013), 'Calculating and reporting effect sizes to facilitate cumulative science: a practical primer for *t*-tests and ANOVAs', *Frontiers of Psychology* **4**(863), 1–12.

Landesinstitut für Schule Bremen (2012), 'Medienbildung. Bildungsplan für Primarstufe, Sekundarstufe I, Sekundarstufe II. Entwurfsfassung 2012.', online. http://www.lis.bremen.de/sixcms/media.php/13/2012_bpmedien_aktuell.36056.pdf

Landow, G. P., ed. (1994), *Hyper / Text / Theory*, John Hopkins University Press, Baltimore and London.

Lascarides, A. and Asher, N. (1993), 'Temporal Interpretation, Discourse Relations, and Common Sense Entailment', *Linguistics and Philosophy* **16**(5), 437–495.

Latour, B. (2005), *Reassembling the Social: An introduction to Actor-Network-Theory*, Oxford University Press, Oxford.

Leander, K. M. and Vasudevan, L. M. (2014), Multimodality and mobile culture, *in* C. Jewitt, ed., 'The Routledge Handbook of Multimodal Analysis', 2 edn, Routledge, London, pp. 152–164.

Leckner, S. (2012), 'Presentation factors affecting reading behaviour in readers of newspaper media: an eye-tracking perspective', *Visual Communication* **11**(2), 163–184.

LeCun, Y., Bengio, Y. and Hinton, G. (2015), 'Deep learning', *Nature* **521**, 436–444.

Lefèvre, P. (2009), 'The conquest of space: evolution of panel arrangements and page layouts in early comics published in Belgium (1880–1929)', *European Comics Art* **2**(2), 227–252.

Lemke, J. (2009), Multimodality, identity and time, *in* C. Jewitt, ed., 'The Routledge Handbook of multimodal analysis', Routledge, London, pp. 140–150.

Lemke, J. L. (1998), Multiplying meaning: visual and verbal semiotics in scientific text, *in* J. Martin and R. Veel, eds, 'Reading science: critical and functional perspectives on discourses of science', Routledge, London, pp. 87–113.

Lemke, J. L. (1999), Typology, topology, topography: genre semantics. MS University of Michigan. http://www-personal.umich.edu/~jaylemke/papers/Genre-topology-revised.htm

Lemke, J. L. (2005), 'Multimedia genre and traversals', *Folia Linguistica* **XXXIX**(1–2), 45–56.

Lewis, D. (1969), *Convention*, Harvard University Press, Cambridge, MA.

Lewis, D. K. (2001), *Reading contemporary picturebooks: picturing text*, Routledge, London.

Lewis, J. P., McGuire, M. and Fox, P. (2007), Mapping the mental space of game genres, *in* 'Proceedings of the 2007 ACM SIGGRAPH symposium on Video games', Sandbox '07, ACM, New York, NY, USA, pp. 103–108. http://doi.acm.org/10.1145/1274940.1274962

Lim, F. V. (2007), The visual semantics stratum: making meaning in sequential images, *in* T. D. Royce and W. L. Bowcher, eds, 'New Directions in the Analysis of Multimodal Discourse', Lawrence Erlbaum Associates, Mahwah, New Jersey, pp. 195–214.

Lima, R. C., de Castro Andrade, R., Monat, A. S. and Spinillo, C. G. (2014), The relation between on-line and print information graphics for newspapers, *in* A. Marcus, ed., 'Design, User Experience, and Usability. User Experience Design for Everyday Life Applications and Services', Vol. 8519 of *LNCS*, Springer, New York, pp. 184–194.

Lin, T., Maire, M., Belongie, S. J., Bourdev, L. D., Girshick, R. B., Hays, J., Perona, P., Ramanan, D., Dollár, P. and Zitnick, C. L. (2014), 'Microsoft COCO: Common Objects in Context', *Computing Research Repository (CoRR)* **abs/1405.0312**.
http://arxiv.org/abs/1405.0312

Loftus, E. F. (1975), 'Leading questions and the eyewitness report', *Cognitive Psychology* **7**, 550–572.

Loschky, L. C., Larson, A. M., Magliano, J. P. and Smith, T. J. (2015), 'What Would Jaws Do? The Tyranny of Film and the Relationship between Gaze and Higher-Level Narrative Film Comprehension', *PLoS One* **10**(11), e142474.

Loschky, L. C., Ringer, R. V., Ellis, K. and Hansen, B. C. (2015), 'Comparing rapid scene categorization of aerial and terrestrial views: A new perspective on scene gist', *Journal of Vision* **15**(6), 1–29.

Lotman, J. M. (1990), *Universe of the mind: A semiotic theory of culture*, I.B. Taurus, London.

Lowie, W. and Seton, B. (2013), *Essential statistics for applied linguistics*, Palgrave Macmillan, Basingstoke.

Lowry, R. (1998-2015), *Concepts and Applications of Inferential Statistics*. Online (last access: 2017.2.5).
http://vassarstats.net/textbook/

Ludwig, L. (2015), *Moving panels: translating comics to film*, Edwardsville, Illinois, Sequart Organization.

Machin, D. (2007), *Introduction to Multimodal Analysis*, Hodder Arnold, London.

Machin, D., ed. (2014), *Visual Communication*, Mouton de Gruyter, Berlin.

Machin, D. and Mayr, A. (2012), *How To Do Critical Discourse Analysis: A Multimodal Introduction*, Sage Publications, London and Thousand Oaks, CA.

Magnani, L. (2001), *Abduction, Reason, and Science. Processes of Discovery and Explanation*, Kluwer Academic/Plenum Publishers, New York.

Maiorani, A. (2007), ''Reloading' movies into commercial reality: A multimodal analysis of The Matrix trilogy's promotional posters', *Semiotica* **166**(1/4), 45–67.

Maiorani, A. (2009), Developing the metafunctional framework to analyse the multimodal hypertextual identity construction: The Lord of the Rings from page to MMORPG, *in* E. Ventola and A. J. M. Guijarro, eds, 'The World Told and the World Shown: Multisemiotic Issues', Palgrave Macmillan, Basingstoke, pp. 220–241.

Mallatratt, S. (1989), *The Woman in Black: A Ghost Play*, French, London.

Manghani, S. (2013), *Image Studies: theory and practice*, Routledge, London.

Manovich, L. (2001), *The language of the new media*, MIT Press, Cambridge, MA.

Manovich, L. (2009), 'The practice of everyday (media) life: From mass consumption to mass cultural production?', *Critical Inquiry* **35**(2), 319–331.

Manovich, L. (2012), How to compare one million images, *in* D. M. Berry, ed., 'Understanding digital humanities', Palgrave Macmillan, Basingstoke, pp. 249–278.

Margolis, E. and Pauwels, L., eds (2011), *The SAGE Handbook of Visual Research Methods*, Sage, London.

Marks, L. U. (2000), *The skin of the film: intercultural cinema, embodiment, and the senses*, Durham and London, Duke University Press.

Martin, J. R. (1985), Process and text: two aspects of human semiosis, *in* J. D. Benson and W. S. Greave, eds, 'Systemic perspectives on discourse, Volume 1; Selected theoretical papers from the ninth International Systemic Workshop', number 15 *in* 'Advances in Discourse processes', Ablex, Norwood, New Jersey, pp. 248–274.

Martin, J. R. (1992), *English text: systems and structure*, Benjamins, Amsterdam.

Martin, J. R. and Rose, D. (2003), *Working with discourse: meaning beyond the clause*, Continuum, London and New York.

Martin, J. R. and Rose, D. (2008), *Genre relations: mapping culture*, Equinox, London and New York.

Martinec, R. (2004), 'Gestures which co-occur with speech as a systematic resource: the realization of experiential meanings in indexes', *Social Semiotics* **14**(2), 193–213.

Martinec, R. and Salway, A. (2005), 'A system for image-text relations in new (and old) media', *Visual Communication* **4**(3), 337–371.

Martinet, A. (1960), *Elements of general linguistics*, Faber and Faber, London.

Mayer, R. E., ed. (2009), *The Cambridge Handbook of Multimedia Learning*, 2 edn, Cambridge University Press, Cambridge.

McAuley, G. (2007), 'State of the Art: Performance Studies', *SemiotiX* **10**.

McAuley, G. (2010), Interdisciplinary Field or Emerging Discipline?: Performance Studies at the University of Sydney, *in* J. McKenzie, H. Roms and C.-L. Wee, eds, 'Contesting Performance: Global Sites of Research', Palgrave Macmillan, pp. 37–50.

McCloud, S. (1994), *Understanding comics: the invisible art*, HarperPerennial, New York.

McCreadie, R., Soboroff, I., Lin, J., Macdonald, C., Ounis, I. and McCullough, D. (2012), On building a reusable Twitter corpus, *in* 'Proceedings of the 35th International ACM SIGIR Conference on Research and Development in Information Retrieval (SIGIR'12)', ACM, pp. 1113–1114.

McGurk, H. and MacDonald, J. (1976), 'Hearing lips and seeing voices', *Nature* **264**(5588), 746–748.

McLuhan, M. (2002 [1964]), The medium is the message, *in* 'Understanding Media: the extensions of man', Routledge Classics, Routledge, London, chapter 1, pp. 7–23.

McNeill, D. (1992), *Hand and Mind: What gestures reveal about thought*, University of Chicago Press, Chicago, Illinois.

McNeill, D. (2000), *Language and Gesture*, Cambridge University Press, Cambridge.

McNeill, D. (2006), Gesture and communication, *in* K. Brown, ed., 'Encyclopedia of Language and Linguistics. Second Edition', Elsevier, Amsterdam, pp. 58–66.

Meier, S. (2014), *Visuelle Stile. Zur Sozialsemiotik visueller Medienkultur und konvergenter Design-Praxis*, transcript, Bielefeld.

Meskin, A. (2007), 'Defining Comics?', *The Journal of Aesthetics and Art Criticism* **65**(4), 369–379.

Messaris, P. (1998), 'Visual aspects of media literacy', *Journal of Communication* **48**(1), 70–80. doi:10.1111/j.1460-2466.1998.tb02738.x

Metz, C. (1966), 'La grande syntagmatique du film narratif', *Communications* **8**, 120–124. Recherches sémiologiques: l'analyse structurale du récit.

Michel, S. (2015), 'Zuschauerkommunikation in sozialen netzwerken: Social TV', *Der Sprachdienst* **2**, 51–67.

Millington, B. (2001), Wagner, Wilhelm Richard, *in* S. Sadie, ed., 'New Grove Dictionary of Music and Musicians', 2 edn, Vol. 26, Macmillan, New York and London, pp. 931–971.

Mitchell, W. (1986), *Iconology: Images, Text, Ideology*, Chicago University Press, Chicago.

Mitchell, W. (1994), *Picture Theory: Essays on verbal and visual representation*, University of Chicago Press, Chicago.

Mitchell, W. (2005), 'There are no visual media', *Journal of visual culture* **4**(2), 257–266.

Mitchell, W. J. T. (1992), 'The pictorial turn', *Artforum* . March.

Mittelberg, I. and Waugh, L. (2014), Gestures and metonymy, *in* C. Müller, A. Cienki, E. Fricke, S. H. Ladewig, D. McNeil and J. Bressem, eds, 'Body – Language – Communication. An International

Handbook on Multimodality in Human Interaction', Vol. 2, Mouton, Berlin/Boston, pp. 1747–1766.

Monaco, J. (2009), *How to read a film: movies, media and beyond*, 30th anniversary edition edn, Oxford University Press, Oxford, U.K.

Mondada, L. (2007), 'Multimodal resources for turn-taking: Pointing and the emergence of possible next speakers', *Discourse Studies* 9(2), 195–226.

Mondada, L. (2014), 'The local constitution of multimodal resources for social interaction', *Journal of Pragmatics* 65, 137–156.

Mondada, L. (2016), Multimodal resources and the organization of social interaction, *in* A. Rocci and L. de Saussure, eds, 'Verbal Communication', Mouton de Gruyter, pp. 329–350.

Moriarty, S. E. (1996), 'Abduction: a theory of visual interpretation', *Communication Theory* 6(2), 167–187.

Moxey, K. (2008), 'Visual studies and the iconic turn', *Journal of Visual Culture* 7, 131–146.

Muckenhaupt, M. (1986), *Text und Bild. Grundfragen der Beschreibung von Text-Bild-Kommunikation aus sprachwissenschaftlicher Sicht*, Tübinger Beiträge zur Linguistik, Narr, Tübingen.

Mukherjee, S. S. and Robertson, N. M. (2015), 'Deep head pose: Gaze-direction estimation in multimodal video', *IEEE Transactions on Multimedia* 17(11), 2094–2107.

Müller, C. (1998), *Redebegleitende Gesten: Kulturgeschichte – Theorie – Sprachvergleich*, Arno Spitz, Berlin.

Müller, C. (2004), Forms and uses of the Palm Up Open Hand: A case study of a gesture family?, *in* C. Müller and R. Posner, eds, 'The Semantics and Pragmatics of Everyday Gestures: The Berlin Conference', Weidler Buchverlag, Berlin, pp. 233–256.

Müller, C. (2008), *Metaphors dead and alive, sleeping and waking: a dynamic view*, Chicago University Press, Chicago.

Müller, C. and Cienki, A. (2009), Words, gestures, and beyond: Forms of multimodal metaphor in the use of spoken language, *in* C. J. Forceville and E. Urios-Aparisi, eds, 'Multimodal Metaphor', Mouton de Gruyter, Berlin/New York, pp. 297–328.

Müller, M. (2012), "Halt's Maul Averell!" — Die Inszenierung multimodaler Interaktion im Comic, *in* D. Pietrini, ed., 'Die Sprache(n) der Comics.', Meidenbauer, München, pp. 75–90.

Müller, M. G. (2007), 'What is visual communication? Past and future of an emerging field of communication research', *Studies in Communication Sciences* 7(2), 7–34.

Müller, M. G. (2011), Iconography and Iconology as a visual method and approach, *in* E. Margolis and L. Pauwels, eds, 'Handbook of Visual Research Methods', Sage, London and Thousand Oaks, CA, pp. 283–297.

Müller, M. G., Kappas, A. and Olk, B. (2012), 'Perceiving press photography: a new integrative model, combining iconology with psychophysiological and eye-tracking methods', *Visual Communication* 11(3), 307–328.

Murray, J. H. (1997), *Hamlet on the Holodeck: The Future of Narrative in Cyberspace*, MIT Press, Cambridge, MA.

Murray, J. H. (2012), *Inventing the Medium. Principles of Interaction Design as a Cultural Practice*, MIT Press, Cambridge, MA.

Neale, S. (2000), *Genre and Hollywood*, Routledge, London.

Neuendorf, K. A. (2002), *The Content Analysis Guidebook*, Sage, London and Thousand Oaks, CA.

Newall, M. (2003), 'A restriction for pictures and some consequences for a theory of depiction', *The Journal of Aesthetics and Art Criticism* 61(4), 381–394.

Nichols, B. (2010), *Introduction to Documentary*, 2 edn, Indiana University Press, Bloomington.

Nielsen, J. (2006), 'F-Shaped Pattern For Reading Web Content', *Jakob Nielsen's Alertbox* . http://www.useit.com/alertbox/reading_pattern.html

Nielsen, J. and Loranger, H. (2006), *Prioritizing Web Usability: the practice of simplicity*, New Riders Publishing, Berkeley, CA.

Nikolajeva, M. and Scott, C. (2001), *How picturebooks work*, Routledge, London.

Norris, S. (2002), 'The implication of visual research for discourse analysis: transcription beyond language', *Visual Communication* 1(1), 97–121.

Norris, S. (2004a), *Analyzing Multimodal Interaction: a Methodological Framework*, Routledge, London and New York.

Norris, S. (2004b), Multimodal discourse analysis: a conceptual framework, *in* P. LeVine and R. Scollon, eds, 'Discourse and technology: multimodal discourse analysis', Georgetown University Press, Washington, D.C., pp. 101–115.

Norris, S. (2011), *Identity in (inter)action. Introducing multimodal (inter)action analysis*, de Gruyter Mouton, Berlin.

Norris, S. (2016a), 'Concepts in multimodal discourse analysis with examples from video conferencing', *Yearbook of the Poznań Linguistic Meeting* 2, 141–165.

Norris, S. (2016b), Multimodal Interaction – Language and Modal Configurations, *in* N.-M. Klug and H. Stöckl, eds, 'Sprache im multimodalen Kontext / Language and Multimodality', number 7 *in* 'Handbooks of Linguistics and Communication Science (HSK)', de Gruyter Mouton, Berlin, pp. 121–144.

Nöth, W. (1997), Can pictures lie?, *in* W. Nöth, ed., 'Semiotics of the Media. State of the Art, Projects, and Perspectives', number 127 *in* 'Approaches to Semiotics', Mouton de Gruyter, Berlin/New York, pp. 133–146.

Ochs, E. (1979), Transcription as Theory, *in* E. Ochs and B. B. Schieffelin, eds, 'Developmental Pragmatics', Academic Press, New York, pp. 43–72.

O'Halloran, K. L. (2015), Multimodal Digital Humanities, *in* P. P. Trifonas, ed., 'International Handbook of Semiotics', Springer, Dordrecht, pp. 389–416.

O'Halloran, K. L., Chua, A. and Podlasov, A. (2014), The role of images in social media analytics: A multimodal digital humanities approach, *in* D. Machin, ed., 'Visual Communication', Mouton de Gruyter, Berlin, pp. 565–588.

O'Halloran, K. L., E, M. K. L. and Tan, S. (2014), Multimodal Analytics: Software and Visualization Techniques for Analyzing and Interpreting Multimodal Data, *in* C. Jewitt, ed., 'The Routledge Handbook of multimodal analysis', 2 edn, Routledge, London, pp. 386–396.

Oliveira, C. (2016), Nexus at the Limits of Possibility: A Few Remarks on the Documentation of Multiplicities via the Curios Case of The Emergence Room #2 Berlin, *in* C. Fernandes, ed., 'Multimodality and Performance', Cambridge Scholars Publishing, Newcastle upon Tyne, pp. 203–218.

O'Toole, M. (1989), Semiotic systems in painting and poetry, *in* C. Poke, R. Russell and M. Falchikov, eds, 'A Festschrift for Dennis Ward', Astra Press, Nottingham.

O'Toole, M. (2011 [1994]), *The language of displayed art*, Routledge, Abingdon, Oxon.

Oviatt, S. and Cohen, P. R. (2015), *The paradigm shift to multimodality in contemporary computer interfaces*, Morgan & Claypool, San Rafael, CA.

Owens, T. (2011), 'Modding the history of science: Values at play in modder discussions of Sid Meier's Civilization', *Simulation & Gaming* 42(4), 481–495.

Page, R., Barton, D., Unger, J. W. and Zappavigna, M. (2014), *Researching Language and Social Media. A Student Guide*, Routledge, New York and London.

Paivio, A. (1986), *Mental Representations: A Dual Coding Approach*, Oxford University Press, London and New York.

Pak, A. and Paroubek, P. (2010), Twitter as a corpus for sentiment analysis and opinion mining, *in* 'Proceedings of the 7th International Conference on Language Resources and Evaluation (LREC'10)', European Language Resources Association, Valletta, Malta, pp. 1320–1326.

Panofsky, E. (1967 [1939]), *Studies in Iconology: Humanistic Themes in the Art of the Renaissance*, 2 edn, Harper & Row, New York. Originally published by the Oxford University Press.

Panofsky, E. (2004 [1934]), Style and medium in the motion pictures, *in* L. Braudy and M. Cohen, eds, 'Film Theory and Criticism', sixth edn, Oxford University Press, Oxford. Originally published 1934, revised in *Critique*, **1**(3): 5–28 (1947).

Pederson, K. and Cohn, N. (2016), 'The changing pages of comics: Page layouts across eight decades of American superhero comics', *Studies in Comics* **7**(1), 7–28.

Pedri, N. (2015), 'Thinking about Photography in Comics', *Image and Narrative* **16**(2), 1–13.

Peirce, C. S. (1931-1958), *Collected Papers of Charles Sanders Peirce*, Harvard University Press, Cambridge, MA. Vols. 1–6, 1931-1935, edited by Charles Hartshorne and Paul Weiss; Vols. 7–8, 1958, edited by Arthur W. Burks.

Peirce, C. S. (1998 [1893–1913]), *The Essential Peirce – Volume 2. Selected Philosophical Writings (1893–1913)*, Indiana University Press, Bloomington.

Pennock-Speck, B. and del Saz-Rubio, M. M., eds (2013), *The Multimodal Analysis of Television Commercials*, University of Valencia Press, Valencia, Spain.

Pentzold, C., Sommer, V., Meier, S. and Fraas, C. (2016), 'Reconstructing media frames in multimodal discourse: the John/Ivan Demjanjuk trial', *Discourse, Context and Media* **12**, 32–39.

Pethő, Ágnes. (2011), *Cinema and intermediality: the passion for the in-between*, Cambridge Scholars Publishing, Newcastle upon Tyne.

Pettersson, R. (2017), Gestalt principles: opportunities for designers, *in* A. Black, P. Luna, O. Lund and S. Walker, eds, 'Information Design: Research and Practice', Routledge, London, pp. 425–434.

Pike, K. L. (1967), *Language in relation to a unified theory of the structure of human behaviour*, 2nd edn, Mouton, The Hague.

Pinheiro, P. O., Collobert, R. and Dollár, P. (2015), Learning to segment object candidates, *in* C. Cortes, N. D. Lawrence, D. D. Lee, M. Sugiyama and R. Garnett, eds, 'Advances in Neural Information Processing Systems 28', Curran Associates, Inc., pp. 1990–1998. http://papers.nips.cc/paper/5852-learning-to-segment-object-candidates.pdf

Pirini, J. (2016), 'Intersubjectivity and Materiality: A Multimodal Perspective', *Multimodal Communication* **5**(1), 1–14.

Pobłoski, K. (2002), 'Becoming-state: The bio-cultural imperialism of Sid Meier's Civilization', *Focaal – European Journal of Anthropology* **39**, 163–177.

Podlasov, A. and O'Halloran, K. L. (2013), Japanese street fashion for young people: A multimodal digital humanities approach for identifying sociocultural patterns and trends, *in* E. Djonov and S. Zhao, eds, 'Critical Multimodal Studies of Popular Discourse', Routledge, New York and London, pp. 71–90.

Poggi, I. (2007), *Mainds, Hands, Face and Body: a Goal and Belief View of Multimodal Communication*, Weidler, Berlin.

Poncin, K. and Rieser, H. (2006), 'Multi-speaker utterances and co-ordination in task-oriented dialogue', *Journal of Pragmatics* **38**, 718–744.

Postema, B. (2013), *Narrative structures in comics: making sense of fragments*, RIT Press, Rochester, New York.

Pratha, N. K., Avunjian, N. and Cohn, N. (2016), 'Pow, Punch, Pika and Chu: The Structure of Sound Effects in Genres of American Comics and Japanese Manga', *Multimodal Communication* **5**(2), 93–109.

Pratt, H. J. (2009), 'Narrative in comics', *The Journal of Aesthetics and Art Criticism* **67**(1), 107–117.

Pratt, H. J. (2012), Making Comics into Film, *in* A. Meskin and R. T. Cook, eds, 'The Art of Comics. A Philosphical Approach', Wiley-Blackwell, Sussex, pp. 147–.

Prince, S. (2007), *Movies and meaning: An introduction to film*, 4th ed. edn, Pearson Allyn and Bacon, Boston, MA.

Pudovkin, V. I. (1926), *Film technique and film acting: the cinema writings of V. I. Pudovkin*, Bonanza Books, New York. Translated by Ivor Montagu. Republished by Sims Press, 2007.

Ramachandran, V. S. and Hubbard, E. M. (2001), 'Psychophysical investigations into the neural basis of synaesthesia', *Proceedings of the Royal Society B: Biological Sciences* **268**, 979–983.

Rambusch, J. and Susi, T. (2008), 'The challenge of managing affordances in computer game play', *Human IT* 9(3), 83–109.

Rayner, K. (1998), 'Eye movements in reading and information processing: 20 years of research', *Psychological Bulletin* **124**(3), 372–422.

Redeker, G. (1996), Coherence and Structure in Text and Discourse, *in* W. Black and H. Bunt, eds, 'Computational Pragmatics: abduction, belief and context', Univesity College Press, London.

Reinhard, C. D. and Olson, C. J., eds (2016), *Making Sense of Cinema: empirical studies into film spectators and spectatorship*, Bloomsbury Academic, New York and London.

Renkema, J. (2006), *Introduction to Discourse Studies*, John Benjamins, Amsterdam.

Renner, K. N. (2001), 'Die Text-Bild-Schere', *Studies in Communication Science* 1(2), 23–44. http://www.scoms.ch/current_issue/abstract.asp?id=55

Rettberg, S. (2015), The American Hypertext Novel, and Whatever Became of It?, *in* H. Koenitz, G. Ferri, M. Haahr, D. Sezen and T. İbrahim Sezen, eds, 'Interactive Digital Narrative. History, Theory and Practice', Routledge, New York and London, pp. 22–35.

Rider, E., Kurtz, S., Slade, D., Longmaid, H., Ho, M.-J., Pun, J.-H., Eggins, S. and Branch, W. (2014), 'The International Charter for Human Values in Healthcare: An interprofessional global collaboration to enhance values and communication in healthcare', *Patient Education and Counselling* **96**(3), 273–280.

Rieser, M. (2015), Artistic Explorations. Mobile, Locative and Hybrid Narrtives, *in* H. Koenitz, G. Ferri, M. Haahr, D. Sezen and T. İbrahim Sezen, eds, 'Interactive Digital Narrative. History, Theory and Practice', Routledge, New York and London, pp. 241–257.

Robin, H. (1993), *The scientific image: from cave to computer*, W.H. Freeman, New York.

Rogers, Y., Sharp, H. and Preece, J. (2011), *Interaction Design: beyond human-computer interaction*, 3 edn, John Wiley & Sons, Chichester.

Rorty, R. (1979), *Philosophy and the Mirror of Nature*, Princeton University Press, Princeton.

Rosand, D. (2002), *Drawing acts: studies in graphic expression and representation*, Cambridge University Press, Cambridge.

Rosch, E. (1978), Principles of categorization, *in* E. Rosch and B. Lloyd, eds, 'Cognition and Categorization', Lawrence Erlbaum, New Jersey.

Rose, G. (2012*a*), *Visual methodologies. An introduction to researching with visual materials*, 3 edn, Sage, London / Thousand Oaks / New Delhi.

Rose, G. (2012*b*), *Visual methodologies. An introduction to researching with visual materials*, 3 edn, Sage, London / Thousand Oaks / New Delhi, chapter 5. Content analysis. Counting what (you think) you see, pp. 81–103.

Rossi, F. and Sindoni, M. G. (2017), The Phantoms of the Opera: Towards a Multi-Dimensional Interpretative Framework of Analysis, *in* M. G. Sindoni, J. Wildfeuer and K. L. O'Halloran, eds, 'Mapping Multimodal Performance Studies', Routledge, London and New York.

Rossman, G. B. and Rallis, S. F. (2012), *Learning in the field: an introduction to qualitative research*, 3 edn, Sage, London and Thousand Oaks, CA.

Rowsell, J. (2013), *Working with Multimodality: Rethinking literacy in a digital age*, Routledge, Abingdon and New York.

Ryan, M.-L. (2003), Cognitive Maps and the Construction of Narrative Space, *in* D. Herman, ed., 'Narrative Theory and the Cognitive Sciences', CSLI, Stanford, CA, pp. 214–242.

Ryan, M.-L. (2004*a*), Introduction, *in* M.-L. Ryan, ed., 'Narrative across Media: The Languages of Storytelling', University of Nebraska Press, Lincoln, pp. 1–40.

Ryan, M.-L., ed. (2004*b*), *Narrative across Media: The Languages of Storytelling*, University of Nebraska Press, Lincoln.

Ryan, M.-L. and Thon, J.-N., eds (2014), *Storyworlds across Media. Toward a Media-Conscious Narratology*, University of Nebraska Press, Lincoln and London.

Sachs-Hombach, K. (2003), *Das Bild als kommunikatives Medium: Elemente einer allgemeinen Bildwissenschaft*, Univ., Habil.-Schr.–Magdeburg, Halem, Köln.

Sachs-Hombach, K., ed. (2001), *Bildhandeln: interdisziplinäre Forschungen zur Pragmatik bildhafter Darstellungsformen*, number 3 *in* 'Reihe Bildwissenschaft', Scriptum-Verlag, Magdeburg.

Sacks, H., Schegloff, E. A. and Jefferson, G. (1974), 'A simplest systematics for the organisation of turn-taking for conversation', *Language* **50**, 696–735.

Sagan, C., Sagan, L. S. and Drake, F. (1972), 'A message from Earth', *Science* **175**(4024), 881–884.

Salter, M. B. (2011), 'The geographical imaginations of video games: Diplomacy, Civilization, America's Army and Grand Theft Auto IV', *Geopolitics* **16**(2), 359–388.

Saussure, F. (1959 [1915]), *Course in General Linguistics*, Peter Owen Ltd., London. edited by Charles Bally and Albert Sechehaye; translated by Wade Baskin.

Schechner, R. (1988), 'Performance Studies: The Broad Spectrum Approach', *The Drama Review* **32**(3), 4–6.

Schechner, R. (2013), *Performance Studies. An Introduction. Third Edition.*, Routledge, London and New York.

Schirra, J. R. and Sachs-Hombach, K. (2007), 'To show and to say: comparing the uses of pictures and language', *Studies in Communication Sciences* **7**(2), 35–62.

Schnettler, B. and Knoblauch, H., eds (2007), *PowerPoint-Präsentationen. Neue Formen der gesellschaftlichen Kommunikations von Wissen*, UVK Verlagsgesellschaft mbH, Konstanz.

Schreier, M. (2012), *Qualitative content analysis in practice*, Sage, London.

Schutz, A. (1962), *Collected Papers I: The problem of social reality*, Nijhoff, The Hague. (Edited by M. Nantanson).

Schwalbe, C. B. (2015), Infographics and interactivity: A nexus of magazine art and science, *in* D. Abrahamson and M. R. Prior-Miller, eds, 'The Routledge Handbook of Magazine Research: The Future of the Magazine Form', Routledge, New York and London, pp. 431–443.

Scollon, R. (2001), *Mediated discourse: the nexus of practice*, Routledge, London.

Scollon, R. and Scollon, S. (2003), *Discourses in Place: Language in the Material World: Reading and Writing in One Community*, Routledge, London.

Scott, K. (2015), 'The Pragmatics of Hashtags: Inference and Conversational Style on Twitter', *Journal of Pragmatics* **81**, 8–20.

Searle, J. R. (1995), *The construction of social reality*, The Free Press, New York.

Senft, T. M. and Baym, N. K. (2015), 'What Does the Selfie Say? Investigating a Phenomenon', *International Journal of Communication* **9**, 1588–1606.

Serafini, F. (2011), 'Expanding perspectives for comprehending visual images in multimodal texts', *Journal of Adolescent and Adult Literacy* **54**(5), 342–350.

Shepherd, M. and Watters, C. (1998), The evolution of cybergenres, *in* 'Proceedings of the 31st Annual Hawaii International Conference on System Sciences', Vol. 2, IEEE Computer Society Press, Los Alamitos, CA, pp. 97–109.
http://csdl.computer.org/comp/proceedings/hicss/1998/8236/02/82360097.pdf

Shifman, L. (2013), 'Memes in a digital world: Reconciling with a conceptual troublemaker', *Journal of Computer-Mediated Communication* **18**(3), 362–377.

Shohat, E. and Stam, R. (1994), *Unthinking Eurocentrism: Multiculturalism and the Media*, Routledge, London.

Short, T. L. (2007), *Peirce's Theory of Signs*, Cambridge University Press, Cambridge.

Siegel, M. (2006), 'Rereading the signs: multimodal transformations in the field of literacy education', *Language Arts* **84**(1), 65–77.

Siever, C. M. (2015), *Mutimodale Kommunikation im Social Web. Forschungsansätze und Analysen zu Text-Bild-Relationen*, Peter Lang, Frankfurt am Main.

Siever, C. M. (2016), Foto-Communitys als multimodale digitale Kommunikationsform, *in* N.-M. Klug and H. Stöckl, eds, 'Sprache im multimodalen Kontext', number 7 *in* 'Handbooks of Linguistics and Communication Science (HSK)', de Gruyter Mouton, Berlin, pp. 455–475.

Simanowski, R., Schäfer, J. and Gendolla, P., eds (2010), *Reading Moving Letters. Digital Literature in Research and Teaching: A Handbook*, transcript, Bielefeld.

Simola, J., Hyönä, J. and Kuisma, J. (2014), 'Perception of visual advertising in different media: from attention to distraction, persuasion, preference and memory', *Frontiers in Psychology* **5**, 1208.

Sina, V. (2014), 'Die Korrelation von Comic und Film. Ein Einblick in die reziproke Entwicklungsgeschichte zweier Medien', *Closure. Kieler e-Journal für Comicforschung* **1**, 99–21.

Sindoni, M. G. (2013), *Spoken and Written Discourse in Online Interactions: A Multimodal Approach*, Vol. 7 of *Routledge Studies in Multimodality*, Routledge, New York and London.

Sindoni, M. G., O'Halloran, K. and Wildfeuer, J. (2016), 'The Expanding Galaxy of Performing Arts: Extending Theories and Questioning Practices', *Social Semiotics* **26**(4), 325–240.

Sindoni, M. G. and Rossi, F. (2016), '"Un nodo avviluppato". Rossini's La Cenerentola as a protoype of multimodal resemiotisation', *Social Semiotics* **26**(4), 385–403.

Sindoni, M. G., Wildfeuer, J. and O'Halloran, K. L., eds (2017), *Mapping Multimodal Performance Studies*, Routledge, London and New York.

Slade, D., Scheeres, H., Manidis, M., Iedema, R., Dunston, R., Stein-Parbury, J., Matthiessen, C., Herke, M. and McGregor, J. (2008), 'Emergency communication: the discursive challenges facing emergency clinicians and patients in hospital emergency departments', *Discourse and Communication* **2**, 271–298.

Sloman, A. (1985), Why we need many knowledge representation formalisms, *in* M.Bramer, ed., 'Rsearch and Development in Expert Systems', Cambridge University Press, Cambridge and New York, pp. 163–183.

Smith, T. J. (2012), 'The Attentional Theory of Cinematic Continuity', *Projections: The Journal for Movies and the Mind* **6**(1), 1–27.

Smith, T. J., Levin, D. T. and Cutting, J. E. (2012), 'A Window on Reality: Perceiving Edited Moving Images', *Current Directions in Psychological Science* **21**(2), 107–113.

Sousanis, N. (2012), 'Comics in the Classroom Guest Post: Nick Sousanis of Teachers College, New York', *Books and Adventures* .

Sowa, T. and Wachsmuth, I. (2009), A Computational Model for the Representation and Processing of Shape in Coverbal Iconic Gestures, *in* K. Coventry, T. Tenbrink and J. Bateman, eds, 'Spatial Language and Dialogue', Oxford University Press, chapter 10, pp. 132–146.

Spitzmüller, J. and Warnke, I. H. (2011), *Diskurslinguistik: eine Einführung in Theorien und Methoden der transtextuellen Sprachanalyse*, De Gruyter, Berlin.

Steedman, M. J. (2002), 'Plans, affordances, and combinatory grammar', *Linguistics and Philosophy* **25**(5–6), 723–753.

Steen, F. and Turner, M. B. (2013), Multimodal Construction Grammar, *in* M. Borkent, B. Dancygier and J. Hinnell, eds, 'Language and the Creative Mind', CSLI Publications, Stanford, CA, pp. 255–274.

Stein, D. and Thon, J.-N., eds (2013), *From comic strips to graphic novels. Contributions to the theory and history of graphic narrative*, number 37 *in* 'Narratologia / Contributions to Narrative Theory', de Gruyter, Berlin and New York.

Stenning, K. and Oberlander, J. (1995), 'A cognitive theory of graphical and linguistic reasoning: logic and implementation', *Cognitive Science* **19**, 97–140.

Stenros, J. (2015), Playfulness, Play, and Games: A Constructionist Ludology Approach, PhD dissertation, University of Tampere.

Stivers, T. and Sidnell, J. (2005), 'Introduction: Multimodal interaction', *Semiotica* **156**(1–4), 1–20.

Stöckl, H. (2004), In between modes: language and image in printed media, *in* E. Ventola, C. Charles and M. Kaltenbacher, eds, 'Perspectives on Multimodality', John Benjamins, Amsterdam, pp. 9–30.

Stöckl, H. (2014), Semiotic paradigms and multimodality, *in* C. Jewitt, ed., 'The Routledge Handbook of multimodal analysis', 2 edn, Routledge, London, pp. 274–286.

Straßner, E. (2002), *Text-Bild-Kommunikation / Bild-Text-Kommunikation*, Niemeyer, Tübingen.

Stukenbrock, A. (2014), 'Pointing to an 'empty' space: *Deixis am Phantasma* in face-to-face interaction', *Journal of Pragmatics* **74**, 70–93.

Suckfüll, M. (2010), 'Films that move us: moments of narrative impact in an animated short film', *Projections* **4**(2), 41–63.

Surridge, K. (2001), 'More than a great poster: Lord Kitchener and the image of the military hero', *Historical Research* **74**(185), 298–313.

Swales, J. M. (1990), *Genre Analysis: English in academic and research settings*, Cambridge University Press, Cambridge.

Taboada, M. and Habel, C. (2013), 'Rhetorical relations in multimodal documents', *Discourse Studies* **15**(1), 65–89.

Tan, S., Wignell, P. and O'Halloran, K. L. (2017), Multimodal Semiotics of Theatrical Performances, *in* M. G. Sindoni, J. Wildfeuer and K. L. O'Halloran, eds, 'Mapping Multimodal Performance Studies', Routledge, London and New York.

Teddlie, C. and Tashakkori, A. (2011), Mixed Methods Research: Contemporary Issues in an Emerging Field, *in* N. K. Denzin and Y. S. Lincoln, eds, 'The Sage handbook of qualitative research', 4 edn, Sage, Los Angeles, pp. 285–299.

Thomas, M. (2014), 'Evidence and circularity in multimodal discourse analysis', *Visual Communication* **13**(2), 163–189.
http://vcj.sagepub.com/content/13/2/163.abstract

Thomas, M., Delin, J. and Waller, R. H. W. (2010), A framework for corpus-based analysis of the graphic signalling of discourse structure, *in* 'Proceedings of Multidisciplinary Approaches to Discourse (MAD 2010)', Moissac, France.

Thon, J.-N. (2016), *Transmedial Narratology and Contemporary Media Culture*, Nebraska Press, Lincoln.

Titzmann, M. (1990), Theoretisch-methodologische Probleme einer Semiotik der Text-Bild-Relationen, *in* W. Harms, ed., 'Text und Bild, Bild und Text: DFG-Symposion 1988', J.B. Metzlersche Verlagsbuchhandlung, Stuttgart, pp. 368–384.

Tomasello, M. (1999), *The Cultural Origins of Human Cognition*, Harvard University Press, Cambridge, MA.

Tomasello, M. (2005), *Constructing a language: a usage-based theory of language acquisition*, Harvard University Press, Cambridge, MA.

Trahndorff, K. F. (1827), *Ästhetik oder Lehre von Weltanschauung und Kunst*, In der Maurerschen Buchhandlung, Berlin.

Tseng, C. (2013), *Cohesion in Film: Tracking Film Elements*, Palgrave Macmillan, Basingstoke.

Tufte, E. R. (1983), *The visual display of quantitative information*, Graphics Press, Cheshire, Connecticut.

Tufte, E. R. (1997), *Visual explanations: images and quantities, evidence and narrative*, Graphics Press, Cheshire, Connecticut.

Tufte, E. R. (2006), *The cognitive style of PowerPoint: pitching out corrupts within*, 2 edn, Graphics Press LLC, Cheshire, Connecticut.

Tversky, B. (2017), Diagrams: cognitive foundations for design, *in* A. Black, P. Luna, O. Lund and S. Walker, eds, 'Information Design: Research and Practice', Routledge, London, pp. 349–360.

Twyman, M. (1979), A schema for the study of graphic language, *in* P. A. Kolers, M. E. Wrolstad and H. Bouma, eds, 'Processing of Visible Language', Vol. 1, Plenum, New York and London, pp. 117–150.

Twyman, M. (1987), A schema for the study of graphic language, *in* O. Boyd-Barrett and B. P., eds, 'Media, Knowledge and Power', Croom Helm, London, pp. 201–225. reprinted version of Twyman (1979).

Urgesi, C., Moro, V., Candidi, M. and Aglioti, S. M. (2006), 'Mapping implied body actions in the human motor system', *Journal of Neuroscience* **26**(30), 7942–7949.

van Leeuwen, T. (1991), 'Conjunctive structure in documentary film and television', *Continuum: journal of media and cultural studies* **5**(1), 76–114.

van Leeuwen, T. (1999), *Speech, Music, Sound*, MacMillan, London.

van Leeuwen, T. (2005*a*), *Introducing social semiotics*, Routledge, London.

van Leeuwen, T. (2005*b*), Multimodality, genre and design, *in* S. Norris and R. Jones, eds, 'Discourse in Action – Introducing Mediated Discourse Analysis', Routledge, London, pp. 73–94.

van Leeuwen, T. (2008), *Discourse and Practice: New Tools for Critical Discourse Analysis*, Oxford University Press, Oxford.

van Leeuwen, T. (2009), Parametric systems: the case of voice quality, *in* C. Jewitt, ed., 'The Routledge Handbook of multimodal analysis', Routledge, London, pp. 68–77.

van Leeuwen, T. and Boeriis, M. (2017), Towards a semiotics of film lighting, *in* J. Wildfeuer and J. A. Bateman, eds, 'Film Text Analysis. New Perspectives on the Analysis of Filmic Meaning', Routledge, New York and Abingdon, pp. 24–45.

van Leeuwen, T. and Djonov, E. (2014), Kinetic Typography: a semiotic exploration, *in* E. Zantides, ed., 'Semiotics and Visual Communication: Concepts and Practices', Cambridge Scholars Publishing, Newcastle, pp. 150–161.

van Leeuwen, T., Djonov, E. and O'Halloran, K. (2013), '"David Byrne really does love Powerpoint": art as research on semiotics and semiotic technology', *Social Semiotics* **23**(3), 409–423.

van Leeuwen, T. and Jewitt, C., eds (2001), *Handbook of visual analysis*, Sage, London.

Varanda, P. (2016), New Media Dance: Where is the Performance?, *in* C. Fernandes, ed., 'Multimodality and Performance', Cambridge Scholars Publishing, Newcastle upon Tyne, pp. 187–202.

Villi, M. (2015), '"Hey, I'm here right now": Camera phone photographs and mediated presence', *Photographies* **8**(1), 3–22.

Voorhees, G. A. (2009), 'I play therefore I am: Sid Meier's Civilization, turn-based strategy games and the *cogito*', *Games and Culture* **4**(3), 254–275.

Wahlster, W., ed. (2006), *SmartKom: Foundations of Multimodal Dialogue Systems*, Cognitive Technologies, Springer, Heidelberg, Berlin.

Waller, R. (1987), The typographical contribution to language: towards a model of typographic genres and their underlying structures, PhD thesis, Department of Typography and Graphic Communication, University of Reading, Reading, U.K.
http://www.robwaller.org/RobWaller_thesis87.pdf

Waller, R. (2012), 'Graphic literacies for a digital age: the survival of layout', *The Information Society: An international journal* **28**(4), 236–252.

Wandtke, T. R., ed. (2007), *The amazing transforming superhero! Essays on the revision of characters in comic books, film and television*, McFarlane & Co., Jefferson, NC and London.

Ward, M. S. (2015), Art in Noise: An Embodied Simulation Account of Cinematic Sound Design, *in* M. Coëgnarts and P. Kravanja, eds, 'Embodied Cognition and Cinema', Leuven University Press, Leuven, pp. 155–186.

Wardrip-Fruin, N. and Harrigan, P., eds (2004), *FirstPerson: New Media as Story, Performance, and Game*, MIT Press, Cambridge, MA.

Ware, C. (2012), *Information Visualization: Perception for Design*, 3 edn, Elsevier, Amsterdam.

Weber, W. (2017), Interactive information graphics: a framework for classifying a visual genre, *in* A. Black, P. Luna, O. Lund and S. Walker, eds, 'Information Design: Research and Practice', Routledge, London, pp. 243–256.

Weber, W. and Wenzel, A. (2013), Interaktive Infografiken: Standortbestimmung und Definition, *in* W. Weber, M. Burmester and R. Tille, eds, 'Interaktive Infografiken', Springer, Berlin, pp. 3–23.

White, P. (2010), 'Grabbing attention: the importance of modal density in advertising', *Visual Communication* **9**(4), 371–397.

Wildfeuer, J. (2014), *Film Discourse Interpretation. Towards a New Paradigm for Multimodal Film Analysis*, Routledge Studies in Multimodality, Routledge, London and New York.

Wildfeuer, J. (2017), From Text to Performance: Discourse. Analytical Thoughts on New Forms of Performances in Social Media, *in* M. Sindoni, J. Wildfeuer and K. O'Halloran, eds, 'Multimodal Perspectives in Performing Arts', Routledge, London and New York.

Wildfeuer, J. and Bateman, J. A. (2014), 'Zwischen *gutter* und *closure*. Zur Interpretation der Leerstelle im Comic durch Inferenzen und dynamische Diskursinterpretation', *Closure. Kieler e-Journal für Comicforschung* **1**, 3–24.
http://www.closure.uni-kiel.de/data/closure1/closure1_wildfeuer_bateman.pdf

Wildfeuer, J. and Bateman, J. A., eds (2017), *Film Text Analysis. New Perspectives on the Analysis of Filmic Meaning*, Routledge, Londpn.

Wildfeuer, J., Schnell, M. W. and Schulz, C. (2015), 'Talking about dying and death: On new discursive constructions of a formerly postulated taboo', *Discourse & Society* **26**(3).

Williamson, J. H. (1986), 'The grid: History, use, and meaning', *Design Issues* **3**(2), 15–30.

Wilson, R. J. (2010), *Introduction to Graph Theory*, 5 edn, Pearson, New York, NY.

Winkler, H. (2008), Zeichenmaschinen: oder warum die semiotische Dimension für eine Definition der Medien unerlässlich ist, *in* S. Münker and A. Roesler, eds, 'Was ist ein Medium?', Suhrkamp Verlag, Frankfurt am Main, pp. 211–222.

Wirth, U. (2005), 'Abductive reasoning in Peirce's and Davidson's account of interpretation', *Semiotica* **153**(1), 199–208.

Wittenburg, P., Brugman, H., Russel, A., Klassmann, A. and Sloetjes, H. (2006), ELAN: a professional framework for multimodality research, *in* 'Proceedings of LREC 2006, Fifth International Conference on Language Resources and Evaluation', pp. 1556–1559.
http://www.lat-mpi.eu/papers/papers-2006/elan-paper-final.pdf

Wodak, R. and Meyer, M., eds (2015), *Methods of Critical Discourse Analysis*, 3 edn, SAGE Publishers, London.

Wolf, M. J. P., ed. (2001), *The Medium of the Video Game*, University of Texas Press, Austin, TX.

Wolf, W. (2003), 'Narrative and narrativity: a narratological reconceptualization and its applicability to the Visual Arts', *Word & Image: A journal of verbal/visual enquiry* **19**(3), 180–197.

Wolf, W. (2005), Metalepsis as a transgeneric and transmedial phenomenon: a case study of the possibilities of 'exporting' narratological concepts, *in* J. C. Meister, T. Kindt, W. Schermus and M. Stein, eds, 'Narrative Beyond Literary Criticism', De Gruyter, Berlin, pp. 83–107.

Wolf, W. and Bernhart, W., eds (2007), *Description in Literature and Other Media*, number 2 *in* 'Studies in Intermediality', Rodopi, Amsterdam and New York.

Woodrow, L. (2014), *Writing about quantitative research in applied linguistics*, Palgrave Macmillan, Basingstoke.

Worth, S. (1982), Pictures can't say ain't, *in* S. Thomas, ed., 'Film / Culture: Explorations of cinema in its social context', The Scarecrow Press, Metuchen, NJ, pp. 97–109.

Yates, J. and Orlikowski, W. J. (2007), The PowerPoint presentation and its corollaries: how genres shape communicative action in organizations, *in* M. Zachry and C. Thralls, eds, 'Communicative Practices in Workplaces and the Professions: Cultural Perspectives on the Regulation of Discourse and Organizations', Baywood Publishing Company, Amityville, New York, pp. 67–91.

Zagal, J. P. and Mateas, M. (2010), 'Time in video games: A survey and analysis', *Simulation & Gaming* **41**(6), 844–868.

Zambrano, R. N. and Engelhardt, Y. (2008), Diagrams for the masses: Raising public awareness – from Neurath to Gapminder and Google Earth, *in* G. Stapleton, J. Howse and J. Lee, eds, 'Diagrams 2008', Vol. 5223 of *LNAI*, Springer-Verlag, Berlin, pp. 282–292.

Zappavigna, M. (2012), *Discourse of Twitter and Social Media*, Continuum, London and New York.

Zappavigna, M. (2015), 'Searchable Talk: The Linguistic Functions of Hashtags', *Social Semiotics* **25**(3), 274–291.

Zappavigna, M. (2016), 'Social media photography: construing subjectivity in Instagram images', *Visual Communication* **15**(3), 271–292.

Zhao, S., van Leeuwen, T. and Djonov, E. (2014), 'The semiotics of texture: from tactile to visual', *Text & Talk* **34**(3), 349–375.

Zhao, X., Salehi, N., Naranjit, S., Alwaalan, S., Voida, S. and Cosley, D. (2013), The Many Faces of Facebook: Experiencing Social Media as Performance, Exhibition, and Personal Archive, *in* 'Proceedings of the SIGCHI Conference on Human Factors in Computing System (SIGCHI'13)', Paris, pp. 1–10.

Index

Aarseth, Espen 45, 99, 105, 106, 110, 362, 366, 372
abduction 61, 222, 335
aesthetics 264
affordance 44, 90, 118
analysis
– chi-square analysis 193
– content analysis 147
– conversation analysis 149, 240
– power analysis 188, 189
– practical analysis 211
– quantitative analysis 364
annotation 146, *see also* transcription, 156, 157
ANVIL 156, 157
API, application programming interface 357, 364
arbitrariness 55
assembly instructions 286–288
audiovisual presentations 340–343

Baldry, Anthony 17, 22, 147, 263
ballet 253, 255
Barthes, Roland 32, 337
Bateman, John A. 44, 63, 107, 153, 165, 223, 263, 287, 303, 329, 337, 338, 376
between-subjects design 182
Bezemer, Jeff 17, 149, 252, 267, 268, 279, 286, 287
blockage 307
body 94, 253
– body movement 239, 248, 257
Bordwell, David 328, 330
Bucher, Hans-Jürgen 17, 68, 108, 159, 215, 286, 340, 341

canvas 87, 89, 101, 213–221, 223, 228, 269, 278, 284, 300, 371
CAQDAS 162
cause and correlation 187
CGI, computer-generated graphics 60
Chomsky, Noam 55
code integration 247
code manifestation 247
coding 150, 151, 199, 201
coherence 273, 308, 321, 348, 367
cohesion 267, 308

Cohn, Neil 25, 93, 123, 186, 296, 301–303, 309, 322
combination 8, 11, 32, 248
comics 275, 287, 295–338
communication 76, 89
– computer-mediated communication 43
– face-to-face communication 243
– non-verbal communication 239
communicative
– communicative action 65
– communicative intention 80
– communicative mode 243
– communicative performance 95
– communicative potential 88
– communicative roles 93
– communicative situation 381
– communicative situations 7, 77–110
community 84
competence 55
complexity 1, 340
compositionality 33, 39
context 15, 55, 88, 132, 144, 149, 215, 222, 242, 300, 340
contingency table 195
conventionality 38
Cook, Roy T. 297, 315, 316, 337
corpora 149, 152–153
correlations 184
cue 293
cultural turn 380
cybergenre 347
cybertext 105

decomposition 265
deep learning 164
descriptive statistics 180
design 90, 224, 227, 321
– graphic design 263–267
– sound design 29
diagrams 279–294
discourse 133–135
– discourse interpretation 222
– discourse relations 267, 291, 308
– discourse semantics 116–123, 275, 281, 308, 315, 321
discourse analysis 243
– critical discourse analysis 50

distributed cognition 88

effect size 189
Eisenstein, Sergei 29
ekphrasis 44
ELAN 156, 157
emoji 7, 356, 361, 364
empiricism 169–210
ensembles, multimodal 7, 9, 17, 19, 70, 81, 99, 223, 224, 279, 295, 340, 354
ergodic 105, 286, 367–369
experiments 151–152
exploration 350
eyetracking 159–162, 284, 307, 338, 342

F-ratio 205
F-value 189
Facebook 355, 356
facial expression 10, 248, 257
family resemblances 297
film 147, 259, 329–332
films
– *Alexander Nevsky (1938)* 29
– concert films 258–260
– *Conte d'été (1996)* 161
– *Die Hard with a Vengeance (1995)* 332
– *Hulk (2003)* 337
– *Inception (2010)* 156
– *Oldboy (2003)* 336
– *Persepolis (2007)* 336
– *Solaris (1972)* 161
– theatre films 258
Fisher's exact test 196
Forceville, Charles 18, 302, 313
frequency 175
Fricke, Ellen 246, 247
function 39, 89

game studies 366
genre 129–131, 213, 264, 296, 315, 342, 372
Gesamtkunstwerk 28
Gestalt 33, 264, 282, 309
gesture 7, 8, 10, 239–248, 341
– PLOH 246
– PUOH 246
Gombrich, Ernst 33, 35, 83, 102
grammar 115
– visual grammar 50
graphic novel 295–338

grid 265, 266, 301, 305, 306, 309, 311, 317, 319

Halliday, M.A.K. 49, 50, 88, 134
hashtag 357, 364
Hiippala, Tuomo 158, 165, 271, 279, 347
holodeck 99, 104, 127
Holsanova, Jana 141, 159, 161, 279, 284, 285, 294
hypertext 45
– literary hypertext 45
hypotheses 141, 271, 321

icon (Peirce) 59, 221, 225, 244
iconic turn 380
iconography 32, 36
index (Peirce) 59, 244
infographics 341
information graphics 279–294, 300
Instagram 362–365
integration 243, 327
inter-rater reliability 198–204
interaction 20, 42, 357
– dialogic interaction 148, 369
– face-to-face interaction 10, 95, 98, 149, 239, 341
– human-computer interaction 41, 64, 88, 356
– moment-by-moment interaction 240
– multimodal interaction 242, 251
– social interaction 240, 360
interaction design 42, 356
Interactive digital narrative 45
interactivity 343, 350
interdependence 17, 312
interface 352, 368, 371, 374
interpretation 86, 308
– discourse interpretation 62, 315
– narrative interpretation 314, 315, 321

Jewitt, Carey 8, 18, 22, 23, 49, 118, 123, 149, 241

Karen M. Cook 370, 372
Kendon, Adam 243
Kress, Gunther 17, 18, 46, 48, 49, 72, 84, 97, 113, 115, 123, 141, 170, 178, 215, 242, 265, 267, 268, 279, 286, 287
Krippendorff, Klaus 143, 192, 200, 203, 210

language 25, 38, 47
– spoken language 8, 247, 319

– written language 8, 255, 279, 295, 297, 319, 362, 364
layout 341, 343, 360
layout space 263–278
Lemke, Jay 16, 129, 279, 367, 372
Lessing, Gotthold Ephraim 44
Likert scales 192
linguistic turn 133, 380
linguistics 8, 47
– media linguistics 68
– systemic-functional linguistics 49
literacy 45, 321–322
– diagrammatic literacy 286
– digital literacy 356
– media literacy 300
– multiliteracy 45, 47
– multimedia literacy 47
– visual literacy 46
literary studies 43, 45
Log-likelihood ratio 197

Machin, David 22, 28, 31, 115, 241, 379
machine learning 164
Mann-Whitney statistical test 191
Manovich, Lev 36, 127, 166, 355, 358, 365
Martin, James R. 129, 272
material 113
– material regularities 83, 87
materiality 24, 63, 89, 254, 263
– tiers of, 215
McGurk effect 114
media
– depictive media 126, 225, 260, 378
– embedded media 127
– media convergence 336–338
– new media 127
media convergence 12, 224–226
medial variant 125
mediated discourse analysis 242
mediatization 66
medium 101–108, 123, 213–217, 299
meme 357, 361
metafunctions 49, 273
metaphor 244, 246, 263, 313
methods 77, 139–168
– computational methods 162–166, 356, 364
– ethnographic methods 144–146, 300
– mixed methods 37, 140
– qualitative methods 140

– quantitative methods 140
mise-en-scène 256
Mitchell, W.J.T. 31, 36, 131, 380
MMORPG 70
mode 27
– mode combinations 128
– mode family 121
– semiotic mode 16, 18, 19, 112–123
module 266
Mondada, Lorenza 22, 144, 241, 242, 252
Müller, Cornelia 36, 243–245, 247
Müller, Marion G. 17, 159
multiliteracy 321
multiplication, semiotic 16, 39
music 28, 29, 226
– music concerts 258–260

narrative 99, 296, 298, 314–315
New London Group 46
non-parametric tests, statistical 191
Norris, Sigrid 18, 22, 149, 215, 242, 274
notation 225
null hypothesis 180

O'Halloran, Kay 165, 166, 356, 357, 364, 372
O'Toole, Michael 48
onomatopoeia 48, 301
opera 28, 253
operationalisation 152, 175, 178

page
– page metaphor 263
– page-flow 270
paradigmatic 54
parametric tests, statistical 191
Peirce, Charles Sanders 52, 56–63, 120
perception 284
performance 55, 85, 251–260, 341
– performance studies 251, 253, 254
performative turn 380
performing arts 252–253
perspective 34
persuasion 239, 296, 342
photograph 7, 10, 279, 280, 297
pictorial turn 380
picturebooks 323
postal model 77
poster 273–278
PowerPoint 340, 343

pragmatics 35, 253
Pragmatism 52, 58
precision 188
program music 29
projection 275, 288
proxemics 239
psychology
– experimental psychology 284
– Gestalt psychology 264

QDAS 162

radical indeterminacy 81
reach, of semiotic modes 119, 122
real estate, screen 353
recall 188
reception studies 140, 145, 338
reflexivity 315
regression line 186
reliability measures
– Scott's π 201
remediation 222
remediatisation 259
resemblance 34
resemiotisation 222
Rose, Gillian 30, 32, 37, 70, 87, 147
RST, Rhetorical Structure Theory 292–294

sample size 189
sampling 142–143
Saussure, Ferdinand de 52, 53
scatter plot 186
self-referentiality 315
semiology *see also* semiotics, 52
semiosis 253
– multimodal semiosis 254
semiotic resource 115, 219
semiotics 25, 51, 78, 88, 253
– experiential semiotics 254
– geosemiotics 242
– relation to material 113
– social semiotics 282, 357
shot 329
sign 54, 78, 81
– sign as symptom 79
– sign vehicle 82
similarity 264
smartphone 7, 355
social media 355–365

social status 84
society 25
software 340, 343
songs
– *Summer in the City (1966)* 333
sound 27
statistics 169–210
Stöckl, Hartmut 22, 51, 125, 126
structural contrast 54
substitutivity 39
sum of squares 173, 183, 185, 202
symbol (Peirce) 60
syntax 283

technology 347, 355
text 52, 131–133
– text-flow 270, 272
textbook 268–273
textuality 131–133, 254
theatre 253–258
Thibault, Paul 17, 22, 147, 263
Threema 361
tiers of materiality 215
time 94
Tinder 362
transcoding 148, 153
transcription 40, 146–151
transience 100, 216, 255, 342, 373, 376
transmedia 45, 260, 314
transmediality 69, 314, 336, 337
triangulation 141, 232
– across disciplines 73, 221
– across media 339
– across methods 159, 221
– across results 141
Tufte, Edward 41, 281, 340
Tukey post hoc test 206
turn 12, 15, 380, 381
– multimodal turn 9
– turn-taking 241
TV 7
Twitter 12, 355, 356
typography 341

ut pictura poesis 44

van Leeuwen, Theo 22, 46, 48, 72, 97, 115, 123,
    141, 170, 178, 215, 242, 265
video games 7, 366–378

visual associations 32
visual experience 35
visual language fluency index 323
visual propositions 35
visualisation 282, 294, 350
VLFI *see* visual language fluency index

Waller, Rob 263, 264, 267, 271
webpages 347
WhatsApp 7, 361
Wildfeuer, Janina 61, 63, 303, 338
within-subjects 182

Yates correction 197

CPSIA information can be obtained
at www.ICGtesting.com
Printed in the USA
BVHW012148010322
630397BV00006B/268